Sustainable Development Goals Series

The **Sustainable Development Goals Series** is Springer Nature's inaugural cross-imprint book series that addresses and supports the United Nations' seventeen Sustainable Development Goals. The series fosters comprehensive research focused on these global targets and endeavours to address some of society's greatest grand challenges. The SDGs are inherently multidisciplinary, and they bring people working across different fields together and working towards a common goal. In this spirit, the Sustainable Development Goals series is the first at Springer Nature to publish books under both the Springer and Palgrave Macmillan imprints, bringing the strengths of our imprints together.

The Sustainable Development Goals Series is organized into eighteen subseries: one subseries based around each of the seventeen respective Sustainable Development Goals, and an eighteenth subseries, "Connecting the Goals," which serves as a home for volumes addressing multiple goals or studying the SDGs as a whole. Each subseries is guided by an expert Subseries Advisor with years or decades of experience studying and addressing core components of their respective Goal.

The SDG Series has a remit as broad as the SDGs themselves, and contributions are welcome from scientists, academics, policymakers, and researchers working in fields related to any of the seventeen goals. If you are interested in contributing a monograph or curated volume to the series, please contact the Publishers: Zachary Romano [Springer; zachary.romano@springer.com] and Rachael Ballard [Palgrave Macmillan; rachael.ballard@palgrave.com].

More information about this series at https://link.springer.com/bookseries/15486

David Moore · Matthias Heilweck ·
Peter Petros

Aquaculture: Ocean Blue Carbon Meets UN-SDGS

Springer

David Moore
Stockport, UK

Matthias Heilweck
Private
Kaysersberg, France

Peter Petros
Kääpä Biotech Oy
Karjalohja, Finland

ISSN 2523-3084 ISSN 2523-3092 (electronic)
Sustainable Development Goals Series
ISBN 978-3-030-94845-0 ISBN 978-3-030-94846-7 (eBook)
https://doi.org/10.1007/978-3-030-94846-7

This Springer imprint is published by the registered company Springer Nature Switzerland AG
The registered company address is: Gewerbestrasse 11, 6330 Cham, Switzerland

Dedicated to our children, and their children, and their children...

We have the chance to turn the pages over
We can write what we want to write
We gotta make ends meet, before we get much older

We're all someone's daughter
We're all someone's son
How long can we look at each other
Down the barrel of a gun?

You're the voice, try and understand it
Make the noise and make it clear, oh, woah
We're not gonna sit in silence
We're not gonna live with fear, oh, woah

[From "You're the Voice", a song released as a single in September 1986. Written by Andy Qunta, Keith Reid, Maggie Ryder and Chris Thompson, and recorded by the Australian singer John Farnham]

https://www.youtube.com/watch?v=tbkOZTSvrHs

Scan the following QR codes

[NOTE: we are not responsible for the content or advertisements of external websites]

1986 ‘official’ version

AND (video made in 2020)

Preface

This book is primarily targeted at everyone who wishes to reverse the greenhouse effect that is the cause of global warming and consequent climate change. From the person in the street, to undergraduate and postgraduate students, senior scientists, decision-makers, opinion formers and political leaders. The book arose as a reaction among the authors to the widely circulated fears about the accumulation of carbon dioxide (CO_2) in our atmosphere and the resultant global warming. Because CO_2 is a greenhouse gas, which does not allow the escape of re-radiated infrared heat energy received from the Sun, events that add greenhouse gases into the atmosphere result in warmer temperatures on Earth and consequential climate change. The Earth has a **Global Carbon Cycle** that maintains a natural balance and acts like a thermostat, helping to keep Earth's temperature relatively stable over long periods of time. This 'thermostat' works over timescales of a few hundred thousand years, so it's a slow part of the overall carbon cycle. But over shorter time periods, say ten thousand to a hundred thousand years, the CO_2 content of the atmosphere, and consequently the temperature of Earth, can quite naturally vary and this is thought to be a contributory cause for the Earth shifting between ice ages and warmer interglacial periods over these timescales. Parts of the carbon cycle may even vary over shorter timescales. The Global Carbon Cycle was almost exactly in equilibrium for several thousand years while humans were evolving and taking their long trek out of Africa. But then industrial humans intervened by burning fossil fuels, thereby returning to the atmosphere CO_2 that the Earth's natural processes had stockpiled in the rocks long before. The rapid pace of the human technological revolution has been imposed upon the slow-paced natural Carbon Cycle, causing such a dramatic increase in atmospheric CO_2 in recent times that, if not corrected, could result in climate change so extreme as to be catastrophic for humanity.

The problem is that continued increase in the amount of CO_2 in the atmosphere will inevitably cause a 'runaway greenhouse' effect that will generate catastrophic increase in the Earth's surface temperatures. This has happened before in the history of the Earth and has been corrected by the Earth's own processes. Perhaps we should look to those natural processes to find the cure for our present predicament.

The earliest part of our planet's history is called the Precambrian. This is an informal unit of geologic time that covers 88% of the Earth's lifetime, extending from the formation of Earth about 4.6 billion years ago to the beginning of the Cambrian Period, about 550 million years ago, which is when the first definitive fossils of hard-shelled creatures are first found in abundance. The life we know about probably arose on Earth after the Moon-forming impact because this made Earth absolutely sterile for a couple of million years, with Earth covered in a deep ocean of magma and enveloped in an atmosphere of rock vapour. When the mantle solidified, steam in the atmosphere condensed and rained out to make hot (about 250° C) salty oceans (salty because NaCl had been an abundant gas in the atmosphere) under an atmospheric pressure about 100 times that of our present atmosphere. This atmosphere was composed mostly of H_2O, CO_2 and N_2, and the impossibly hot conditions on the surface would have lasted while Earth's CO_2 remained in the atmosphere. But a thick primordial CO_2 atmosphere, a liquid water ocean and a fresh basalt mantle crust are a highly reactive trinity. It is thought that the CO_2 reacted with the newly formed rocks of the seafloor forming carbonates that were subducted into the mantle over a period of 20 to 100 million years. Removal of CO_2 from the atmosphere allowed the Earth's surface to cool so much that ice covered the ocean and 'Snowball Earth' resulted.

So, this was Earth's first experience of the runaway greenhouse even though the Sun at this distant time radiated much less than it does today. It is also an indicator for a way of dealing with our present-day CO_2 excess, and what we call the *industrial engineering* approaches to *carbon capture and storage* (described in Chap. 7) all seek to react captured atmospheric CO_2 with deep mantle rocks. In the present day, a carbon sequestration solution that is fast gaining traction among wealthier nations is the application of CO_2 capture processes to flue gases of power plants, and other heavy industries like cement and steel producers. This is certainly a promising technology for aggressive emission reduction of CO_2, but the high energy requirement of the carbon capturing process (its energy penalty) and consequent high infrastructure cost for long-term storage of CO_2 limits the impact of this industrial technology and threatens to create further problems for the future. So, if technology might be too expensive and disruptive, what about *bio*technology?

From discussions aimed at finding some way of combating climate change, proposals have been made to develop *biological* methods that would pull carbon dioxide out of Earth's atmosphere and sequester it in some way on a long-term basis. One frequently recommended approach is to remove CO_2 from the atmosphere with activities such as reforestation and changing forest management and agricultural practices to enhance soil carbon storage. However, it is also noted that such activities would limit land for food production and negatively affect biodiversity. Furthermore, decay of dead wood and fallen leaves in natural forests releases huge quantities of CO_2 and other greenhouse gases back into the atmosphere, even in the same year the carbon was sequestered. Several recent studies indicate that massive tree planting is not the panacea that many people believe (and hope); and putting such plans into effect could do more harm than good to our environment.

In this book we suggest that we should look to the oceans for a solution and properly harness the ability of marine calcifier organisms (molluscs, crustacea, corals, foraminifera and coccolithophore algae) to remove *permanently* CO_2 from the atmosphere into solid (crystalline) $CaCO_3$. This also has an ancient historical precedent, as we explain in Chap. 6. At intervals over the past 500 million years the fossil record shows that the distant ancestors of today's marine calcifiers had the physiological tools to cope with both acidified oceans and great excesses of atmospheric CO_2 and still create vast remains of shells made from crystalline $CaCO_3$. These organisms have dealt with excess atmospheric CO_2 before; we should enable them to do that again.

Our suggestion is that we should apply this calcifier physiology to solving our present problem with excess atmospheric CO_2 by cultivating the calcifiers on a massive global scale to sequester that excess atmospheric CO_2 into the ocean's sediments. We know that some marine scientists are unconvinced that shell biomineralisation is effective in carbon *sequestration*, but we believe and demonstrate here (Chaps. 1, 2 and 6) that the scientific evidence shows it is an *effective carbon sink* providing overall CO_2 budgets are considered rather than individual reactions. We also emphasise that this $CaCO_3$ not only sequesters atmospheric carbon but has the bio-circular economic potential for use as a sustainable biomaterial in a wide variety of different ways, and that the activity has enormous potential for sustainable aquaculture, conservation and restitution of marine ecosystems (Chaps. 3–6).

We argue that if the level of finance and global effort that are readily foreseen for forest management and flue gas treatments were to be applied to expansion of shellfish and coccolithophore cultivation around the world, significant amounts of carbon dioxide could be permanently removed from the atmosphere within the timescale that is currently envisaged for carbon capture by afforestation. The overwhelming advantage of our action plan (Chap. 8) is that *the excess atmospheric CO_2 released by our use of fossil fuels will be returned to the place it belongs—as a present-day fossil safe to the distant future*. With the additional advantages of improved natural capital value (including food security), and ecosystem services (many of the organisms involved are natural habitat engineers). Further, as a nature-based solution, there is a minimum of hard infrastructure and consequently faster implementation (we could start tomorrow) and lower investment risk (many of the organisms that will sequester carbon in their shells are saleable food-animals). Carbon sequestration through shellfish cultivation is much more permanent, being secured for geological periods of time, rather than for the few years or decades secured by planting trees or by industrial carbon capture and storage, both of which can only be considered as temporary solutions. So, we suggest cultivating shellfish *for their shells*.

Trees are widely cultivated for the timber they produce, and many people now expect this timber to sequester carbon from the atmosphere. But a timber carbon sink is only temporary; all timber degrades, being digested by fungi, microbes and arthropods that return its carbon to the atmosphere as respiratory CO_2 over a timescale of tens to hundreds of years at best. In contrast, half the body weight of shellfish is a crystalline calcium carbonate shell made from atmospheric CO_2. The shell lasts forever—well, it lasts for hundreds of millions of years. Hundreds of millions of years sequestration of today's excess atmospheric carbon dioxide. That's worth cultivating.

Stockport, UK — David Moore
Kaysersberg, France — Matthias Heilweck
Karjalohja, Finland — Peter Petros
December 2021

Contents

1 Diagnosing the Problem 1
David Moore, Matthias Heilweck, and Peter Petros
1.1 In this Chapter... 1
1.2 A Plain Language Guide to the Earth's Carbon Cycle 1
1.3 The Denial of Anthropogenic CO_2-Driven Climate Change 6
1.4 The Global Carbon Cycle 7
1.5 The Likely Effects of Climate Change 10
1.6 Climate Change and What We Might Do About It. 18
1.7 Blue Carbon on the High Seas and the United Nations Sustainable Development Goals 21
References. 30

2 Cultivate Shellfish to Remediate the Atmosphere 35
David Moore, Matthias Heilweck, and Peter Petros
2.1 In this Chapter... 35
2.2 Plant Trees for the Intrinsic Value of Forests 35
2.3 Cultivate Shellfish: Save the Atmosphere 41
2.4 There's a Lot of Shell in Shellfish 42
2.5 Shellfish Shell is Mineralised Carbon Dioxide from the Atmosphere 44
2.6 Shellfish Shell is Not Digested and is Chemically Stable for Geological Periods of Time 46
2.7 Change the Paradigm 46
2.8 Additional Benefits 49
2.9 The Shellfish Cultivation Industry Offers Unique Opportunities for Limiting Climate Change and Enhancing Conservation Strategies 56
References. 59

3 Aquaculture: Prehistoric to Traditional to Modern 65
David Moore and Matthias Heilweck
3.1 In this Chapter... 65
3.2 Our Primeval Shellfish-Eating Tradition 65
3.3 Ancient Clam Gardening 68
3.4 What is Aquaculture? 70

3.5 Fish and Shellfish in Recorded History 73
3.6 Nineteenth Century Oyster and Scallop Dredging on Both Sides of the Atlantic 78
3.7 Bringing the Oyster Back to the Table 84
3.8 Present-Day Aquaculture 88
References 91

4 The High Seas Solution 97
Matthias Heilweck
4.1 In this Chapter... 97
4.2 The Context 97
4.3 A Novel Idea 98
4.4 Why Should It Be Done? 99
4.5 How Are We Going to Achieve It? 103
4.6 Where Could This Take Place? 108
4.7 What Shall We Raise There? 110
4.8 My Vision about the High Seas 116
4.9 My Dream of a Shared Half World 119
References 127

5 Farming Giant Clams in 2021: A Great Future for the 'Blue Economy' of Tropical Islands 131
David Moore
5.1 In this Chapter... 131
5.2 Introducing Giant Clams 131
5.3 Biology and History of Giant Clams 133
5.4 Giant Clams and Coral Reefs 137
5.5 Giant Clam Cultivation and Restocking 141
5.6 Giant Clam Aquaculture is More Than Just Science! 147
5.7 More a Matter of Vision and Governance Than Either Science or Art 149
References 150

6 Coccolithophore Cultivation and Deployment 155
David Moore
6.1 In this Chapter... 155
6.2 Introducing Calcifying Algae 155
6.3 The Nature, Biology and Ecology of Coccolithophores 156
6.4 Why Coccolithophores Could Be Good for Us and Our Planet 164
6.5 Coccolithophore Cultivation 166
6.6 Large-Scale Cultivation of Coccolithophores 167
6.7 What We Might Do 171
References 172

7 Comparing Industrial and Biotechnological Solutions for Carbon Capture and Storage 177
Peter Petros and David Moore
7.1 In this Chapter... 177
7.2 Solutions for Atmospheric Carbon Reduction 177
7.3 The Carbon Market 178
7.4 Industrial Carbon Dioxide Capture, Utilisation and Storage (CCUS) 180
7.5 Carbon Emissions Reduction Targets 194
7.6 The Comparison with 'Soft' (Nature-Based) Carbon Sequestration 194
7.7 'Soft' Carbon Sequestration Solutions (Nature-Based) 196
7.8 Coastal Blue Carbon 200
7.9 The Ultimate Blue Carbon: The Oceans' Calcifiers 204
7.10 Sustainability Assessment of CCS Methods 206
7.11 Conclusions 209
References 210

8 What Should Be Done 217
David Moore, Matthias Heilweck, and Peter Petros
8.1 In this Chapter... 217
8.2 Executive Summary of Chaps. 1–7 218
8.2.1 Chapter 1. Diagnosing the Problem by David Moore, Matthias Heilweck and Peter Petros 218
8.2.2 Chapter 2. Cultivate Shellfish to Remediate the Atmosphere by David Moore, Matthias Heilweck and Peter Petros 219
8.2.3 Chapter 3. Aquaculture: Prehistoric to Traditional to Modern by David Moore and Matthias Heilweck 219
8.2.4 Chapter 4. The High Seas Solution by Matthias Heilweck 220
8.2.5 Chapter 5. Farming Giant Clams in 2020: A Great Future for the 'Blue Economy' of Tropical Islands by David Moore 220
8.2.6 Chapter 6. Coccolithophore Cultivation and Deployment by David Moore 221
8.2.7 Chapter 7. Comparing Industrial and Biotechnological Solutions for Carbon Capture and Storage by Peter Petros and David Moore . . . 221
8.2.8 Summary of Recommendations 221
8.2.9 The Action Plan 222
8.2.10 Immediate Activity (Assuming Global Funding and Programme Management Are Both in Place) 222
8.2.11 Waste Shells? 223
8.2.12 Infrastructural Activity Designed to Change the Paradigm 224

8.3 Specific Recommendations 224
8.3.1 Seamount Installations and Factory Ships 224
8.3.2 Factory Ships. 226
8.3.3 Coral Reef Restoration 226
8.4 FAQs—Frequently Asked Questions and Online Comments. 227

Index 243

Abbreviations

AAIW	Antarctic Intermediate Water
BBC	British Broadcasting Corporation
BBNJ	Biodiversity Beyond areas of National Jurisdiction (a UN Intergovernmental Conference)
BCSGA	British Columbia Shellfish Growers' Association
BECCS	Bioenergy with Carbon Capture & Storage
BSF	Black Soldier Fly
C2ES	Center for Climate and Energy Solutions
CAT	Climate Action Tracker (website)
CCD	Calcite Compensation Depth
CCS	Carbon dioxide Capture & Storage
CCU	Carbon dioxide Capture & Utilisation
CCUS	Carbon dioxide Capture, Utilisation & Storage
CDM	Clean Development Mechanism
CDR	Carbon Dioxide Removal
CH_4	Methane
CITES	Convention on International Trade in Endangered Species (of Wild Fauna and Flora)
CLCS	Commission on the Limits of the Continental Shelf
CO_2	Carbon dioxide
COI	Cytochrome oxidase subunit 1
COP21	United Nations Conference of the Parties, Paris 2015
COP26	United Nations Conference of the Parties, Glasgow 2021
COP27	United Nations Conference of the Parties, Sharm al-Sheikh, Egypt 2022
CPRE	UK's Countryside Charity (formerly Council for the Protection of Rural England)
CRFM	Caribbean Regional Fisheries Mechanism
DAC	Direct Air Capture
DAI	Dangerous Anthropogenic Interference
DCM	Deep Chlorophyll Maximum
DHA	Docosahexaenoic acid
DMS	Dimethyl sulfide
EEA	European Environment Agency
EEZ	Exclusive Economic Zone
EPA	Eicosapentaenoic acid

ETS	Emission Trading System
EU	European Union
FAO	Food and Agriculture Organization of the United Nations
GCC	Global Carbon Cycle
GCP	Global Carbon Project
GFANZ	Glasgow Financial Alliance for Net Zero
GHG	Greenhouse gas
GMO	Genetically Modified Organism
HCO_3^-	Bicarbonate ion
HFC	Hydrofluorocarbon
IEA	International Energy Agency
IGC	Intergovernmental Conference
IMTA	Integrated Multi-Trophic Aquaculture
IPBES	Intergovernmental Science-Policy Platform on Biodiversity and Ecosystem Services
IPCC	Intergovernmental Panel on Climate Change
ISA	International Seabed Authority
IUCN	International Union for Conservation of Nature
LCA	Life Cycle Assessment
LC-PUFA	long chain omega-3 polyunsaturated fatty acids
LED	Light emitting diode
MPA	Marine Protected Area
MSI	Mariculture Sustainability Index
Mya or MYA	Million years ago
N_2O	Nitrous oxide
NASA	US National Aeronautics and Space Administration
NASEM	US National Academies of Sciences, Engineering, and Medicine
NB-NET	Nature-Based Negative Emissions Technology
NbS	Nature-based solution
NCP	Nature's Contributions to People
NCS	Natural Climate Solution
NET	Negative Emissions Technology
NMFS	US National Marine Fisheries Service
NOAA	US National Oceanic & Atmospheric Administration
NORA	Native Oyster Restoration Alliance
OAE	Oceanic Anoxic Event
ONS	UK Office for National Statistics
PBR	Photobioreactor
PETM	Palaeocene/Eocene Thermal Maximum Warming Event
PFC	Perfluorocarbon
POP	Persistent Organic Pollutant
PSA	Perpetual Salt Fountain
RAS	Recirculating Aquaculture System
RFC	Reason For Concern
RSM	Restorative Shellfish Mariculture
SCE	Shallow Coastal Ecosystems

SCUBA	Self-Contained Underwater Breathing Apparatus
SeaWIFS	Sea-Viewing Wide Field-of-View Sensor
SPM	Summary for Policymakers
SSP	Shared Socio-economic Pathway
TMDC	Tridacna Mariculture Development Center
T-OAE	Toarcian Oceanic Anoxic Event
UKNEA	UK National Ecosystem Assessment (website)
UNCLOS	UN Convention on the Law Of the Sea
UNEP	UN Environment Programme
UNESCO	UN Educational, Scientific and Cultural Organization
UNFCCC	UN Framework Convention on Climate Change
UN-SDGS	UN Sustainable Development Goals
URL	Uniform Resource Locator
US DOE	US Department of Energy
WCS	Wildlife Conservation Society
WEF	World Economic Forum
WFP	World Food Program
WMO	World Meteorological Organization
WWF	World Wide Fund for Nature

Diagnosing the Problem 1

David Moore, Matthias Heilweck,
and Peter Petros

1.1 In this Chapter...

We give a plain language guide to the Earth's carbon cycle by briefly summarising the observations and origins of increased levels of greenhouse gases, mainly CO_2 but including CH_4 and N_2O, in our present-day atmosphere. They are increased in the sense that they have not occurred naturally in the Earth's atmosphere at any time during the past 420,000 years. The only tenable explanation for our atmosphere's present state is that it is the consequence of mankind's excessive use of fossil fuels since the Industrial Revolution onwards. Something that has been described as a planetary-scale experiment in which humans return to the atmosphere and oceans the concentrated organic carbon that had previously been stored in sedimentary rocks for many hundreds of millions of years. We deal with the arguments that deny the truth of anthropogenic CO_2-driven climate change, then illustrate the Earth's global carbon cycle. Explaining how it was almost exactly in equilibrium for several thousand years while humans were evolving, before industrial humans intervened. We describe how the excess greenhouse gas emissions are projected to change the global climate over this century and beyond, and discuss 'dangerous anthropogenic interference' (DAI), 'reasons for concern' (RFCs) and climate tipping points. Finally, we give a short account of the various improved management, engineering and natural climate solutions advocated to increase carbon storage and/or avoid greenhouse gas emissions across global forests, wetlands, grasslands, agricultural lands and industry, and indicate how they are discussed in our later chapters, and we will propose an alternative natural carbon sink that is currently greatly undervalued and underdiscussed.

1.2 A Plain Language Guide to the Earth's Carbon Cycle

The birth of the Industrial Revolution is marked by the invention of the first practical steam engine by Thomas Newcomen in 1712; his 'Atmospheric Steam Pump Engine', was installed at a coalmine at Dudley Castle in Staffordshire, England; working day and night the Engine raised 120 gallons of water every minute from a depth of 156 feet. Newcomen engines were expensive, rugged and reliable but were extremely inefficient. By the time Newcomen died on 5 August 1729 there were at least 100 of his engines in Britain and across Europe.

In 1764, James Watt was commissioned to repair a Newcomen steam engine and found ways to make it much more efficient. Five years later, Watt was granted his first British patent for the unique design of his new steam engine; this was the design that set the world in motion with steam powered railway locomotives and steam ships, and went on to power the textile mills that brought the Industrial Revolution into full activity in the 1760s. By the turn of the century, 1800, about 10 million tonnes of coal had already been mined, and burned, in Britain (see Table 1.1).

D. Moore et al., *Aquaculture: Ocean Blue Carbon Meets UN-SDGS*,
Sustainable Development Goals Series, https://doi.org/10.1007/978-3-030-94846-7_1

Table 1.1 Explanation of some units used in atmosphere science

It's a mass weighing a ton, or is it a tonne, or a long ton, or even a short ton?	
The British Imperial ton, also known as the Long Ton or Displacement Ton, is the name for the unit called 'the ton' in the avoirdupois system of weight measurements, as standardised in the thirteenth century to be equal to 2,240 pounds. The UK adopted the metric system in 1985 and most Commonwealth countries followed British practice	=1,016.0469088 kg
The short ton is commonly used in the United States and was formalised in 1832 to be equivalent to 2,000 pounds. In the United States it is known simply as a common ton	=907.18474 kg
In the metric system the mass of one cubic metre of pure water at 4 °C is specified as being 1,000 kg, and is called the tonne (referred to as metric ton in the United States). A tonne is equivalent to 2,204.6 pounds	=1,000 kg
The International Bureau of Weights and Measures adopted the symbol 't' for the tonne (it is a symbol, not an abbreviation). You will encounter several ways of defining the very large masses dealt with in atmospheric and geophysical sciences; a few equivalencies are shown below	
1,000 t (1×10^3 t) = 1 kt (kilotonne)	=1 Gg (gigagram) or 10^9 g
1 million t (1×10^6 t) = 1 Mt (megatonne)	=1 Tg (teragram) or 10^{12} g
1 billion t (1×10^9 t) = 1 Gt (gigatonne)	=1 Pg (petagram) or 10^{15} g
You will also encounter very low concentrations of materials, in the atmosphere and elsewhere, expressed in the units 'ppm' (parts per million) and 'ppb' (parts per billion). These are ratios that describe how much of substance X is present in mixture M by expressing it in the form '10 parts of X in one million parts of M' (which would be written simply as '10 ppm'). This saves writing strings of zeros because 10 ppm = 0.001%, or 0.00001 as a fraction of one. Obviously, 10 ppb = 0.01 ppm = 0.000001% = 0.00000001%. These ratios may refer to the relative volumes of gases or liquids (usually written as 'ppmv', meaning 'parts per million by volume'); or weights (masses) of components of a dry mix (usually written as 'ppmw', meaning 'parts per million by weight'); or both, as for example, when a solid material is added to a solution (usually written as 'ppm w/v', meaning 'parts per million, weight into volume'). This last one is particularly convenient for making up dilute chemical solutions because 'one milligram of dry substance per litre of solution' (1 mg l^{-1}) = 1 ppm	

In the nineteenth century, and for many years subsequently, coal was king, steam power its agent and the iron foundry the maker of industry in Britain and, increasingly, around the world.

This is when appreciable amounts of carbon dioxide (CO_2) began to be added to the atmosphere through the combustion of ***fossil*** fuels (coal, oil and gas). The rate of combustion has continually increased with the passing of time, so that, by 2019, global carbon emissions from fossil fuels (and including cement production) reached an estimated mass of CO_2 of 36.8 Gt (=gigatonne, see Table 1.1).

The last ice age ended about 12,000 years ago and the period since then (called the **Holocene**) has featured relative stability in both climate and atmospheric gas concentrations over most of that time. The compositions of really ancient atmospheres are deduced from isotope ratios in geological samples (Zahnle et al. 2010). For more recent times the composition can be measured directly in ice and still frozen bubbles of gas in ice cores removed from the polar ice sheets of the Arctic and Antarctic or high mountain glaciers. Glacial ice is formed from the gradual accumulation of annual layers of snow, so the upper layers are the most recent and layers are successively older the deeper you go. A really deep-drilled ice core can contain layers of ice formed thousands of years ago that has remained frozen and undisturbed until the core was cut. Core drilling at Vostok station in East Antarctica extended the ice record of atmospheric composition and climate over the past four glacial–interglacial cycles and revealed that atmospheric levels of the two important greenhouse gases, CO_2 and methane, of the present-day have not been experienced by the atmosphere at any time during the past 420,000 years (Petit et al. 1999).

Other ice core data have revealed that levels of CO_2 (at about 280 ppm by volume) and CH_4 (at about 650 ppb by volume) in the atmosphere, as well as another greenhouse gas, nitrous oxide (N_2O), have been relatively constant for the past two thousand years (Fig. 1.1). As Fig. 1.1 shows, levels of all three gases started to increase rapidly about 200 years ago, and the increases in these three greenhouse gases are the primary cause of the warming of the Earth's averaged temperature by more than 1 °C over the past century.

Importantly, the rate of increase of atmospheric CO_2 over the past 70 years is nearly 100 times greater than that at the end of the last ice age. Such abrupt changes in the atmospheric levels of CO_2 have never before been seen (Fig. 1.2) and must be caused by human activities (that is, they are anthropogenic).

Ice core data reveal other significant changes in the atmosphere during the last 200 years or so, particularly in the Northern Hemisphere. For example, the ice itself reveals increases in the amounts of nitrate and sulphate, which, like the greenhouse gases, are also produced ultimately from the combustion of fossil fuels. These constituents are the key components of acid rain and, indeed, data from the same ice cores also reveal an increase in acidity (Geng et al. 2014).

William H. Brune (Distinguished Professor of Meteorology, PennState College of Earth and Mineral Sciences), in his 2020 online course METEO 300: *Fundamentals of Atmospheric Science* website (https://www.e-education.psu.edu/meteo300/node/606) describes the situation this way:

> … As fossil fuel emissions have increased over recent decades, so has the growth rate of atmospheric CO_2, as indicated by the concave-upward curvature in Fig. 1.1. The growth rate has approximately doubled from about 1 ppmv per year in the 1960s to about 2 ppmv per year in the 2000s (Fig. 1.2). According to the Global Carbon Project [https://www.globalcarbonproject.org/], 86% of the anthropogenic CO_2 emissions during 2009–2018 were from fossil fuel burning and 14% were from land-use change (e.g., deforestation).
>
> However, CO_2 injected to the atmosphere from human activity does not stay there. 44% of the

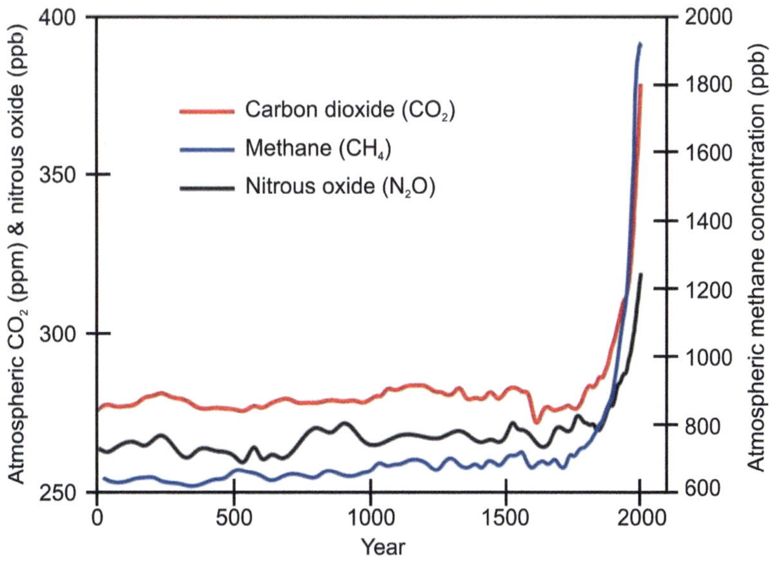

Fig. 1.1 Atmospheric concentrations of carbon dioxide, methane and nitrous oxide between the years zero AD and 2000 AD. Data derived from the IPCC Report AR4 (2007); Forster et al. 2007. Figure redrawn after a graphic from PennState College of Earth and Mineral Sciences, METEO 300 Fundamentals of Atmospheric Science by Brune (2020) (https://www.e-education.psu.edu/meteo300/node/606)

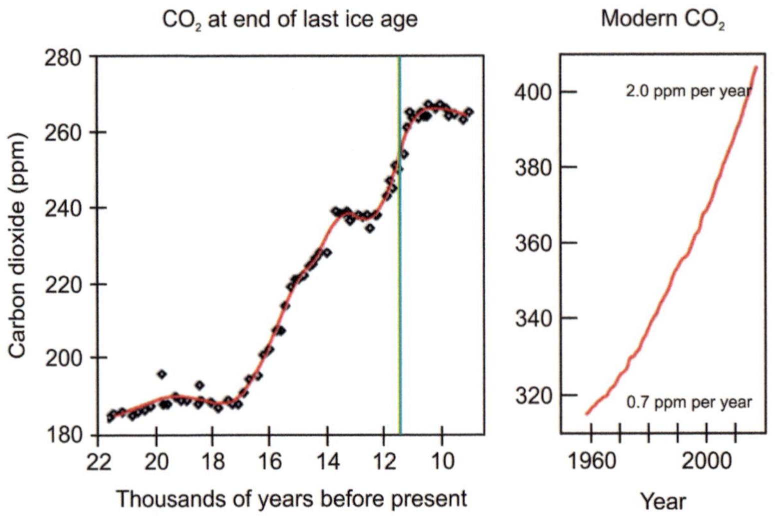

Fig. 1.2 Atmospheric content of CO_2 since the end of the last ice age. The figure on the left shows the CO_2 atmospheric concentration (in ppm) from the end of the last ice age to the present day. The figure on the right shows the atmospheric CO_2 content over the most recent 60 years. The vertical green line on the lefthand figure corresponds, as closely as can be achieved at this scale, to a period of 60 years similar to that depicted in the righthand figure for modern times. This serves to show that the tremendously rapid rise in atmospheric CO_2 concentration we are experiencing in our lifetimes is totally unprecedented in the last 22,000-year history of planet Earth. Redrawn after a figure in the World Meteorological Organization's *WMO Greenhouse Gas Bulletin,* issue No. 13 (2017) (https://ane4bf-datap1.s3-eu-west-1.amazonaws.com/wmocms/s3fs-public/ckeditor/files/GHG_Bulletin_13_EN_final_1_1.pdf)

> emissions from human activity during 2009–2018 accumulated in the atmosphere, 29% were absorbed by terrestrial ecosystems, 23% were absorbed by the ocean, and 4% is unaccounted for. Superimposed on the accelerating trend over the past few decades is an annual cycle in which CO_2 declines during Northern Hemisphere summer and rises during most of the rest of the year. This cycle reflects photosynthesis (an atmospheric CO_2 sink) and respiration (an atmospheric CO_2 source) of terrestrial ecosystems in the Northern Hemisphere, where most land is present. Note that the current increase to above 400 ppm now extends well above any other time in, at least, the past 800,000 years when CO_2 varied only between about 180 and 280 ppm by volume … (Brune 2020; https://www.e-education.psu.edu/meteo300/node/606).

No natural cause for these concentration increases has been found, instead these unnaturally rapid changes in the composition of the atmosphere over the past several decades primarily reflect changes in human activity. These include enhanced deforestation and agriculture, but the changing atmosphere is mainly caused by the burning of fossil fuels; the so-called 'fossil fuel emissions' resulting from using coal, oil and natural gas to release their energy content for our transport, industrial and domestic activities.

Although 'an atmospheric hypothesis' of the Earth's glacial periods possibly being due to the concentration of CO_2 in the atmosphere was framed by Chamberlin (1899), the first Swedish Nobel laureate, the physical chemist, Svante August Arrhenius, made the earliest quantification of the contribution of CO_2 to the greenhouse effect by deduction from observational data, and was also the first to speculate about whether variations in atmospheric concentration of CO_2 might contribute to long-term variations in climate (Arrhenius 1896). This notion had a fairly chequered history for a while, because the role of water vapour in absorption of infrared radiation in the lower atmosphere was given more prominence. Improvements in measurement of the absorption spectra of gases, though, enabled Callendar (1949) to restate the theory of the contribution of CO_2 to the greenhouse effect in these terms:

> … this theory depends on the fact that, whereas carbon dioxide is almost completely transparent to solar radiation, it is particularly opaque to the heat [infrared radiation] which is radiated back to space from the earth. In this way it (the CO_2) acts as a heat trap, allowing the temperature near the earth's surface to rise above the level it would attain if there were no carbon dioxide in the air … Callendar (1949).

That quotation states the fundamental essence of the meaning of 'greenhouse gas'.

Subsequently, Revelle and Suess (1957) stated the consequential impact of that greenhouse gas in very direct terms by describing a ***planetary-scale experiment*** in which mankind is:

> … returning to the atmosphere and oceans the concentrated organic carbon [previously] stored in sedimentary rocks over hundreds of millions of years … (Revelle and Suess 1957).

Revelle and Suess (1957) also demonstrated, by comparing $^{14}C/^{12}C$ and $^{13}C/^{12}C$ carbon isotope ratios in wood and in marine material that the average lifetime of a CO_2 molecule in the atmosphere before it dissolves into the sea is of the order of 10 years. It follows that most of the CO_2 released by fossil fuel combustion since the beginning of the industrial revolution must have been absorbed by the oceans. They concluded (in 1957) that "… the increase of atmospheric CO_2 from this cause is at present small but may become significant during future decades if industrial fuel combustion continues to rise exponentially …" Unfortunately, fossil fuel combustion has further intensified since then and the CO_2 concentration in the atmosphere has risen steadily, and it is still rising. Except for a one-year reduction in 2008/2009, every year of the twenty-first century has seen a year-on-year increase in anthropogenic CO_2 emissions (MacDowell et al. 2017). The latest data we can find are CO_2 measurements by the Scripps Institution of Oceanography and the National Oceanic and Atmospheric Administration (NOAA) which show that the amount of CO_2 in the air in May 2020 reached the monthly average value of **417 ppm** (source: https://www.washingtonpost.com/). This value is the highest atmospheric concentration observed in human history, and is probably the highest reached at any time in the last 3 million years.

McKinley et al. (2020) state that

> … The ocean has absorbed the equivalent of 39% of fossil carbon emissions since 1750, significantly modulating the growth of atmospheric CO_2 and the associated climate change …. If emissions continue to accelerate, this sink is expected to grow … (McKinley et al. 2020, and references therein).

These authors show that two processes external to the ocean are sufficient to explain major variability of the ocean carbon sink in recent decades. First, the global-scale reduction in the ocean carbon sink in the 1990s can be attributed to slowed growth rate of atmospheric CO_2 level, followed by recovery of the sink after 2001 due to acceleration of atmospheric CO_2 growth. Second, the timing of global sink variability in the 1990s is explained as a global response to the 1991 eruption of Mount Pinatubo in the Philippines, on 15 June 1991, which was the second-largest volcanic eruption of the twentieth century. They conclude that the most important control on the average magnitude of the ocean carbon sink is the variability in the growth rate of atmospheric CO_2 levels. This implies that if future fossil fuel emissions can be cut sufficiently to reduce growth of atmospheric CO_2, the ocean sink will act as a buffer, be reduced immediately "… and substantially mitigate atmospheric carbon accumulation for the next several centuries …" (McKinley et al. 2020).

Rapidly increasing atmospheric levels of CO_2 and other greenhouse gases are the atmospheric drivers of climate change because they can generate unpredictable changes in the climate system leading to severe ecological and economic disruptions. And so, our diagnosis is that human activities in the recent past have released into the atmosphere such quantities of greenhouse gases that were previously locked into fossilised rock strata that the resultant climate change will inevitably cause damaging disruption to future human activities. Table 1.2 lists selected URLs and hyperlinks to other reliable sources of information.

Table 1.2 URLs and hyperlinks to other reliable sources of information

The Carbon Cycle at NASA's Earth Observatory at this URL: http://www.earthobservatory.nasa.gov/Features/CarbonCycle/?src=eoa-features
Download the US DOE Report of 2008, *Carbon Cycling and Biosequestration*: *Report from the March 2008 Workshop*, DOE/SC-108, U.S. Department of Energy Office of Science; free download from https://doi.org/10.2172/948438
Earth System Research Laboratories' *Global Monitoring Laboratory* (U.S. Department of Commerce, National Oceanic & Atmospheric Administration) (https://www.esrl.noaa.gov/gmd/outreach/behind_the_scenes/gases.html)
RealClimate is a commentary site on climate science by working climate scientists for the interested public and journalists (http://www.realclimate.org/index.php/archives/2018/01/the-global-co2-rise-the-facts-exxon-and-the-favorite-denial-tricks/)
The Global Carbon Project (GCP) integrates knowledge of greenhouse gases for human activities and the Earth system (https://www.globalcarbonproject.org/)
World Meteorological Organization *WMO Greenhouse Gas Bulletin*. See issue No. 13 of 30 October 2017 (https://library.wmo.int/doc_num.php?explnum_id=4022)
Carbon Dioxide Measurements of the Scripps Institution of Oceanography CO_2 Program (https://www.scrippsco2.ucsd.edu/)
Greenpeace: *Nine ways humans have altered Earth's Holocene climate* https://www.greenpeace.org/international/story/22792/
PennState College of Earth and Mineral Sciences, METEO 300 *Fundamentals of Atmospheric Science*, by William H. Brune (2020) https://www.e-education.psu.edu/meteo300/node/606
BBC World Service **podcasts** entitled *The Climate Question* reports on why we find it so hard to save our own planet, and how we might change that. Regular presenters Neal Razzell in Canada and Graihagh Jackson in London are be joined each week by a range of BBC specialists around the world including Julie Yoon in Seoul; Janhavee Moole in India; Resty Woro Yuniar in Indonesia and science and political experts like the BBC's environment correspondents Navin Singh Kudka and Justin Rowlatt and by climate experts, politicians, campaigners and influencers. View all episodes at https://www.bbc.co.uk/programmes/w13xtvb6/episodes/downloads

All of the facts that lead to our diagnosis are well known and easy to understand. But our interpretation of those facts is often challenged by those wishing to play down the role of human activities in causing dangerous CO_2-increases.

1.3 The Denial of Anthropogenic CO_2-Driven Climate Change

The fact is that the carbon dioxide greenhouse gas that we blame for climate warming represents only 0.04% of the total gases in our atmosphere. And another fact that weighs heavily with those wishing to deny that human activities cause climate change is that the overwhelming majority of the CO_2 that is emitted, day by day, into the atmosphere comes from natural geological and biological sources, such as volcanoes or decomposition processes in nature or the aerobic respiration of all the living things on the planet. The anthropogenic contribution of CO_2 is (still) not much more than 5% of the atmosphere's total CO_2 burden; so, the anthropogenic CO_2 content in the air that we breathe is only 0.002%.

Written like this, these undeniable facts do seem to provide reason for those who deny the validity of the claims of the world's scientists that human activities are causing dangerous CO_2-increases, arguing instead that the human contribution to the emissions of CO_2 in the air we breathe is too small to cause the dramatic changes the scientists are warning us all about; it's all down to Nature's natural carbon cycle they say. This, though, is pure mischief. Because, written like this, there is another undeniable and crucial fact that this denial does not consider, which is that the anthropogenic release of previously fossilised carbon from coal, petroleum and natural gas is a net addition to the natural carbon cycling of the present-day global atmosphere. To explain what we mean, we must examine the normal scheme of things by finding out about the global carbon cycle.

1.4 The Global Carbon Cycle

The chemistry of carbon is the chemistry of life on Earth. Carbon compounds make up the bodies of all the Earth's living organisms, provide the nutrients and energy that sustains them, and deliver the energy that fuels our global economy. And the carbon compounds that are emitted into the atmosphere regulate the temperature of the Earth through their activity as greenhouse gasses.

Most of the carbon on Earth is stored in rocks and sediments; with the rest being located in the ocean, the atmosphere and in all those living organisms. These are the reservoirs through which carbon atoms are continually recycled. Living organisms have a high turnover of carbon, but do not make any net addition of CO_2 to the atmosphere. Non-photosynthetic organisms use the carbon compounds of their food to make their own biomass and although the digestion of food releases CO_2 back to the atmosphere, through respiration, the growth of their biomass in life represents a net removal of carbon from the atmosphere, but when they die, the decomposition of their bodies releases all their carbon back to the atmosphere. This is the regular biological cycle: remove carbon from the atmosphere to build live biomass and then return that carbon to the atmosphere after death. It's a cycle that applies equally to photosynthetic organisms from the most archaic of photosynthetic bacteria to the most stately of forest trees. Of course, the difference is that photosynthesis enables these organisms to make a much greater net removal of carbon from the atmosphere as they turn CO_2 into nutrient sugars. But this remains true for photosynthetic organisms only as long as the sun shines and they remain alive. At night these organisms also respire, thus returning carbon to the atmosphere, and when they die their biomass also rots, returning all their carbon to the atmosphere.

The Earth's global carbon cycle was almost exactly in equilibrium before industrial humans intervened; which is evident from the constancy of the CO_2 concentration in the air for several thousand years while humans were evolving (Fig. 1.2). There are various reservoirs or sinks, some of which have short lifetimes (like the three score years and ten of humans), others have long lifetimes (like the hundred million-year-old geological limestone strata, or the equally old coal measures and deep reserves of petroleum and natural gas). Carbon flows between the reservoirs, shifting carbon out of one reservoir by putting more carbon into another reservoir. This is the exchange that is called **The Global Carbon Cycle** (Fig. 1.3).

In the long term, the carbon cycle maintains a natural balance that avoids all of Earth's carbon being dumped into the atmosphere or being stockpiled entirely in rocks. Because CO_2 is a greenhouse gas, which does not allow escape of re-radiated infrared, this balance acts like a thermostat, helping to keep Earth's temperature relatively stable over long periods of time. Any changes that put greenhouse gases into the atmosphere (Fig. 1.1) result in warmer temperatures on Earth. This thermostat works over a timescale of at least a few hundred thousand years, so it's a slow part of the overall carbon cycle. But over shorter time periods, say ten thousand to a hundred thousand years, the CO_2 content of the atmosphere, and consequently the temperature of Earth, can quite naturally vary (Fig. 1.2), and this is thought to be a contributory cause for the Earth shifting between ice ages and warmer interglacial periods over these timescales. Parts of the carbon cycle may even vary over shorter timescales.

For example, seasonal variation in the CO_2 concentration of the atmosphere is consistently measured by stations of the global CO_2 measurement network, such as the Mauna Loa Observatory of the Scripps Institution of Oceanography, in Hawaii (view the current year's data at https://www.esrl.noaa.gov/gmd/ccgg/trends/). Seasonal variation is mainly due to seasonal changes over the year in the forests of the land masses of the northern hemisphere;

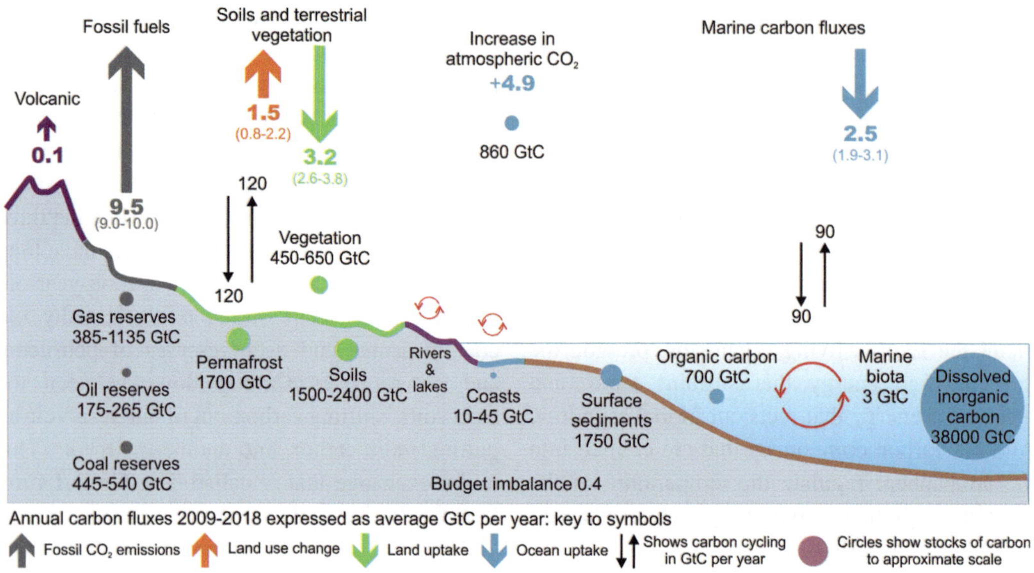

Fig. 1.3 The global carbon cycle. Schematic representation of the overall global carbon cycle emphasising those caused by anthropogenic activities. Data cover the decade 2009–2018. The key to symbols below the graphic shows the meaning of the arrows and units; large bold numerals indicate the mean annual total of carbon emitted or stocked in GtC yr^{-1}, with the statistical range of the estimates (±one standard deviation) shown below. Uncertainty in the atmospheric CO_2 growth rate is very small (±0.02 GtC yr^{-1}) and is neglected for the figure. An overall budget imbalance of 0.4 GtC yr^{-1} is due to overestimated emissions and/or underestimated sinks. The anthropogenic perturbations ***are additional to*** the Earth's ***natural active carbon cycle***; with fluxes (vertical bidirectional arrows) and stocks (annotated circles) shown across the figure. Redrawn after a figure in Friedlingstein et al. (2019)

spring and summer drawdown of CO_2 for plant growth, followed by emission of CO_2 from autumn and winter decay and digestion of shed flowers, fruit, leaves and branches.

Detailed quantifications of carbon fluxes and reservoirs, such as those shown in Fig. 1.3, are the starting points for the myths of the climate change deniers and global warming sceptics. The myths that deny the facts that human activities are causing climate change are not just an argumentative mischief because when those sceptics are in government and responsible for environmental regulations that scale back or eliminate climate mitigation measures, our climate disaster which is on the horizon can be brought even closer (view the *Climate Deregulation Tracker* of the Sabin Center for Climate Change Law at Columbia Law School, New York, at this URL: https://climate.law.columbia.edu/content/climate-reregulation-tracker).

The first, and major, myth is based on the ***true*** observations that although the great majority of the CO_2 emitted every day into the atmosphere is the result of natural phenomena, specifically, respiration of live organisms and decomposition of dead ones; only a few per cent of the total result from human activities like burning fossil fuels, making cement from fossilised limestone and forest clearing and forest burning for agricultural expansion. The myth is that this few per cent of anthropogenic CO_2 emissions must therefore be irrelevant. This is the big, spurious, totally missing the point sceptic myth. The point that is being missed is that ***human activities like burning fossil fuels and making cement from fossilised limestone are making a net addition of CO_2 to the present-day atmosphere*** by releasing today carbon that was removed from the atmosphere long, long ago. The majority emitters, respiration and decomposition, are merely **recycling** atmospheric CO_2. By which we mean that the food that you respire today (releasing CO_2 in the process) was made by the organisms that became your food using CO_2 drawn down from the atmosphere earlier the same year; so, you are recycling it back to the atmosphere, you are not making a net addition to the atmosphere. Similarly, decomposition of the biomass of an organism that dies today will return to the atmosphere (as CO_2) the carbon of which it was made when alive using CO_2 drawn down from the atmosphere in its recent past. Again, there is no net addition to the atmosphere.

Another climate sceptic myth, is that the recent increase in concentration of CO_2 in the atmosphere is derived from volcanic emissions. This cannot be so because the total volcanic emissions can be measured to be about 0.1 Gt of carbon per year, compared to the anthropogenic emissions from fossil fuel burning alone of 9.5 Gt of carbon per year (Fig. 1.3). Total anthropogenic emissions (which include our damage to forest ecosystems) are now more than a hundred times greater than those from volcanoes. The volcanic emissions are important for long-term changes of atmospheric CO_2 levels over timescales of millions of years, but not over a few decades as we are experiencing (Fig. 1.2).

There is ***yet another denier myth***, that the oceans are the cause of the atmospheric CO_2 increase. This also ignores the rapid timescale of the rise in atmospheric CO_2 levels we are experiencing now because it depends on the variation in CO_2 levels during the Earth's glacial cycles. It is certainly true that during ice ages greater concentrations of CO_2 are dissolved into the oceans and there is correspondingly less in the atmosphere. It is also true that as the ice retreats and the world warms at the end of the glacial cycle that the CO_2 is returned to the atmosphere from the oceans. But this is a cycle that takes place over timescales of many thousands or millions of years; it is a fallacy to claim that the same natural phenomenon is happening today. Indeed, direct measurements completely dispose of this misconception. The upper ocean has been mapped and documented in detail by countless ship surveys that have demonstrated that ***today's oceans absorb CO_2 and do not release any***. The increase in CO_2 concentration in the upper ocean is itself a serious environmental problem because CO_2 dissolved in water forms carbonic acid. Consequently, rising CO_2 concentrations lead to acidification of the oceans, which has significant, and mostly adverse, ecological effects.

It's almost not even worth discussing the ***final climate sceptics myth***, which blames the world's forests for most of the increase in atmospheric CO_2—it is too foolish to contemplate. But, for the sake of completion, this particular fallacy puts the blame on the world's forests because of their undeniable emission of CO_2 by the regular decay of their shed foliage and dead wood. By looking at this emission in isolation, these climate sceptics ignore the fact that the CO_2 emitted during the decay of leaves and dead wood is merely returning to the atmosphere the CO_2 that was removed from it to make those leaves and that wood in the first place. This natural activity of the forest (and other vegetation) is one of the carbon cycles that contribute to the Global Carbon Cycle (Fig. 1.3). To break that cycle and force the forests to really make a contribution to our accumulating atmospheric CO_2, you would have to clear-cut the trees and burn them, replacing the long-lived, carbon sequestering, forest trees with transient pasture grasses or oil-producing monocultures. Now, who would be misguided enough to do that?

Ruling out the denial myths this way, we are left with the uncomfortable conclusion (already stated above) that the rise we are measuring in the atmospheric concentration of the greenhouse gas CO_2 has just one cause, which is our profligate use of ***fossil*** fuels. We are driving, flying, heating and cooking on gas towards our own extinction.

1.5 The Likely Effects of Climate Change

We are already experiencing the climatic effects of the increase in CO_2 concentration in the atmosphere, but the potential future effects of global climate change can be calculated from our understanding of the physical processes, and/or estimated from knowledge of the Earth's climate history. Both come to the conclusion that the average global warming due to the increase in CO_2 to date, is expected to be about +1 °C. This corresponds exactly to the measured observations of global warming (Fig. 1.4).

As we have shown above, there is no natural explanation for this, meaning that the best estimate for the anthropogenic share of global warming since 1950 is 100%.

This climate change has already had noticeable effects on our environment. Glaciers have dwindled, some have disappeared, winter ice on rivers, lakes and in polar waters is breaking up earlier, and continued melting of polar ice will only accelerate sea level rise, a gloomy prospect for coastal communities. And we mean coastal communities like Tokyo, New York, Shanghai, Kolkata, Dhaka, Osaka, Mumbai, Bangkok, Guangzhou, Shenzen and Miami; all of which appear among the Top 20 cities expected to be exposed to climate-change-induced coastal flooding by the 2070s (OECD 2010; Nicholls et al. 2011; and view this 2019 UN News report at https://news.un.org/en/story/2019/09/1047392).

The previous paragraph suggests a bleak future caused by climate change, but ecologists around the world are ***already*** recording lengthening of summer seasons and drastic changes in the distribution ranges of fungi, plants and animals, including widening host ranges of disease and pest organisms. We are all aware of an

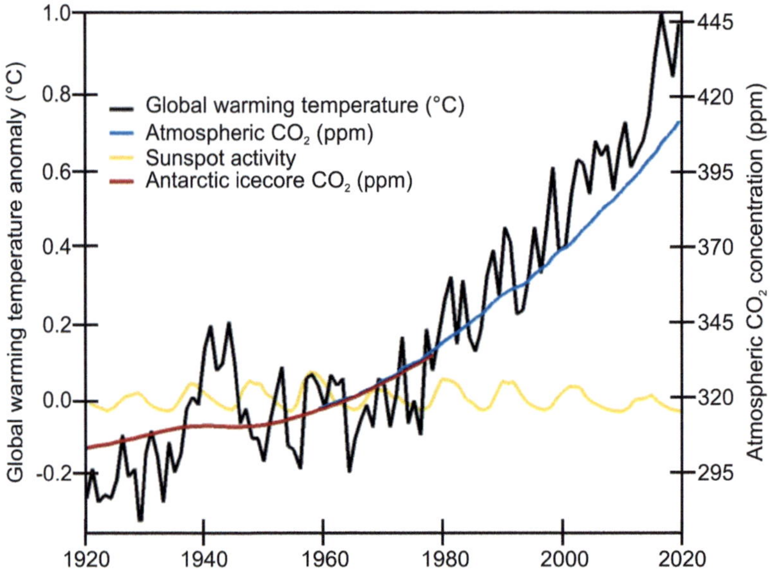

Fig. 1.4 Evolution of global temperature (black), atmospheric CO_2 concentration (blue), CO_2 concentration in air trapped in Antarctic ice cores (magenta) and solar activity (yellow) over the 100 years from 1920 to 2020. Temperature and CO_2 are scaled relative to each other as the physically expected CO_2 effect on the climate predicts (that is, the best estimate of climate sensitivity). The sunspot activity curve shows average number of sun spots per year; its amplitude is scaled from the observed correlation of solar and temperature data. Data taken from the website *RealClimate: Climate Science From Climate Scientists* (http://www.realclimate.org/). This graphic was produced using the climate widget at this URL: http://www.herdsoft.com/climate/widget/. 1920 was chosen as the start date as it is the start of the dominance of the internal combustion engine in transport on land, sea and air. At the start of the First World War, horse-drawn transport dominated, but by the end of that war motorised transport dominated. You can create a version of this graph for yourself, covering years of your own choice with the widget at http://herdsoft.com/climate/widget/

increase in the number, duration and intensity of extreme weather events caused by the greater amounts of energy that are now being trapped in the atmosphere, and the great majority of the world's scientists agree on the hazards that will come if atmospheric CO_2 levels are allowed to rise even more (Randers 2012).

For example, the Intergovernmental Panel on Climate Change (**IPCC**) includes over 1,300 scientists from around the world and forecast in their reports that global temperatures will continue to rise for decades to come, due to the greenhouse gases produced by human activities (IPCC 2007, 2013; Forster et al. 2007; Stocker et al. 2013 (all available free online)). According to the IPCC, the extent of climate change effects on individual regions will vary between regions, and over time, and with the ability of different community and environmental structures to adapt to, or even mitigate the changes.

The IPCC reports further predict that increases in global mean temperature of 1–3 °C above 1990 levels will produce beneficial impacts in some regions and harmful ones in others. Net annual costs will increase over time as global temperatures increase. The IPCC states that, "... Taken as a whole, the range of published evidence indicates that the net damage costs of climate change are likely to be significant and to increase over time ...".

And a lot of evidence ***has*** been published in the last decade or so, which we cannot review here, so, rather than repeat other summaries we will refer to just two more (Melillo et al. 2014; Wuebbles et al. 2017 [both available free online]), which together amount to over 1,000 pages of well documented projections. These are the Third and Fourth Reports of the *US National Climate Assessment*, which summarise the impacts of climate change on the United States, now and in the future. These reports were produced by a team of more than 300 experts guided by a 60-member ***Federal Advisory Committee***, and were extensively reviewed by the public and independent experts, including federal agencies and a panel from the ***US National Academy of Sciences*** (but if you would rather get your information from videos or podcasts, take a look at those listed in Table 1.3).

Restricting myself to just the headline statements in these *National Climate Assessment* reports, some of the long-term effects of global climate change in the United States are projected to be as follows:

- The global climate is projected to continue to change over this century and beyond. The magnitude of climate change beyond the next few decades depends primarily on the amount of heat-trapping gases emitted globally, and how sensitive the Earth's climate is to those emissions.
- Temperatures will continue to rise but this "... will not be uniform or smooth across the country or over time ..."
- Frost-free seasons (and growing seasons) will lengthen; these have been "... increasing nationally since the 1980s, with the largest increases occurring in the western United States, affecting ecosystems and agriculture. Across the United States, the growing season is projected to continue to lengthen..." by a month or more, if heat-trapping gas emissions continue to increase.
- "... Average US precipitation has increased since 1900, but some areas have had increases greater than the national average, and some areas have had decreases. More winter and spring precipitation is projected for the northern United States, and less for the Southwest, over this century ..."
- "... Droughts in the [US] Southwest and heat waves (periods of abnormally hot weather lasting days to weeks) everywhere [in the US] are projected to become more intense, and cold waves less intense everywhere ... Summer temperatures are projected to continue rising, and a reduction of soil moisture, which exacerbates heat waves, is projected for much of the western and central US in summer. By the end of this century, what have been once-in-20-year extreme heat days (one-day events) are projected to occur every two or three years over most of the nation ..."

Table 1.3 Some ***YouTube videos*** that describe the climate and climate change

A Brief History of CO_2 Emissions, a video illustrating the history of CO_2 emissions by the Potsdam Institute for Climate Impact Research (PIK) and the Urban Complexity Lab: https://www.youtube.com/results?search_query=EQ7S0D1iucY
What is Climate Change?—Start Here. The hard facts about global warming from Al Jazeera English (https://www.youtube.com/watch?v=dcBXmj1nMTQ)
Climate Change 101 with National Geographic's Bill Nye, explains what causes climate change, how it affects our planet, why we need to act promptly to mitigate its effects, and how each of us can contribute to a solution (https://www.youtube.com/watch?v=EtW2rrLHs08)
A new high-resolution computer model created by NASA shows CO_2, the greenhouse gas driving global warming, in 2014, 'the warmest year ever recorded' (https://www.youtube.com/watch?v=fJ0o2E4d8Ts&t=7s)
Carbon Brief is a UK-based website covering the latest developments in climate science, climate policy and energy policy. In this video, Dr Glen Peters explains why global CO_2 emissions rose in 2019 (https://www.youtube.com/watch?v=_hE-gGauVDg)
Carbon dioxide emissions inventory for commercial aviation. Video of highlights from a September 2019 paper that details calendar year 2018, presented by one of the paper's co-authors, Brandon Graver (https://www.youtube.com/watch?v=oAkvaDwjsc0&t=95s)
UN Secretary-General António Guterres warns of the threat posed by climate change, in a major address in 2018 (https://www.youtube.com/watch?v=VNe-jBVij-g&t=3s)
Word artist Prince Ea makes a powerful case for protecting the planet, and challenges the human race to create a sustainable future in this short film in the National Geographic Short Film Showcase. Winner of the Film4Climate competition organised by the Connect4Climate Program of the World Bank (https://www.youtube.com/watch?v=B-nEYsyRlYo)
Climate science explained in 60 s by the Royal Society of London and the US National Academy of Sciences (https://www.youtube.com/watch?v=n4e5UPu1co0&t=33s)
How does the climate system work? An animation to explain how the climate system works by the UK Met Office (https://www.youtube.com/watch?v=lrPS2HiYVp8)
The jet stream and how it affects the major climate patterns of the world. The effects of climate change on climate patterns and how the jet stream plays a major role in those changes by Oregon State University (https://www.youtube.com/watch?v=ifkc_NNufT4)
Jakarta: A warning? (BBC World Service podcast broadcast 11 January 2021). Sea level rise caused by climate change means that Jakarta is on fast forward when it comes to flooding. Will the city cope? What can we learn from it? (TheClimateQuestion-20210110-JakartaAWarning.mp3) download from https://www.bbc.co.uk/programmes/w3ct0xb7
A year to save the world. (BBC World Service podcast broadcast 14 January 2021). Is 2021 the make-or-break year for our environment? (TheClimateQuestion-20210103-AYearToSaveTheWorld.mp3) download from https://www.bbc.co.uk/programmes/w3ct0xb6
A degree away from carnage. (BBC World Service podcast broadcast 30 November 2020). What do rising temperatures mean for our world? (TheClimateQuestion-20201129-ADegreeAwayFromCarnage.mp3) download from https://www.bbc.co.uk/programmes/w3ct0xb1
The Science Show is broadcast by the Australian Broadcasting Corporation's Radio National. It has featured several aspects of climate change and its impacts from bush fires to ocean fish migrations. View the range of available **podcasts** at: https://www.abc.net.au/radionational/programs/scienceshow/

- Hurricanes will become stronger and more intense. "… The intensity, frequency and duration of North Atlantic hurricanes, as well as the frequency of the strongest (Categories 4 and 5) hurricanes, have all increased since the early 1980s … Hurricane-associated storm intensity and rainfall rates are projected to increase as the climate continues to warm …"
- Sea level will rise by 1–8 feet by the end of the twenty-first century. "Global sea level has risen by about 8 inches since reliable record keeping began in 1880. It is projected to rise

another 1–8 ft by 2100. This is the result of added water from melting land ice and the expansion of seawater as it warms … In the next several decades, storm surges and high tides could combine with sea level rise and land subsidence to further increase flooding in many regions. Sea level rise will continue past 2100 because the oceans take a very long time to respond to warmer conditions at the Earth's surface. Ocean waters will therefore continue to warm and sea level will continue to rise for many centuries at rates equal to or higher than those of the current century …"

- "… The Arctic Ocean is expected to become essentially ice free in summer before mid-[21st]-century …"

The Third (Melillo et al. 2014) and Fourth (Wuebbles et al. 2017) *National Climate Assessment Reports* predict the following ***regional*** effects on the United States:

- "… **Northeast**. Heat waves, heavy downpours and sea level rise pose growing challenges to many aspects of life in the Northeast. Infrastructure, agriculture, fisheries and ecosystems will be increasingly compromised. Many states and cities are beginning to incorporate climate change into their planning …"
- "… **Northwest**. Changes in the timing of streamflow reduce water supplies for competing demands. Sea level rise, erosion, inundation, risks to infrastructure and increasing ocean acidity pose major threats. Increasing wildfire, insect outbreaks and tree diseases are causing widespread tree die-off …"
- "… **Southeast**. Sea level rise poses widespread and continuing threats to the region's economy and environment. Extreme heat will affect health, energy, agriculture and more. Decreased water availability will have economic and environmental impacts"
- "… **Midwest**. Extreme heat, heavy downpours and flooding will affect infrastructure, health, agriculture, forestry, transportation, air and water quality, and more. Climate change will also exacerbate a range of risks to the Great Lakes …"
- In the **Southwest**, increased heat and drought, linked to climate change, have already increased wildfire occurrences, while declining water supplies and insect outbreaks have reduced agricultural yields, and "… health impacts in cities due to heat, and flooding and erosion in coastal areas are additional concerns …".

A major concern about climate change is that tiny perturbations in critical thresholds may cause irreversible changes in the climate system that could dramatically alter the Earth's planetary environment as we know it (McCarthy et al. 2001). The United Nations Framework Convention on Climate Change (UN 1992), in Article 2, obligates signatory nations to stabilise greenhouse gas (GHG) concentrations in the atmosphere at a level that "… would prevent ***dangerous anthropogenic interference*** (DAI) with the climate system…" (Mann 2009). McCarthy et al. (2001) identified a number of ***reasons for concern*** (RFCs) (and see Smith et al. 2009). These are points-of-no-return, which, once exceeded, plunge the world into new dynamics. They have been defined over recent years as ***tipping points*** (Lenton et al. 2008; IPCC 2014) Among the tipping points that are most discussed are (Fig. 1.5; Russill and Nyssa 2009; Lenton et al. 2019; Randers and Goluke 2020):

- The Arctic sea ice melts.
- Greenland becomes ice-free.
- The West Antarctic ice sheet disintegrates.
- Siberian permafrost thaws.
- The Amazon rain forest dies back due to drought and fires.
- Boreal forests suffer damaging fires and new pests and diseases.

The greatest fear is that these tipping points, singly or in combination, could cause runaway climate change, contributing to mass extinction of species (not excluding humans), dramatic sea level rise, extensive droughts and the transformation of forests into vast grasslands. The evolution of life on Earth has been interrupted by several mass extinction events in the past 500

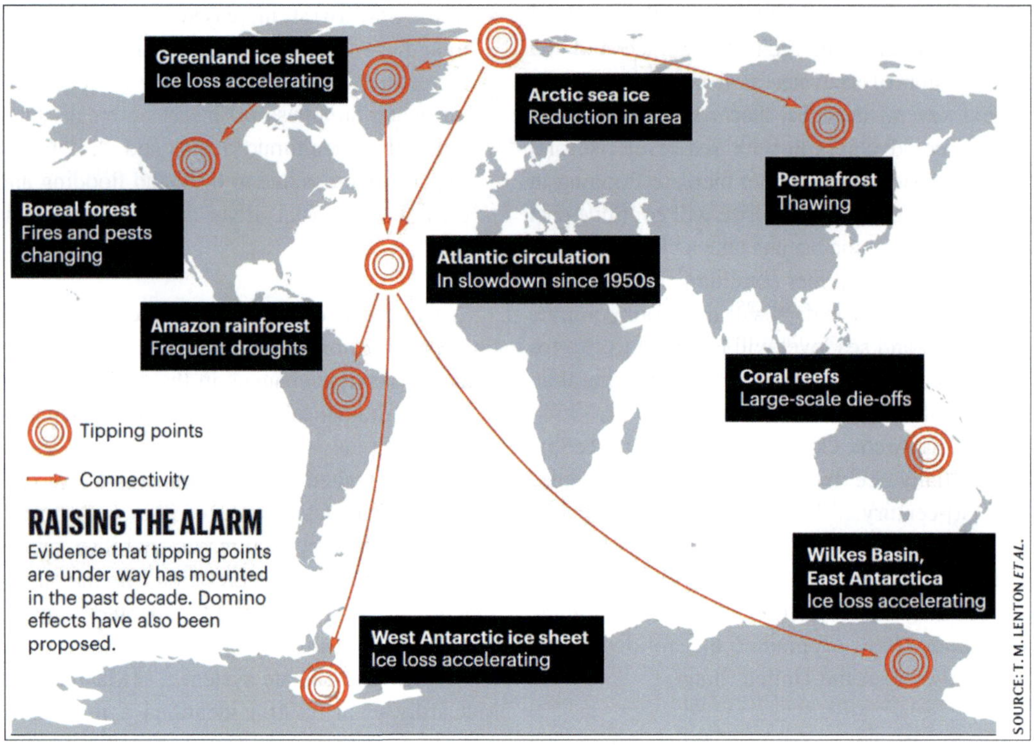

Fig. 1.5 Raising the alarm. Potential tipping points in the climate system (from Lenton et al. 2008, 2019)

million years of Earth's geological history. It is widely accepted that these catastrophic ecological crises were triggered by the effects of severe changes in atmospheric chemistry resulting in greenhouse gas-generated global warming (Knoll et al. 1996; Mayhew et al. 2008; Gehler et al. 2016). At least some of the mass extinctions were associated with super-volcanic eruptions. In the present day, anthropogenic CO_2 emissions are more than 100 times greater than all of Earth's volcanoes combined. Human activity is today's super-volcano, emitting greenhouse gases at a greater rate than the mega-eruptions of our geological past; those that extinguished so many living organisms of their time.

Lenton et al. (2019) state that (the emphasis is ours):

> … In our view, the evidence from tipping points alone suggests that we are in a state of ***planetary emergency***: both the risk and urgency of the situation are acute … We argue that the intervention time left to prevent tipping could already have shrunk towards zero, whereas the reaction time to achieve net zero emissions is 30 years at best. Hence we might already have lost control of whether tipping happens. A saving grace is that the rate at which damage accumulates from tipping - and hence the risk posed - could still be under our control to some extent. ***The stability and resilience of our planet is in peril***. International ***action*** - not just words - must reflect this.

The above are projections for the future, but in 2021 the Meteorological Office in the UK reported that we are ***already*** suffering the disruptive effects of climate change even as this book goes to press. The truth of this claim is substantiated by the catalogue of extreme weather events already experienced across the planet during 2021, which include: scorching heatwaves, often followed by massive wildfires in Canada, the United States, Italy, Turkey, Greece and Russia, while unusually heavy rains triggering catastrophic flash flooding have hit western Europe, western India, Korea and China (view: https://www.theweek.co.uk/news/environment/953574/worlds-most-extreme-weather-events-2021). And the projections keep getting worse. Tellman et al. (2021)

used satellite imaging to reveal the proportion of the world's population exposed to floods, demonstrating that from 2000 to 2015 this proportion increased by 20–24% (ten times higher than previous estimates). Their view is that "… Climate change projections for 2030 indicate that the proportion of the population exposed to floods will increase further". Flooding impairs land use for agriculture but the Oxfam report, *Tightening the Net: Net zero climate targets—implications for land and food equity* (Sen and Dabi 2021) independently argues that because of "… an explosion in demand for land…" large-scale afforestation could increase food prices by about 80% by 2050 (see also the IPCC *Special Report Climate Change and Land* IPCC 2019). Sen and Dabi (2021) go on to state that net zero climate targets "… could end up being a dangerous distraction that could delay the rapid reductions in emissions that high-emitting countries and companies need to make if we are to avoid catastrophic climate breakdown".

We have to hope that the approval by 195 member ***governments*** of the IPCC to the Working Group I contribution to the Sixth Assessment Report '*AR6 Climate Change 2021: The Physical Science Basis*', which was released on 9 August 2021 (see https://www.ipcc.ch/report/ar6/wg1/; IPCC 2021) will lead to the rapid reductions around the globe in greenhouse gas emissions that are now both urgent and essential. This 4000-page report was published just prior to, and intended to inform, the United Nation's *Conference of the Parties* (COP26) climate change conference held in Glasgow, UK, in November 2021. Keynote findings and projections in the *Summary for Policymakers* (SPM) of this report are essential reading for all concerned about climate change (which surely must encompass all of humanity). We quote these keynote findings here:

> …This SPM provides a high-level summary of the understanding of the current state of the climate, including how it is changing and the role of human influence, the state of knowledge about possible climate futures, climate information relevant to regions and sectors, and limiting human-induced climate change…:

A. **The Current State of the Climate**

A.1 It is unequivocal that human influence has warmed the atmosphere, ocean and land. Widespread and rapid changes in the atmosphere, ocean, cryosphere and biosphere have occurred.

A.2 The scale of recent changes across the climate system as a whole and the present state of many aspects of the climate system are unprecedented over many centuries to many thousands of years.

A.3 Human-induced climate change is already affecting many weather and climate extremes in every region across the globe. Evidence of observed changes in extremes such as heatwaves, heavy precipitation, droughts, and tropical cyclones, and, in particular, their attribution to human influence, has strengthened since AR5.

A.4 Improved knowledge of climate processes, paleoclimate evidence and the response of the climate system to increasing radiative forcing gives a best estimate of equilibrium climate sensitivity of 3 °C with a narrower range compared to AR5.

B. **Possible Climate Futures**

B.1 Global surface temperature will continue to increase until at least the mid-century under all emissions scenarios considered. Global warming of 1.5 °C and 2 °C will be exceeded during the 21st century unless deep reductions in CO_2 and other greenhouse gas emissions occur in the coming decades.

B.2 Many changes in the climate system become larger in direct relation to increasing global warming. They include increases in the frequency and intensity of hot extremes, marine heatwaves, and heavy precipitation, agricultural and ecological droughts in some regions, and proportion of intense tropical cyclones, as well as reductions in Arctic sea ice, snow cover and permafrost.

B.3 Continued global warming is projected to further intensify the global water cycle, including its variability, global monsoon precipitation and the severity of wet and dry events.

B.4 Under scenarios with increasing CO_2 emissions, the ocean and land carbon sinks are projected to be less effective at slowing the accumulation of CO_2 in the atmosphere.

C. **Climate Information for Risk Assessment and Regional Adaptation**

C.1 Natural drivers and internal variability will modulate human-caused changes, especially at regional scales and in the near term, with little effect on centennial global warming. These modulations are important to consider in planning for the full range of possible changes.

C.2 With further global warming, every region is projected to increasingly experience concurrent and multiple changes in climatic impact-drivers. Changes in several climatic impact-drivers would be more widespread at 2 °C compared to 1.5 °C global warming and even more widespread and/or pronounced for higher warming levels.

C.3 Low-likelihood outcomes, such as ice sheet collapse, abrupt ocean circulation changes, some compound extreme events and warming substantially larger than the assessed very likely range of future warming cannot be ruled out and are part of risk assessment.

D. **Limiting Future Climate Change**

D.1 From a physical science perspective, limiting human-induced global warming to a specific level requires limiting cumulative CO_2 emissions, reaching at least net zero CO_2 emissions, along with strong reductions in other greenhouse gas emissions. Strong, rapid and sustained reductions in CH_4 emissions would also limit the warming effect resulting from declining aerosol pollution and would improve air quality.

D.2 Scenarios with very low or low GHG emissions (SSP1-1.9 and SSP1-2.6) [SSP = ***Shared Socio-economic Pathway***; a concept adapted from Lund et al. (2020)] lead within years to discernible effects on greenhouse gas and aerosol concentrations, and air quality, relative to high and very high GHG emissions scenarios (SSP3-7.0 or SSP5-8.5). Under these contrasting scenarios, discernible differences in trends of global surface temperature would begin to emerge from natural variability within around 20 years, and over longer time periods for many other climatic impact-drivers (high confidence)" (IPCC 2021). The above ***keynote findings and projections*** are quoted from the *Summary for Policymakers* of AR6 (SPM) which can be downloaded from https://www.ipcc.ch/report/ar6/wg1/downloads/report/IPCC_AR6_WGI_Headline_Statements.pdf.

BBC News reported on 10 November 2021 (https://www.bbc.co.uk/news/science-environment-59220687) that "...Despite pledges made at the climate summit COP26, the world is still nowhere near its goals on limiting global temperature rise..." The report was based on a new analysis carried out by the ***Climate Action Tracker*** (CAT) website https://www.climateactiontracker.org/, which makes independent scientific assessments of government climate actions, measuring them against the globally agreed Paris Agreement reached at COP 21 in Paris, in 2015 of "...holding warming well below 2 °C, and pursuing efforts to limit warming to 1.5 °C". Despite these pledges made in 2015, and the level of global warming in 2021 being quoted as 1.1 °C, the world is still nowhere near its goals on limiting global temperature rise. The CAT analysis (in 2021) calculated that the world is heading for 2.4 °C of warming, a far higher temperature than was internationally decided in the Paris Agreement. The BBC report commented that 2021's COP26 "…has a massive credibility, action and commitment gap…" to close. Nevertheless, the UN Secretary-General's statement on the conclusion of the UN Climate Change Conference COP26 (https://unfccc.int/news/secretary-general-s-statement-on-the-conclusion-of-the-un-climate-change-conference-cop26#:~:text=A%20climate%20of%20mistrust%20is,No%20more%20IOUs), while recognising that "…The approved texts are a compromise…" went on to identify several areas of progress, including commitments:

- to end deforestation (over 100 countries approved the ***Glasgow Leaders Declaration on Forests and Land Use***, which aims to halt deforestation by 2030) (https://www.ukcop26.org/glasgow-leaders-declaration-on-forests-and-land-use/).
- To drastically reduce methane emissions (over 100 countries agreed to a 30% reduction in methane emissions by 2030 under the ***Global Methane Pledge*** (https://www.ukcop26.org/world-leaders-kick-start-accelerated-climate-action-at-cop26/), an initiative launched by the United States and European Union. Methane is over 80 times more powerful than CO_2 at warming the climate, but only remains in the lower atmosphere for about 12 years because it is oxidised to CO_2 and H_2O in the high atmosphere. Reducing the amount of methane that human activities add to the atmosphere can buy time for us to deal with CO_2.
- To mobilise private finance around net zero. The ***Glasgow Financial Alliance for Net Zero (GFANZ)***, commits over $130 trillion of private capital to transforming the economy for net zero (https://assets.bbhub.io/company/sites/63/2021/11/GFANZ-Progress-Report.pdf). These commitments, from over 450 firms across 45 countries, can deliver the estimated $100 trillion of finance needed for net zero over the next three decades.
- More than 40 countries signed an agreement at COP26 to ***phase out*** coal in electricity generation during the 2030s and 2040s, while public finance institutions committed to ending international public support for coal in the energy sector by the end of 2022. Signatories include some of the world's biggest coal burners (Canada, Poland, Vietnam, South Korea, Ukraine and Indonesia) (https://www.ukcop26.org/end-of-coal-in-sight-at-cop26/). Unfortunately, before the final ***Glasgow Climate Pact*** could be agreed by all delegations (https://www.unfccc.int/sites/default/files/resource/cop26_auv_2f_cover_decision.pdf), India and China with other coal-dependent developing nations, rejected the ***phase out*** clause, and this was amended to ask countries to ***phase down*** their coal use.
- Reaffirm resolve towards the 1.5 degree goal. The unanimously agreed ***Glasgow Climate Pact*** kept the goal of limiting global warming to 1.5 °C alive by finalising, with agreement on carbon markets and transparency, the outstanding elements of the Paris Agreement (https://www.ukcop26.org/cop26-keeps-1-5c-alive-and-finalises-paris-agreement/).
- Boost climate ***finance*** to help countries adapt to climate change (called 'adaptation' in UN-speak) and strengthen support for vulnerable countries suffering from irreparable climate damage (https://www.unep.org/news-and-stories/story/what-does-cop26-mean-adaptation).
- For the first time encourage International Financial Institutions to consider climate vulnerabilities in concessional financial and other forms of support, including Special Drawing Rights. US$413 million was pledged for the most vulnerable countries at COP26 [https://unfccc.int/news/us-413-million-pledged-for-most-vulnerable-countries-at-cop26].

Returning to the words of António Guterres, the Secretary-General of the United Nations, his closing statement described these actions as "…welcome steps, but they are not enough…" and went on to say "Adaptation [to climate change] isn't a technocratic issue, it is life or death…We have another climate crisis today. A climate of mistrust is enveloping our globe. Climate action can help rebuild trust and restore credibility. That means finally delivering on the $100 billion climate finance commitment to developing countries. No more IOUs. It means measuring progress, updating climate plans every year and raising ambition. I will convene a global stocktaking summit at the heads of state level in 2023 … We are in the fight of our lives. Never give up. Never retreat. Keep pushing forward. I will be with you all the way. COP 27 starts now".

The 27th session of the *Conference of the Parties* (COP 27) is currently scheduled for November 2022 in Sharm al-Sheikh, Egypt. We hope that promotion of calcifier cultivation finds a place on its agenda. We know it makes sense.

None of the above makes particularly comfortable reading (especially so for my children and grandchildren) because it makes the point very starkly that nobody escapes, everybody suffers, and time is so short we must do something about it, ***NOW***. The strapline for the *Comment* article in the journal *Nature* by Girardin et al. (2021) is: "*Analysis suggests that to limit global temperature rise, we must slash emissions and invest now to protect, manage and restore ecosystems and land for the future*".

1.6 Climate Change and What We Might Do About It

There are also a great many published resources that deal with potential methods of mitigation of global warming and climate change. Griscom et al. (2017) made a comprehensive analysis of 20 conservation, restoration and/or improved land management natural climate solutions; these being actions that increase carbon storage and/or avoid greenhouse gas emissions across global forests, wetlands, grasslands and agricultural lands. They showed that most such actions, when implemented effectively, offer additional benefits such as water filtration, flood risk reduction, improved soil health, improved habitat biodiversity and enhanced climate resilience, and concluded that.

> … existing knowledge … provides a robust basis for immediate global action to improve ecosystem stewardship as a major solution to climate change.…

We will discuss some of these additional benefits in Chaps. 2 and 3 (Moore et al., 2021b).

Here, we will bring attention to the first critical assessment by an intergovernmental body in almost 15 years of the "… status and trends of the natural world, the social implications of these trends, their direct and indirect causes, and, importantly, the actions that can still be taken to ensure a better future for all…" was published in 2019 by IPBES (the Intergovernmental Science-Policy Platform on Biodiversity and Ecosystem Services) (IPBES 2019). The ***key messages*** of this report are stated to be:

- Nature and its vital contributions to people, which together embody biodiversity and ecosystem functions and services, are deteriorating worldwide.
- Direct and indirect drivers of change have accelerated during the past 50 years. The rate of global change in nature during the past 50 years is unprecedented in human history.
- Goals for conserving and sustainably using nature and achieving sustainability cannot be met by current trajectories, and goals for 2030 and beyond may only be achieved through transformative changes across economic, social, political and technological factors.
- Nature can be conserved, restored and used sustainably while other global societal goals are simultaneously met through urgent and concerted efforts fostering transformative change. Societal goals, including those related to food, water, energy, health and the achievement of human well-being for all, mitigating and adapting to climate change and conserving and sustainably using nature, can be achieved in sustainable pathways through the rapid and improved deployment of existing policy instruments and new initiatives that more effectively enlist individual and collective action for transformative change.

These key messages reflect the facts that the biosphere, on which we all depend, is being changed to an unprecedented degree across all spatial scales causing diversity within and between species and within and between ecosystems to decline more rapidly during the past 50 years than at any time in human history. Biodiversity and ecosystem functions are being reduced so rapidly that they seriously threaten our ability to achieve already agreed societal and environmental goals for biodiversity and sustainability, as well as goals specified in the Paris Agreement of 2015. Sustainable development is often inhibited by current political and governance structures and fundamental structural change is called for to transform the public and private sectors to achieve sustainability at local, national, and global levels. Particularly "…a commitment to mutually supportive international goals and targets, supporting actions by indigenous peoples and local communities at the local level, new frameworks for private sector investment and innovation, inclusive and adaptive governance approaches and arrangements, multi-sectoral planning, and strategic policy mixes…".

The 2019 report of the US National Academies of Sciences, Engineering, and Medicine entitled ***Negative Emissions Technologies and Reliable Sequestration: A Research Agenda*** (NASEM 2019) will be used as the basis for further discussion of options for removing CO_2 from the atmosphere and sequestering it reliably. The *Committee on Developing a Research Agenda for Carbon Dioxide Removal and Reliable Sequestration*, which produced this report, was created to recommend a detailed research development plan for what are known as **negative emissions technologies**, or NETs. NETs are technologies that remove and sequester CO_2 from the atmosphere with the intention of mitigating climate change. NETs have previously received less attention than technologies aimed at reducing the level of future CO_2 emissions by reducing fossil fuel consumption, though this requires massive deployment of low-carbon technologies and agricultural land use change between now and 2050.

Deploying NETs may be less expensive and less disruptive than reducing some emissions, such as a substantial portion of agricultural and land use emissions and some transportation emissions. NETs are envisaged by this Committee to:

- use biological processes to produce energy from biomass, while capturing and storing the resulting CO_2 emissions, and increase carbon stocks in soils, forests and wetlands by pro-active conservation. Unfortunately, our present forests are already suffering from the effects of the climate changes that have already occurred. Many forested areas are dying due to drought, often amplified by more devastating wildfires, and virulent, newly emerged and invasive pests and diseases (Demeude and Gadault 2020). The threat to forests is worldwide and, in many cases, can be traced to invasions of non-native bark and ambrosia beetles which carry symbiotic fungi to feed their larvae within galleries they bore into the tree. It is the sudden appearance of pathogenicity in the fungus that is the new and currently uncontrollable threat to forest ecosystems, and fruit and timber industries, around the globe. Triggered by climate change, some invasive bark and ambrosia beetle/fungus symbioses are shifting from non-pathogenic saprotrophy in their native ranges to a prolific tree-killing in invaded ranges (Moore et al. 2020). We cannot rely on forests to mitigate the effects of climate change; they're dying because of it!
- use chemical processes to capture CO_2 directly from the air and then sequester it in geologic reservoirs,
- enhance geologic processes that capture CO_2 from the atmosphere and permanently bind it with rocks (quoted from NASEM 2019).

The summary of this report lists a number of conclusions that outline the main thrust of the research agenda it goes on to develop, and which we quote directly below because they quantify the task ahead:

- **Conclusion 1**: Negative emissions technologies are best viewed as a component of a ***mitigation portfolio***, rather than a way to decrease atmospheric concentrations of carbon dioxide only after anthropogenic emissions have been eliminated. Indeed a different publication concludes that any attempt to solve the global climate change problem must be based on a portfolio approach that incorporates a full spectrum of strategies based on nature-based solutions, ***and*** alternative energy contributions ***and*** industrial mitigation (Anderson et al. 2019). In her article about direct air capture on the iNews website, Madeleine Cuff (Cuff 2020) points out that while trees can absorb CO_2, there isn't enough land on the planet to create a carbon sink of trees the size humanity needs. Cuff's solution is to turn to "... giant machines that can suck CO_2 out of the atmosphere ...". Our solution is to make more sustainable use of the other 70% of the planet, its ***oceans***.

- **Conclusion 2**: Four negative emissions technologies are ready for large-scale deployment: afforestation/reforestation, changes in forest management, uptake and storage by agricultural soils, and bioenergy with carbon capture and storage (BECCS). These NETs have low to medium costs (\$100/t CO_2 or less) and substantial potential for safe scale-up from current deployment.
- **Conclusion 3**: Current negative emissions technologies with direct costs that do not exceed \$100/t CO_2 can be safely scaled up to capture and store substantial amounts of carbon, but significantly less than ~1 Gt/y CO_2 in the United States and ~10 Gt/y CO_2 globally. These levels represent a substantial fraction of the total emissions of ~6.5 Gt CO_2 [emitted] in the United States and more than 50 Gt CO_2 [emitted] globally, but they may be difficult to attain because they require unprecedented rates of adoption of agricultural soil conservation practices, forestry management practices and waste biomass capture.
- **Conclusion 4:** If the goals for climate and economic growth are to be achieved, negative emissions technologies will likely need to play a large role in mitigating climate change by removing ~10 Gt/y CO_2 globally by mid-century and ~20 Gt/y CO_2 globally by the century's end.

The recent report published by EEA, the European Environment Agency (Davis et al. 2021) points out that:

> Climate change, biodiversity loss and degradation of ecosystems are interdependent and pose significant societal challenges, threatening economic and social stability, public health and well-being. The World Economic Forum considers extreme weather- and climate-related events and biodiversity loss to be among the five most imminent global risks (WEF 2020). Fighting climate change and preventing ecosystem degradation and biodiversity loss are highly interdependent, requiring increased coherency between their respective policy agendas and actions.

Yet, the NET solutions the report advocates seem to be rather limited in scope and aimed at ***adapting*** to climate change and reducing the risk of disasters. The report describes restoration of nature in river valleys and uplands to reduce downstream flooding risks and using natural vegetation in coastal regions to stabilise coastlines. Greening of urban spaces and renovation of canals and rivers are promoted to increase "... resilience to heatwaves and brings additional health and wellbeing benefits", but the report also observes that "re-forestation is increasingly used for storing carbon". In such examples NETs are becoming very similar to **NbSs**, which are ***Nature-based solutions***, these being:

> ... solutions to societal challenges that involve working with nature to deliver benefits for people and biodiversity. They include the protection, restoration or management of natural and semi-natural ecosystems; the sustainable management of productive land and seascapes; or the creation of novel ecosystems such as urban 'green infrastructure'. Well-designed NbS can contribute to tackling climate change and biodiversity loss ..." (quoted from https://www.nbsguidelines.info/ and see https://www.naturebasedsolutionsinitiative.org/what-are-nature-based-solutions/).

At the 2016 *World Conservation Congress*, members of the ***International Union for Conservation of Nature*** (IUCN) adopted a resolution which defined, for the first time, the use of nature for simultaneous benefits to biodiversity and human well-being. According to ***that*** resolution, ***Nature-based Solutions*** (NbS) are defined as:

> … actions to protect, sustainably manage and restore natural or modified ecosystems that address societal challenges effectively and adaptively, simultaneously providing human well-being and biodiversity benefits …

Descriptions of NbS are accompanied by a set of core principles for successfully implementing and upscaling NbS to address the ongoing degradation of natural resources that adversely affects both biodiversity and human well-being (Cohen-Shacham et al. 2016, 2019). The first version of the ***Global Standard for Nature-based Solutions*** was launched in July 2020 to assure funders, investors and decision-makers that NbS initiatives they support are effective and scalable (view https://www.iucn.org/theme/nature-based-solutions/resources/iucn-global-standard-nbs). Discussion of NbS to cope with the environmental challenges we currently face has become increasingly popular in recent years, particularly in suggestions involving people working with nature in response to the climate crisis. The concept is being proposed as a mechanism to achieve the transformative changes needed to establish more resilient and sustainable landscapes (Palomo et al. 2021; Welden et al. 2021). Palomo et al. (2021) focussed on mountain regions because of their vulnerability to climate change and the many different nature-based solutions already implemented there. They confirmed the potential of NbS for transformative change and showed that most NbS "… are based on four elements with transformation potential: nature's values, knowledge types, community engagement, and nature management practices". They also maintain "… that NbS are as much 'people based' as 'nature based'".

Milner-Gulland et al. (2021) discuss a framework they call the ***Mitigation and Conservation Hierarchy*** that is intended to mitigate and compensate the impacts on biodiversity due to commercial developments. Their approach features four distinct steps:

- **refrain**, involves avoiding negative impacts on nature as far as possible;
- **reduce**, involves minimising damage to nature where it cannot be completely avoided;
- **restore**, involves remediating any immediate damage to nature;
- **renew**, involves investing in revitalising nature.

They recognise that because 2021 begins the UN Decade on Ecosystem Restoration ***and*** the decade of action on the UN Sustainable Development Goals "… there has never been a more important time for ***ambitious action*** for biodiversity …" (quoted from Milner-Gulland et al. 2021; with our emphasis).

While all such approaches will undoubtedly ***help*** to build a sustainable economy and deliver many benefits, such as reduced greenhouse gas emissions, reduced pressures on biodiversity, and improved human health and well-being, we have to say that some major opportunities to achieve these desirable aims are being overlooked in all these reports.

1.7 Blue Carbon on the High Seas and the United Nations Sustainable Development Goals

The authors of this book do not in any way disagree with the findings of the NASEM (2019) or IUCN (Cohen-Shacham et al. 2016) reports, nor any of the other publications mentioned above, and the only fault we find with the recommendations of the European Environment Agency, (Davis et al. 2021) or any of the other proposals is that they are ***not sufficiently ambitious***; we should be seeking to **solve and remove** the problems we face rather than simply **adapt** to them. Because, with a little more imaginative use of calcifying organisms in the 70% of the world's surface that is covered in ocean, we really do have the opportunity to ***solve our present-day problems with the atmosphere, the climate and our food supply and our struggling ecosystems***. But it will require a great deal of effort and extraordinary ambition.

In the contents of this book, we develop arguments for alternative biotechnologies which we believe have not been considered in NASEM (2019), or indeed elsewhere, although Gattuso et al.

(2021) conclude "… Ocean-based NETs are uncertain but potentially highly effective. They have high priority for research and development …".

The specific technologies considered by NASEM (2019) are listed below.

Coastal Blue Carbon, namely, the "… land use and management practices that increase the carbon stored in living plants or sediments in mangroves, tidal marshlands, seagrass beds, and other tidal or salt-water wetlands. These approaches are sometimes called 'blue carbon' even though they refer to coastal ecosystems instead of the open ocean …" The report does point out that the committee's initial task statement (or 'job description') was to focus exclusively on ***near-shore coastal*** NETs despite the recognition that oceanic options for CO_2 removal and sequestration, which fall outside the scope of its task, could sequester an enormous amount of CO_2. We wish to remedy this exclusion. Furthermore, the ***Carnegie Climate Governance Initiative*** report (Mace et al. 2021) emphasises that ***reducing emissions alone will not be sufficient*** to limit the climate temperature increase to the 1.5 °C agreed in 2015 by parties to the United Nations Framework Convention on Climate Change (UNFCCC). Instead, CO_2 will also need to be ***removed from the atmosphere, on a scale never previously attempted***. This report finds that though a number of reporting rules and accounting practices are already in place with direct applicability to the implementation of carbon dioxide removal options, many governance gaps remain.

The central thrust of the argument presented in the book you are reading now is that the physiological chemistry of a few types of ocean creatures, the ***calcifiers of the coasts and open seas***, (coccolithophore algae, corals, crustacea and molluscs) enables them to extract CO_2 from the atmosphere and ***sequester it permanently as crystalline*** $\boldsymbol{CaCO_3}$. Our contention is that High Seas Blue Carbon promises to conserve and sustainably use the oceans, and its proper implementation and management can provide food for our growing population as well as restoring the atmosphere to its pre-Industrial-Revolution state.

The overwhelming advantages of calcifying organisms in this respect derives from their long evolutionary history (Moore 2021). We discuss this in Chap. 6 but the essence of the story is that when the first precellular living things evolved they employed calcium ions to carry signals in many different processes. When all those processes were finally brought together in the first proper cells it became essential for these to develop precise control over their internal Ca^{2+} levels. Subsequently, at several to many times during the Earth's history the seas have become calcium-rich and in those calcium-rich waters the cells were in danger of having their calcium-control mechanisms over-stretched. While some cells coped with this by evolving improved calcium-control, the calcifiers followed a different evolutionary pathway to detoxify the calcium by reacting it with a waste product of their metabolism (CO_2) to make $CaCO_3$ shells, and by so doing they solved everybody else's 'excess calcium' problem. We should stress that using $CaCO_3$ this way was a specific evolutionary innovation, and was far from an inevitable way to provide protection, which is the other function of these shells. Any fungus could make chitin reinforced with melanin for protection, any plant could make cellulose+lignin, and animals could make chitin and/or keratin and/or collagen, and even bone (which is a calcium+phosphate salt). So, calcifying organisms evolved in the distant past to detoxify the excess ***calcium*** in their environment, and we could harness them today to detoxify excess $\boldsymbol{CO_2}$ in our environment. The calcifying organisms have a good track record for environmental engineering, while industrial humans have a good track record for getting things done quickly. Together we could work the miracle of curing our atmosphere.

Cultivation of coccolithophores, corals, crustacea and molluscs on a massive scale would make a huge and continued ameliorative contribution to the planetary ecosystem; we discuss these suggestions in Chaps. 2–6 (inclusive) which follow, and which join a growing chorus of believers in the value of this argument. But there are non-believers and before going further we wish to address one of our *Frequently Asked*

Questions, which is "*How do you overcome the issue of shell making being a CO_2 production and not a sink?*" We can all agree where we start any discussion of marine calcification—with the **calcification reaction** that proceeds according to the following scheme:

$$2HCO_3^- + Ca^{2+} \leftrightarrow CaCO_3 + CO_2 + H_2O \text{ [reaction 1]} \quad (1.1)$$

This calcifying reaction scheme shows that two bicarbonate ions (which originally were both derived from the atmosphere, photosynthetic fixation of atmospheric CO_2 being the only source of metabolic carbon) react with a Ca^{2+} ion and one of them is precipitated as $CaCO_3$, and the other released as CO_2.

It is the ongoing ***interpretation*** of this by marine chemists "***as a net return of CO_2 to the atmosphere***" that is worrying, because it seems to us to be illogical and, in the real world, simply wrong. The implication of this interpretation that shellfish offer no net removal of carbon from the atmosphere starts to worry us at the lunch table, when we discard all those $CaCO_3$ shells that are left after our meal of *moules marinière* (see Figs. 1.3 and 1.4 and the Section 'There's a lot of shell in shellfish' in Chap. 2). Mulling this over after lunch, we realised that there are **five** major scientific reasons for describing this interpretation as illogical or even simply wrong, so we decided to ***audit*** what is usually described as the ***Blue Carbon Account***.

1. The **first** fault in logic arises from concentrating on the disposition of the product CO_2, rather than the product $CaCO_3$. In shallow waters, where shellfish are cultivated, $CaCO_3$ is essentially **insoluble** and totally **stable** (limestone). Consequently, this reaction removes from any further chemistry or biochemistry one of the two initial reactant bicarbonate ions. The source of both of those bicarbonate ions is atmospheric CO_2; either through carbon dioxide reacting with water to form carbonic acid, or from metabolism of food-derived organic carbon (ALL of which on this planet is derived from photosynthetic fixation of atmospheric CO_2). Hence, **reaction** (1.1) can be expressed as 2 atmospheric carbons + calcium ↔ one ***precipitated*** carbon + one potentially ***atmospheric*** carbon. It is simply **arithmetically and metabolically** wrong to claim this as a ***net return of CO_2 to the atmosphere***.
2. The **second** illogicality is the failure to give that little double headed arrow in the middle of the reaction scheme its due weight. **Reaction** (1.1) **is reversible**. Paragraph 1 refers to the chemistry of shallow waters. In the deep ocean, shells of dead calcareous plankton and other calcifiers occur throughout the water column ***above*** the Calcite Compensation Depth (CCD) (Bickert 2009). This is located at ocean depths of about 3,500 to 5,000 m and separates calcareous from noncalcareous sediments, with the 'calcareous ooze', which accretes into a type of limestone or chalk, being restricted roughly to the shallower half of the deep-sea floor. This is because **calcium carbonate solubility increases dramatically** with ***depth*** and ***pressure*** and at the depth of the CCD all calcium carbonate dissolves to form bicarbonate ions according to this equation:

$$CaCO_3 + CO_2 + H_2O \leftrightarrow Ca^{2+} + 2HCO_3^- \quad (1.2)$$

Note that CO_2 **is taken up** in this reaction and **the carbonate ion ($-CO_3^-$) remains intact**. If the seabed is above the CCD, bottom sediments consist of calcareous ooze. If the exposed seabed is below the CCD, $CaCO_3$ will dissolve before reaching this depth, preventing deposition of calcareous sediment, and the sea floor sediment will be a layer of siliceous ooze or abyssal clay (Berger 2016). You will also note that **reaction** (1.2) (dissolution at high hydrostatic pressure) ***is the exact reverse*** of **reaction** (1.1) (calcification). ***The atmosphere is not directly involved in either reaction direction***, this calcification/dissolution reaction being a balanced oceanic CO_2-cycle that depends on water depth. In the extreme case the solubilised bicarbonate ions will be carried by the global thermohaline circulation and could take a thousand years to surface and interact again with the atmosphere.

3. **Thirdly**, the phrase "[...calcification is...] a net return of CO_2 ***to the atmosphere***" is a major error of cell biology, as it ignores most of what we know about living things on this planet. Calcium storage organelles ('acidocalcisomes' in which Ca and P accumulate) are common to ***all eukaryotic organisms***, having a key role in calcium signalling and cellular calcium homeostasis. Calcifying coccolithophores have similar organelles that transfer calcium in vesicles, containing calcium-loaded particles, that fuse with the coccolith vesicle in which coccolith calcification occurs (Gal et al. 2018). Thus, to be returned to the sea water (one step prior to the atmosphere), CO_2 molecules produced by **reaction** (1.1) would have to be ***transported*** across at least two cell membranes. Yet in all calcifying cells this CO_2 ***will*** dissolve in the first aqueous compartments it encounters in a matter of seconds (Mitchell et al. 2010) becoming a bicarbonate ion which is a candidate for another round of calcification, and in coccolithophores, will most likely be harvested for photosynthesis. There will be **no return** of CO_2 ***to the atmosphere***. Mussels on their rocky shore would fizz like sparkling wine if that really happened.
4. **Fourthly**, although there is a dearth of data bearing on 'shellfish for carbon sequestration', there is some which should not be ignored. Tamburini et al. (2020) made a life cycle assessment of Manila Clam (*Ruditapes philippinarum*) farming in the Po River Delta, located in the northwestern Adriatic Sea (although Manila clam farming is a quantitatively important and valuable form of aquaculture production worldwide). This life cycle assessment found that: annual production of 1,000 kg fresh ready-to-sell clams sequestered in their shells 444.55 kg of CO_2, 1.54 kg of nitrogen and 0.31 kg of phosphorus y^{-1}. As the title of this article proclaims: this LCA "proves that Manila Clam farming … is a fully sustainable aquaculture practice ***and a carbon sink***" (the emphasis is ours) and the data are there for all to see.
5. **Fifthly**, any auditor worth his salt will establish an ***audit trail***, which is a sequential record of the history and details around an event. Reaching back into human history, we find that intact shellfish shells are excavated regularly from the middens associated with coastal communities of early humans, from around 120,000 years ago (Moore et al. 2021a, b), while intact shellfish shells abound in deepwater cores of ancient coastal sediments of hundreds of thousands of years ago. Further back in time our audit trail of the calcification reaction reveals the power of ***biogenic*** carbonate in the deeper history of planet Earth as illustrated by the global paleoceanographic reorganisations of carbonate accumulation and dissolution from the Cretaceous to the Miocene (between 125 and 9 million years ago) (van Andel et al. 1975, 1977; Preiß-Daimler et al. 2021). Sedimentary limestone rocks derive **all** their calcium carbonate from the ***biological activities*** of bryozoa, corals, crinoids, microscopic algae, foraminifera in the plankton and/or benthos of the day, as well as shellfish shells. Even chemical precipitation, which is an important method by which limestones form, depends on solution of ***biologically produced*** $CaCO_3$ as water currents agitate grains of sand and shell fragments together. Calcium carbonate is essentially insoluble in ***surface*** sea waters today, so warm, shallow waters can be saturated with $CaCO_3$, which recrystallises as aragonite on nuclei formed from shell fragments and builds up in concentric layers to form small multilayered spheres called ooids. And remember the fossils from really deep time? Ammonites (**65–240 million years ago**), trilobites (**520 million years ago**), brachiopods (**550 million years ago**), shellfish all. At intervals over the past 500 million years the fossil record clearly demonstrates that the distant ancestors of today's marine

calcifiers had the ***physiological tools*** to cope with both **acidified oceans** and great **excesses of atmospheric CO_2** and still create vast remains of shells, reefs and carbonaceous ooze made from crystalline $CaCO_3$. These organisms have dealt with excess atmospheric CO_2 before; we should enable them to do that again.

Ultimately, the CO_2 for the shell of shellfish and those other calcifiers comes from the atmosphere, but carbonates in shells are neither digested nor degraded. The shells of dead shellfish are chemically stable for geological periods of time, so in reality, this CO_2 is **removed** from the atmosphere, ***permanently***.

Misunderstanding these points lead Munari et al. (2013) to conclude that "… shell formation in cultivated shellfish cannot be part of carbon trading systems". As we show in Chap. 2, the later analyses of Alonso et al. (2021) dispel this myth of calcifier physiology. An even more important point, also discussed in Chap. 2, is that as far as the aquaculture fishery ***industry*** is concerned, the diesel fuel consumption of diesel powered fishing vessels and the electricity consumption of onshore industrial plant (refrigeration, road transport, etc.) are ***the*** major contributors to the environmental burden (the carbon cost) of cultivating calcifiers like bivalves. This makes any adverse contribution of the respiratory flux of CO_2 to production of the shell that might be suggested even more irrelevant.

Another commonly held view, which we believe to be an oversimplification, is that due to the rising levels of anthropogenic CO_2 in the world's oceans the pH of seawater is being reduced and ocean acidification is predicted to have a harmful effect on the physiology of calcifying organisms in present and future oceans (Kroeker et al. 2013). However, such a prediction seems to us to ignore several important ***facts*** about the last 4 billion years of biological evolution (expanding on some of the comments in Point 3, above), namely:

- selectively permeable phospholipid bilayer membranes isolating the cell from its environment and compartments within the cell;
- ion-specific transporters across those membranes;
- the coupling of proton transporters to energy generation;
- the coupling of proton transporters to the co-transport of other ions, metabolic substrates and metabolic wastes the coupling of proton transporters to the co-transport of other ions, metabolic substrates and metabolic wastes (a case in point being that Foraminifera actively pump protons out from the site of calcification which is therefore surrounded by a low (acidic) external pH ***of their own making*** (Kawahata et al. 2019)). Foraminifera are amoeba-like, single-celled protists that secrete a protective shell (called a 'test' because it is intracellular). The most primitive tests are made from cemented sand grains, but most are made of calcite or aragonite crystals. Fossilised tests are found in sediments as old as the earliest Cambrian (about 545 Mya) and planktonic and benthic foraminifera are still abundant today, living in marine and brackish waters.

We find no surprise, therefore, in the conclusions of Connell et al. (2017), who tested the effects of ocean acidification on a calcifying gastropod herbivore in a volcanic CO_2 vent ecosystem with CO_2 levels close to those predicted for the world's future oceans:

> …We conclude that the effect of ocean acidification on algae (primary producers) can have a strong, indirect positive influence on the abundance of some calcifying herbivores, which can overwhelm any direct negative effects Connell et al. (2017).

The review paper entitled *Rebuilding marine life* (Duarte et al. 2020) indicates that achieving the UN's Sustainable Development Goal 14 (to conserve and sustainably use the oceans, seas and marine resources for sustainable development)

> … will require rebuilding the marine life-support systems that deliver the many benefits that society receives from a healthy ocean …". But they finally conclude that "… Rebuilding marine life represents a doable Grand Challenge for humanity, an ethical obligation and a smart economic objective to achieve a sustainable future ….

In the opinion of Duarte et al. (2020), recovery rates seen in past studies of conservation interventions suggest that:

> … substantial recovery of the abundance, structure and function of marine life could be achieved by 2050, if major pressures — including climate change — are mitigated ….

And in their letter to the journal *Science*, Gordon et al.(2020) assert that:

> … Marine restoration projects are undervalued …". In their final paragraph they conclude that "… [marine] restoration projects could help maintain species survival and ecosystem services, ultimately providing humanity with the breathing space to stabilize the climate … (Gordon et al. 2020).

We have discussed above the denial of ***anthropogenic CO_2-driven*** climate change, but alongside the chorus of denials of its cause, the climate ***has been changing*** and not for the better. Over the past few years, climate change has been exposing the real world to a catalogue of sudden environmental pressures. Climate change results in an increasing frequency of extreme, even disastrous, weather events on a global scale. Similarly, by altering the conditions of established natural habitats, climate change drives major redistributions of ecological communities and species of organisms. Titley et al. (2021) point out that changing distributions of species ranges stretch across political and national borders and:

> … By mapping transboundary range shifts globally, … highlight regions where international cooperation may be most useful for conservation and where border barriers may be most detrimental … [and] … by modeling the climatic niches of terrestrial mammals and birds globally, show that projected species loss under climate change is greatest in countries with weaker governance and lower Gross Domestic Product, with loss of mammal species projected to be greater in countries with lower CO_2 emissions (Titley et al. 2021).

Essentially the same general conclusion applies to the incidence of extreme weather events. The significance of this is that as evidence mounts that climate change ***HAS*** been driven by anthropogenic release of fossil CO_2 into the atmosphere by the industrial nations of the world there is a need for an appropriate transnational legal framework to ensure application of the principle that 'the polluter pays' for repair and restitution of the damages caused.

Lawsuits seeking compensation for climate-related losses and to compel governments to reduce their GHG emissions have proliferated around the world, though most claims have been unsuccessful because judicial treatment of scientific evidence lags behind the state-of-the-art in climate science (Stuart-Smith et al. 2021). Not surprisingly, in recent years we have seen:

(a) a steady increase in the number of climate change laws introduced by legislations around the world (view these URLs: https://www.lse.ac.uk/granthaminstitute/; https://climate-laws.org/; https://climate.law.columbia.edu/; https://www.theccc.org.uk/the-need-to-act/a-legal-duty-to-act/; https://ec.europa.eu/clima/policies/eu-climate-action/law_en; https://www.bbc.com/future/article/20200706-the-law-that-could-make-climate-change-illegal).

(b) An increasing number of publications attempting to guide lawyers and legal scholars to resolve the legal issues raised by 'legally disruptive' climate change within existing legal regimes (e.g. Meltz 2014; Fisher et al. 2017).

The legal and legislative issues raised by national and international remedial and conservation programmes are important aspects of such programmes, but they are beyond our expertise, and we are unable to discuss them further. Here, we concentrate on activities that will contribute to effective, permanent and sustainable removal of carbon from the atmosphere.

Terrestrial carbon removal and sequestration is usually understood to mean land use and management practices such as afforestation and reforestation, changes in forest management, or changes in agricultural practices, that enhance carbon storage in agricultural soils. This is possibly the most conventional aspect, as photosynthetic carbon capture by trees and other green plants is widely considered to be an effective strategy to limit the rise of CO_2 concentrations in the atmosphere by sequestering carbon in the plant body. The *Intergovernmental Panel on Climate Change* Special Report of 2018 (Masson-Delmotte et al. 2019) suggested that an increase of 1 billion hectares of forest will be necessary to limit global warming to 1.5 °C by 2050.

In the same publication year, Bastin et al. (2019) mapped the global potential tree coverage and estimated that the world's terrestrial ecosystems could support an additional 0.9 billion (0.9×10^9) hectares of continuous forest (corresponding to more than a 25% increase in presently forested area) and that such a change has the potential to cut the atmospheric carbon pool by about 25%. We all like trees and we are in favour of planting more of them, but there are negative aspects to these estimations that indicate that the value of Green Carbon as a means of sequestering carbon from the atmosphere on the long-term basis required for full and lasting benefit has been seriously overestimated (Moore et al. 2021a). All of this is discussed in Chap. 2.

It is the ***certainty*** and ***permanence*** of the removal of CO_2 from the atmosphere that would make biotechnology using calcifying organisms so attractive. NASEM (2019) notes that terrestrial options and the few ***coastal*** blue carbon options they consider are reversible if the carbon sequestering practices are not maintained. Forested land could be cleared again, but the reversion to intensive tillage would reverse any gains in soil carbon sequestration achieved by the afforestation. Similarly, restored coastal wetland could be drained again for agricultural use, losing any advantage gained by the wetland restoration.

> … Although temporary CO_2 storage will have some climate benefit, scientific and economic requirements to ensure the permanence of storage within ecosystems are substantial ….

But while we would offer easily cultivated calcifying organisms as candidates to provide these benefits, NASEM (2019) offers only bioenergy with carbon capture and sequestration (BECCS), direct air capture and carbon mineralisation.

Bioenergy with carbon capture and sequestration (BECCS). Energy production using plant biomass to produce electricity using liquid fuels (derived from plant oils), and/or heat by direct burning effectively only recycles today's CO_2 back to the atmosphere (in contrast to fossil fuels, which make a net increase of ***ancient*** CO_2 to today's atmosphere. If combined with capture and sequestration of any CO_2 produced when using the bioenergy the whole process can provide a net reduction of CO_2 in the atmosphere.

Direct air capture. Uses chemical processes that capture CO_2 from ambient air and concentrate it, so that it can be injected into a storage reservoir. Terlouw et al. (2021) have assessed different direct air carbon capture and storage systems that require low-carbon electricity and heat sources for the CO_2 capture process and demonstrate negative greenhouse gas emissions in all cases. We also note with interest the carbon-neutral jet fuel ('*syngas*') produced by hydrogenation of direct air capture carbon dioxide using sustainable renewable hydrogen and solar energy (Yao et al. 2020).

Carbon mineralisation. In which CO_2 from the atmosphere forms a chemical bond with reactive rocks, like mantle peridotite and basaltic lava, both at the surface (ex situ) where CO_2 in ambient air is mineralised on exposed rock, and in the subsurface (in situ) where concentrated CO_2 streams are injected into rocks to mineralise in the pores. This might employ supercritical CO_2 in deep sedimentary geological formations. CO_2 usually behaves as a gas in air at standard temperature and pressure, or as a solid called dry ice when cooled and/or pressurised sufficiently. ***Supercritical*** $\boldsymbol{CO_2}$ is a ***fluid state phase*** that occurs when CO_2 is held at or above its critical temperature and critical pressure [view YouTube video at https://www.youtube.com/watch?v=-gCTKteN5Y4].

These last three mechanisms of CO_2 capture and sequestration are discussed in Chap. 7, and in Chap. 8 we outline ***our*** recommendations for action.

The *World Resources Institute*'s 2019 report by its High Level Panel for a Sustainable Ocean Economy entitled "*The Ocean as a Solution to Climate Change: Five Opportunities for Action*" (Hoegh-Guldberg, et al. 2019) pointed out that the oceans are a dominant feature of our planet and emphasised that "… while much of recent attention is focused on the problems that the ocean faces, the ocean is also a source of potential solutions and innovation … [to] … provide opportunities in the fight against climate change".

The five areas of ocean-based climate action to mitigate greenhouse gas emissions considered by Hoegh-Guldberg et al. (2019) are:

- Ocean-based renewable energy (defined [their Table 1, p. 22] as scaling up fixed and floating offshore wind turbine installations and extracting energy from ocean waves, tides, currents, salinity and temperature differences, and floating photovoltaic solar energy).
- Ocean-based transport (reducing emissions from shipping).
- Coastal and marine ecosystems, such as restoration and protection of mangroves (Su et al. 2021), salt marshes (Raw et al. 2021), seagrass meadows (Kerninon et al. 2021) and seaweed habitats (Riquet et al. 2021).
- The ocean-based food system (defined as reducing emissions from fishing vessels and aquaculture in general and switching sources of protein in human diets away from intensive carbon emission land-based sources of protein, notably beef and lamb, in favour of low-carbon ocean-based food from the sea).
- Carbon storage in the seabed (defined as "Geological storage offshore of captured CO_2 in the seabed").

We agree with the mantra "A healthy ocean is critical to achieving global targets to limit climate change" (website homepage: http://www.oceanpanel.org/climate) and the only fault we find with Hoegh-Guldberg et al. (2019) is the lack of ambition, which it shares with the other reports we have discussed above. We emphasise, again, that humanity should seek to solve and remove the problems we face rather than simply adapt to those problems.

Importantly, though, Hoegh-Guldberg et al. (2019) discuss in detail how ocean-based interventions of the sorts they describe can contribute to the targets in 16 of the 17 United Nations Sustainable Development Goals (see their Figure ES-5, page 12).

Given the **UN-SDGS** news headlines ***Sustainable Blue Foods are Vital to Global Food Security*** (https://www.youtube.com/watch?v=-gCTKteN5Y4), of 5 May 2021, and ***The Ocean Race launches Relay4Nature***, of 7 May 2021, (https://www.theoceanrace.com/en/news/12688_The-Ocean-Race-launches-Relay4Nature.html), we wish to emphasise that all we describe above as being discussed later in ***this book*** align well with the ***United Nations Sustainable Development Goals***, specifically **SDG-2** (Zero hunger), **SDG-3** (Good health and well-being; which we think are encompassed by the healthy food, reef-building and pollution-filtration services of cultivated shellfish), **SDG-12** (Responsible consumption and production), **SDG-13** (Climate action) and **SDG-14** (Life below water) (Table 1.4).

Our core belief is that humankind must look to the oceans for a solution to climate change resulting from excess CO_2 in the atmosphere, and that marine calcifiers (coccolithophores, foraminifera, molluscs, crustacea, corals) are the tools that would provide that solution. But in most of the above discussion we have put practical aspects to one side in order to make the essential broader points about the science, governance and politics involved in any attempts to repair our atmosphere. In later chapters we will develop ***actions plans*** (for a summary and synthesis, see Chap. 8) that are potentially both practical and effective. If you think we are being over-ambitious with our claims for the potential benefits of massively increasing global aquaculture production, then read on!

Table 1.4 Blue carbon on the high seas and the United Nations' sustainable development goals*

2 ZERO HUNGER	**Goal 2** End hunger, achieve food security and improved nutrition and promote sustainable agriculture (*to which **we** would add* '**and aquaculture**') (https://sdgs.un.org/goals/goal2) • Target 2.1: By 2030, end hunger and ensure access by all people, in particular the poor and people in vulnerable situations, including infants, to safe, nutritious and sufficient food all year round • Target 2.2: By 2030, end all forms of malnutrition, including achieving, by 2025, the internationally agreed targets on stunting and wasting in children under 5 years of age, and address the nutritional needs of adolescent girls, pregnant and lactating women and older persons • Target 2.3: By 2030, double the agricultural productivity and incomes of small-scale food producers, in particular women, indigenous peoples, family farmers, pastoralists and fishers, including through secure and equal access to land, other productive resources and inputs, knowledge, financial services, markets and opportunities for value addition and non-farm employment • Target 2.4: By 2030, ensure sustainable food production systems and implement resilient agricultural practices that increase productivity and production, that help maintain ecosystems, that strengthen capacity for adaptation to climate change, extreme weather, drought, flooding and other disasters and that progressively improve land and soil quality • Target 2.a: Increase investment, including through enhanced international cooperation, in ***… aquaculture in general and at all scales from small communities to large industries***
3 GOOD HEALTH AND WELL-BEING	**Goal 3** Ensure healthy lives and promote well-being for all at all ages (https://sdgs.un.org/goals/goal3)
12 RESPONSIBLE CONSUMPTION AND PRODUCTION	**Goal 12** Ensure sustainable consumption and production patterns (https://sdgs.un.org/goals/goal12) • Target 12.2: By 2030, achieve the sustainable management and efficient use of natural resources • Target 12.8: By 2030, ensure that people everywhere have the relevant information and awareness for sustainable development and lifestyles in harmony with nature • Target 12.a: Support developing countries to strengthen their scientific and technological capacity to move towards more sustainable patterns of consumption and production • Target 12.c: Rationalize inefficient fossil-fuel subsidies that encourage wasteful consumption by removing market distortions, in accordance with national circumstances, including by restructuring taxation and phasing out those harmful subsidies, where they exist, to reflect their environmental impacts, taking fully into account the specific needs and conditions of developing countries and minimizing the possible adverse impacts on their development in a manner that protects the poor and the affected communities
13 CLIMATE ACTION	**Goal 13** Take urgent action to combat climate change and its impacts (https://sdgs.un.org/goals/goal13) • Target 13.1: Strengthen resilience and adaptive capacity to climate-related hazards and natural disasters in all countries • Target 13.2: Integrate climate change measures into national policies, strategies and planning • Target 13.3: Improve education, awareness-raising and human and institutional capacity on climate change mitigation, adaptation, impact reduction and early warning • Target 13.b: Promote mechanisms for raising capacity for effective climate change-related planning and management in least developed countries and small island developing States, including focusing on women, youth and local and marginalized communities. Acknowledging that the United Nations Framework Convention on Climate Change is the primary international intergovernmental forum for negotiating the global response to climate change

(continued)

Table 1.4 (continued)

14 LIFE BELOW WATER	**Goal 14** Conserve and sustainably use the oceans, seas and marine resources for sustainable development (https://sdgs.un.org/goals/goal14) • Target 14.2: By 2020, sustainably manage and protect marine and coastal ecosystems to avoid significant adverse impacts, including by strengthening their resilience, and take action for their restoration in order to achieve healthy and productive oceans • Target 14.3: Minimize and address the impacts of ocean acidification, including through enhanced scientific cooperation at all levels • Target 14.6: By 2020, prohibit certain forms of fisheries subsidies which contribute to overcapacity and overfishing, eliminate subsidies that contribute to illegal, unreported and unregulated fishing and refrain from introducing new such subsidies, recognizing that appropriate and effective special and differential treatment for developing and least developed countries should be an integral part of the World Trade Organization fisheries subsidies negotiation • Target 14.7: By 2030, increase the economic benefits to Small Island developing States and least developed countries from the sustainable use of marine resources, including through sustainable management of fisheries, aquaculture and tourism • Target 14.a: Increase scientific knowledge, develop research capacity and transfer marine technology, taking into account the Intergovernmental Oceanographic Commission Criteria and Guidelines on the Transfer of Marine Technology, in order to improve ocean health and to enhance the contribution of marine biodiversity to the development of developing countries, in particular small island developing States and least developed countries • Target 14.c: Enhance the conservation and sustainable use of oceans and their resources by implementing international law as reflected in UNCLOS, which provides the legal framework for the conservation and sustainable use of oceans and their resources, as recalled in paragraph 158 of *The Future We Want* (which is the outcome document of the United Nations *Conference on Sustainable Development* Rio de Janeiro, Brazil, 20–22 June 2012 (https://sustainabledevelopment.un.org/content/documents/733FutureWeWant.pdf)

*View the 17 Sustainable Development Goals of the United Nations at this URL: https://sdgs.un.org/goals

References

Alonso AA, Álvarez-Salgado XA, Antelo LT (2021) Assessing the impact of bivalve aquaculture on the carbon circular economy. J Clean Prod 279: article 123873. https://doi.org/10.1016/j.jclepro.2020.123873

Anderson CM, DeFries RS, Litterman R, Matson PA, Nepstad DC, Pacala S, Schlesinger WH, Shaw MR, Smith P, Weber C, Field CB (2019) Natural climate solutions are not enough. Science 363:933–934. https://doi.org/10.1126/science.aaw2741

Arrhenius SA (1896) On the influence of carbonic acid in the air upon the temperature of the ground. Philos Mag J Sci 41:237–276. https://www.rsc.org/images/Arrhenius1896_tcm18-173546.pdf

Bastin J-F, Finegold Y, Garcia C, Mollicone D, Rezende M, Routh D, Zohner CM, Crowthe TW (2019) The global tree restoration potential. Science 365:76–79. https://doi.org/10.1126/science.aax0848

Berger WH (2016) Calcite compensation depth (CCD). In: Harff J, Meschede M, Petersen S, Thiede J (eds) Encyclopedia of marine geosciences. Encyclopedia of earth sciences series. Springer, Dordrecht. https://doi.org/10.1007/978-94-007-6238-1_47

Bickert T (2009) Carbonate compensation depth. In: Gornitz V (ed) Encyclopedia of paleoclimatology and ancient environments. Encyclopedia of earth sciences series. Dordrecht: Springer. https://doi.org/10.1007/978-1-4020-4411-3_33

Callendar GS (1949) Can carbon dioxide influence climate? Weather 4:310–314. https://doi.org/10.1002/j.1477-8696.1949.tb00952.x

Chamberlin TC (1899) An attempt to frame a working hypothesis of the cause of glacial periods on an atmospheric basis. J Geol **7**:545–584. https://www.jstor.org/stable/30055497

Cohen-Shacham E, Janzen C, Maginnis S, Walters G (2016) Nature-based solutions to address global societal challenges. A report produced for the IUCN Commission on Ecosystem Management (CEM) and IUCN World Commission on Protected Areas (WCPA). Gland, Switzerland: IUCN. ISBN: 9782831718125. https://doi.org/10.2305/IUCN.CH.2016.13.en

Cohen-Shacham E, Andrade A, Dalton J, Dudley N, Jones M, Kumar C, Maginnis S, Maynard S, Nelson CR, Renaud FG, Welling R, Walters G (2019) Core principles for successfully implementing and upscaling Nature-based Solutions. Environ Sci Policy 98:20–29. https://doi.org/10.1016/j.envsci.2019.04.014

Connell SD, Doubleday ZA, Hamlyn SB, Foster NR, Harley CDG, Helmuth B, Kelaher BP, Nagelkerken I, Sarà G, Russell BD (2017) How ocean acidification can benefit calcifiers. Curr Biol 27:R95–R96. https://doi.org/10.1016/j.cub.2016.12.004

Cuff M (2020) Direct air capture: Giant machines that can suck CO_2 out of the atmosphere could help control pollution levels. iNews Website. https://www.inews.co.uk/news/analysis/direct-air-capture-giant-machines-co2-pollution-levels-775208

Davis M, Vandewalle M, Förster J, Medri S, Marttunen M, Veerkamp C, Picatoste JR, Castellari S, Mysiak J, Zandersen M (2021) Nature-based solutions in Europe: Policy, knowledge and practice for climate change adaptation and disaster risk reduction. EEA Report No 01/2021. Luxembourg: Publications Office of the European Union. ISBN: 9789294803627. https://doi.org/10.2800/919315. https://www.eea.europa.eu/publications/nature-based-solutions-in-europe

Demeude H, Gadault T (2020) Le cri de détresse des forêts françaises confrontées aux crises écologiques. The GoodPlanet Info magazine. https://www.goodplanet.info/2020/11/26/le-cri-de-detresse-des-forets-francaises-confrontees-aux-crises-ecologiques/

Duarte CM, Agusti S, Barbier E, Britten GL, Castilla JC, Gattuso JP, Fulweiler RW, Hughes TP, Knowlton N, Lovelock CE, Lotze HK, Predragovic M, Poloczanska E, Roberts C, Worm B (2020) Rebuilding marine life. Nature 580:39–51. https://doi.org/10.1038/s41586-020-2146-7

Fisher E, Scotford E, Barritt E (2017) The legally disruptive nature of climate change. Mod Law Rev 80:173–201. https://doi.org/10.1111/1468-2230.12251

Forster P, Ramaswamy V, Artaxo P, Berntsen T, Betts R et al (2007) Changes in atmospheric constituents and in radiative forcing. In: Solomon S, Qin D, Manning M, Chen Z, Marquis M, Averyt KB, Tignor M, Miller HL (eds) Climate Change 2007: The physical science basis. contribution of working group I to the fourth assessment report of the intergovernmental panel on climate change, IPCC AR4. Cambridge University Press, Cambridge, UK and NY, USA

Friedlingstein P, Jones MW, O'Sullivan M, Andrew RM, Hauck J et al (2019) Global carbon budget 2019. Earth Syst Sci Data 11:1783–1838. https://doi.org/10.5194/essd-11-1783-2019

Gal A, Sorrentino A, Kahil K, Pereiro E, Faivre D, Scheffel A (2018) Native-state imaging of calcifying and noncalcifying microalgae reveals similarities in their calcium storage organelles. Proc Natl Acad Sci USA 115:11000–11005. https://doi.org/10.1073/pnas.1804139115

Gattuso JP, Williamson P, Duarte CM, Magnan AK (2021) The potential for ocean-based climate action: net negative emissions and beyond. Front Clim. https://www.frontiersin.org/articles/10.3389/fclim.2020.575716/full

Gehler A, Gingerich PD, Pack A (2016) Temperature and atmospheric CO_2; concentration estimates through the PETM using triple oxygen isotope analysis of mammalian bioapatite. Proc Natl Acad Sci USA 113:7739–7744. https://doi.org/10.1073/pnas.1518116113

Geng L, Alexander B, Cole-Dai J, Steig EJ, Savarino J, Sofen ED, Schauer AJ (2014) Nitrogen isotopes in ice core nitrate linked to anthropogenic atmospheric acidity change. Proc Natl Acad Sci USA 111:5808–5812. https://doi.org/10.1073/pnas.1319441111

Girardin CAJ, Jenkins S, Seddon N, Allen M, Lewis SL, Wheeler CE, Griscom BW, Malhi Y (2021) Nature-based solutions can help cool the planet—if we act now. Nature 593:191–194. https://doi.org/10.1038/d41586-021-01241-2

Gordon TAC, Radford AN, Simpson SD, Meekan MG (2020) Marine restoration projects are undervalued. Science 367:635–636. https://doi.org/10.1126/science.aba9141

Griscom BW, Adams J, Ellis PW, Houghton RA, Lomax G, Miteva DA et al (2017) Natural climate solutions. Proc Natl Acad Sci USA 114:11645–11650. https://doi.org/10.1073/pnas.1710465114

Hoegh-Guldberg O, Caldeira K, Chopin T, Gaines S, Haugan P et al (2019) The ocean as a solution to climate change: five opportunities for action. Report of the high level panel for a sustainable ocean economy. World Resources Institute, Washington, DC. https://www.oceanpanel.org/climate

IPBES (2019) The global assessment report on biodiversity and ecosystem services. In: Díaz S, Settele J, Brondízio ES, Ngo HT, Guèze M, Agard J et al (eds) Summary for policymakers. Published by the Intergovernmental Science-Policy Platform on Biodiversity and Ecosystem Services. IPBES Secretariat. Bonn, Germany, 56 p. ISBN No: 9783947851133. https://www.ipbes.net/sites/default/files/inline/files/ipbes_global_assessment_report_summary_for_policymakers.pdf

IPCC (2007) Summary for policymakers, in climate change 2007: impacts, adaptation and vulnerability. Contribution of Working Group II to the Fourth Assessment Report of the Intergovernmental Panel on Climate Change. Cambridge University Press. , Cambridge, UK, p. 17. https://www.ipcc.ch/report/ar4/wg2/

IPCC (2013) Summary for policymakers, in climate change 2013: the physical science basis. In: Stocker TF, Qin D, Plattner G-K, Tignor M, Allen SK, Boschung J, Nauels A, Xia Y, Bex V, Midgley PM (eds) Contribution of working group I to the fifth assessment report of the intergovernmental panel on climate change. Cambridge University Press, Cambridge, UK and New York, USA. https://www.ipcc.ch/report/ar5/wg1/

IPCC (2014) Summary for policymakers. In: Field CB, Barros VR, Dokken DJ, Mach KJ, Mastrandrea MD, Bilir TE, Chatterjee M, Ebi KL, Estrada YO, Genova RC, Girma B, Kissel ES, Levy AN, MacCracken S, Mastrandrea PR, White LL (eds) Climate change 2014: impacts, adaptation, and vulnerability. Part A: Global and sectoral aspects. Contribution of working group II to the fifth assessment report of the intergovernmental panel on climate change. Cambridge University Press, Cambridge, UK and New York, USA, pp 1–32. https://doi.org/10.1017/CBO9781107415379

IPCC (2019) Climate change and land: an IPCC special report on climate change, desertification, land degradation, sustainable land management, food security, and greenhouse gas fluxes in terrestrial ecosystems. In: Shukla PR, Skea J, Calvo Buendia E, Masson-Delmotte V, et al (eds). Published for the Intergovernmental Panel on Climate Change. https://www.ipcc.ch/srccl/

IPCC (2021) Climate change 2021: the physical science basis. Contribution of working group I to the sixth assessment report of the intergovernmental panel on climate change. In: Masson-Delmotte V, Zhai P, Pirani A, Connors SL, Péan C, Berger S, Caud N, Chen Y, Goldfarb L, Gomis MI, Huang M, Leitzell K, Lonnoy E, Matthews JBR, Maycock TK, Waterfield T, Yelekçi O, Yu R, Zhou B (eds). Cambridge University Press. https://www.ipcc.ch/report/ar6/wg1/downloads/report/IPCC_AR6_WGI_Headline_Statements.pdf; https://www.ipcc.ch/report/ar6/wg1/downloads/report/IPCC_AR6_WGI_SPM_final.pdf; https://www.ipcc.ch/report/ar6/wg1/downloads/report/IPCC_AR6_WGI_Full_Report.pdf

Kawahata H, Fujita K, Iguchi A, Inoue M, Iwasaki S et al (2019) Perspective on the response of marine calcifiers to global warming and ocean acidification—behavior of corals and foraminifera in a high CO_2 world "hot house". Progr Earth Planet Sci 6(5):37p. https://doi.org/10.1186/s40645-018-0239-9

Kerninon F, Payri CE, Le Loc'h F, Alcoverro T, Maréchal J-P, Chalifour J et al (2021) Selection of parameters for seagrass management: towards the development of integrated indicators for French Antilles. Mar Pollut Bull 170, article number 112646. https://doi.org/10.1016/j.marpolbul.2021.112646

Knoll AH, Bambach RK, Canfield DE, Grotzinger JP (1996) Fossil record supports evidence of impending mass extinction. Science 273:452–457. https://doi.org/10.1126/science.273.5274.452

Kroeker KJ, Kordas RL, Crim R, Hendriks IE, Ramajo L, Singh GS, Duarte CM, Gattuso JP (2013) Impacts of ocean acidification on marine organisms: quantifying sensitivities and interaction with warming. Glob Change Biol 19:1884–1896. https://doi.org/10.1111/gcb.12179

Lenton TM, Held H, Kriegler E, Hall JW, Lucht W, Rahmstorf S, Schellnhuber HJ (2008) Tipping elements in the Earth's climate system. Proc Natl Acad Sci USA 105:1786–1793. https://doi.org/10.1073/pnas.0705414105

Lenton TM, Rockström J, Gaffney O, Rahmstorf S, Richardson K, Steffen W, Schellnhuber HJ (2019) Comment: Climate tipping points—too risky to bet against. Nature 575:592–595. https://doi.org/10.1038/d41586-019-03595-0

Lund MT, Aamaas B, Stjern CW, Klimont Z, Berntsen TK, Samset BH (2020) A continued role of short-lived climate forcers under the shared socioeconomic pathways. Earth Syst Dyn 11:977–993. https://doi.org/10.5194/esd-11-977-2020

MacDowell N, Fennell PS, Shah N, Maitland GC (2017) The role of CO_2 capture and utilization in mitigating climate change. Nat Clim Chang 7:243–249. https://doi.org/10.1038/nclimate3231

Mace MJ, Fyson CL, Schaeffer M, Hare WL (2021) Governing large-scale carbon dioxide removal: are we ready?—An update. Carnegie Climate Governance Initiative (C2G), New York, USA. https://www.climateanalytics.org/media/c2g2-2018-cdr-governance-1_1_1.pdf

Mann ME (2009) Defining dangerous anthropogenic interference. Proc Natl Acad Sci USA 106:4065–4066. https://doi.org/10.1073/pnas.0901303106

Masson-Delmotte V, Zhai P, Pörtner H-O, Roberts D, Skea J et al (eds) (2019) IPCC, 2018: Global warming of 1.5°C. An IPCC special report on the impacts of global warming of 1.5°C above pre-industrial levels and related global greenhouse gas emission pathways, in the context of strengthening the global response to the threat of climate change, sustainable development, and efforts to eradicate poverty. https://www.ipcc.ch/sr15/; https://www.ipcc.ch/sr15/download/

Mayhew PJ, Jenkins GB, Benton TG (2008) A long-term association between global temperature and biodiversity, origination and extinction in the fossil record. Proc R Soc B Biol Sci 275:47–53. https://doi.org/10.1098/rspb.2007.1302

Meltz R (2014) Climate change and existing law: a survey of legal issues past, present and future. Congressional Research Service Report no. R42613. https://www.fas.org/sgp/crs/misc/R42613.pdf

McCarthy JJ, Canziani OF, Leary NA, Dokken DJ, White KS (2001) Climate change 2001: impacts, adaptation, and vulnerability. A report of working group II of the intergovernmental panel on climate change. Cambridge University Press, Cambridge UK & New York, USA, 1042 p. ISBN: 0521807689. https://www.ipcc.ch/report/ar3/wg2/

McKinley GA, Fay AR, Eddebbar YA, Gloege L, Lovenduski NS (2020) External forcing explains recent decadal variability of the ocean carbon sink. AGU Adv 1, article e2019AV000149. https://doi.org/10.1029/2019AV000149

Melillo JM, Richmond TC, Yohe GW (eds) (2014) Climate change impacts in the United States: the third national climate assessment. U.S. Global Change Research Program, Washington, DC, USA, 841 p. U.S. Government Printing Office. ISBN: 9780160924026. https://doi.org/10.7930/J0Z31WJ2. https://nca2014.globalchange.gov/

Milner-Gulland EJ, Addison P, Arlidge WNS, Baker J, Booth H et al (2021) Four steps for the Earth: mainstreaming the post-2020 global biodiversity framework. One Earth 4(1):75–87. https://doi.org/10.1016/j.oneear.2020.12.011

Mitchell MJ, Jensen OE, Cliffe K, Maroto-Valer MM (2010) A model of carbon dioxide dissolution and mineral carbonation kinetics. Proc R Soc A Math Phys Eng Sci 466:1265–1290. https://doi.org/10.1098/rspa.2009.0349

Moore D (2021) Saving the planet with appropriate biotechnology: 4. Coccolithophore cultivation and deployment/Salvando el planeta con biotecnología apropiada: 4. Cultivo de cocolitóforos e implementación. Mexican J Biotechnol 6:129–155. https://doi.org/10.29267/mxjb.2021.6.1.129

Moore D, Heilweck M, Petros P (2021a) Saving the planet with appropriate biotechnology: 1. Diagnosing the Problems/Salvando el planeta con biotecnología apropiada: 1. Diagnóstico de los problemas. Mexican J Biotechnol 6:1–30. https://doi.org/10.29267/mxjb.2021.6.1.1

Moore D, Heilweck M, Petros P (2021b) Saving the planet with appropriate biotechnology: 2. Cultivate Shellfish to Remediate the Atmosphere/Salvando el planeta con biotecnología apropiada: 2. Cultivar mariscos para remediar la atmósfera. Mexican J Biotechnol 6:31–91. https://doi.org/10.29267/mxjb.2021.6.1.31

Moore D, Robson GD, Trinci APJ (2020) 21st century guidebook to fungi, 2nd edn. Cambridge University Press, Cambridge, UK

Munari C, Rossetti E, Mistri M (2013) Shell formation in cultivated bivalves cannot be part of carbon trading systems: a study case with *Mytilus galloprovincialis*. Mar Environ Res 92:264–267. https://doi.org/10.1016/j.marenvres.2013.10.006

NASEM (2019) Negative emissions technologies and reliable sequestration: a research agenda. National Academies of Sciences, Engineering, and Medicine of the United States of America. The National Academies Press, Washington, DC. https://doi.org/10.17226/25259

Nicholls RJ, Marinova N, Lowe JA, Brown S, Vellinga P, de Gusmão D, Hinkel J, Tol RSJ (2011) Sea-level rise and its possible impacts given a 'beyond 4°C world' in the twenty-first century. Philos Trans R Soc A Math Phys Eng Sci 369:161–181. https://doi.org/10.1098/rsta.2010.0291

OECD (2010) Cities and climate change. OECD Environment Working Paper No 1 (ENV/WKP(2007)1). OECD Publishing, Paris, France. https://doi.org/10.1787/9789264091375-en

Palomo I, Locatelli B, Otero I, Colloff M, Crouzat E et al (2021) Assessing nature-based solutions for transformative change. One Earth 4(5):730–741. https://doi.org/10.1016/j.oneear.2021.04.013

Petit JR, Jouzel J, Raynaud D, Barkov NI, Barnola J-M et al (1999) Climate and atmospheric history of the past 420,000 years from the Vostok ice core Antarctica. Nature 399:429–436. https://doi.org/10.1038/20859

Preiß-Daimler I, Zarkogiannis SD, Kontakiotis G, Henrich R, Antonarakou A (2021) Paleoceanographic perturbations and the marine carbonate system during the middle to late Miocene carbonate crash—a critical review. Geosciences 2021 11(2), article 94. https://doi.org/10.3390/geosciences11020094

Randers J (2012) 2052: A global forecast for the next forty years. Chelsea Green Publishing Co., Hartford, Vermont, USA, 376 p. ISBN: 9781603584210

Randers J, Goluke U (2020) An earth system model shows self-sustained melting of permafrost even if all man-made GHG emissions stop in 2020. Sci Rep 10, article number 18456. https://doi.org/10.1038/s41598-020-75481-z

Raw JL, Adams JB, Bornman TG, Riddin T, Vanderklift MA (2021) Vulnerability to sea-level rise and the potential for restoration to enhance blue carbon storage in salt marshes of an urban estuary. Estuar Coast Shelf Sci 260, article number 107495. https://doi.org/10.1016/j.ecss.2021.107495

Revelle R, Suess HE (1957) Carbon dioxide exchange between atmosphere and ocean and the question of an increase of atmospheric CO_2 during the past decades. Tellus 9:18–27. https://doi.org/10.1111/j.2153-3490.1957.tb01849.x

Riquet F, De Kuyper CA, Fauvelot C, Airoldi L, Planes S, Fraschetti S et al (2021) Highly restricted dispersal in habitat-forming seaweed may impede natural recovery of disturbed populations. Sci Rep 11, article number 16792. https://doi.org/10.1038/s41598-021-96027-x

Russill C, Nyssa Z (2009) The tipping point trend in climate change communication. Glob Environ Chang 19:336–344. https://doi.org/10.1016/j.gloenvcha.2009.04.001

Sen A, Dabi N (2021) Tightening the net: net zero climate targets—implications for land and food equity. Oxfam GB for Oxfam International. ISBN: 9781787487796. 42 p. https://doi.org/10.21201/2021.7796. https://oxfamilibrary.openrepository.com/handle/10546/621205

Smith JB, Schneider SH, Oppenheimer M, Yohe GW, Hare W, Mastrandrea MD, Patwardhan A, Burton I, Corfee-Morlot J, Magadza CHD, Füssel H-M, Pittock AB, Rahman A, Suarez A, van Ypersele J-P (2009) Assessing dangerous climate change through an update of the Intergovernmental Panel on Climate Change (IPCC) "reasons for concern." Proc Natl Acad Sci USA 106:4133–4137. https://doi.org/10.1073/pnas.0812355106

Stocker TF, Qin D, Plattner G-K, Tignor M, Allen SK, Boschung J, Nauels A, Xia Y, Bex V, Midgley PM (eds) (2013) Summary for policymakers, in climate change 2013: the physical science basis. Contribution of working group I to the fifth assessment report of the intergovernmental panel on climate change. Cambridge University Press, Cambridge, UK and New York, USA. https://www.ipcc.ch/report/ar5/wg1/

Stuart-Smith RF, Otto FEL, Saad AI, Lisi G, Minnerop P, Lauta KC, van Zwieten K, Wetzer T (2021) Filling the evidentiary gap in climate litigation. Nat Clim Change. https://doi.org/10.1038/s41558-021-01086-7

Su J, Friess DA, Gasparatos A (2021) A meta-analysis of the ecological and economic outcomes of mangrove restoration. Nat Commun 12, article number 5050. https://doi.org/10.1038/s41467-021-25349-1

Tamburini E, Turolla E, Fano EA, Castaldelli G (2020) Sustainability of Mussel (*Mytilus galloprovincialis*) farming in the Po River Delta, Northern Italy, based on a life cycle assessment approach. Sustainability 12, article number 3814. https://doi.org/10.3390/su12093814

Tellman B, Sullivan JA, Kuhn C, Kettner AJ, Doyle CS, Brakenridge GR, Erickson TA, Slayback DA (2021) Satellite imaging reveals increased proportion of population exposed to floods. Nature 596:80–86. https://doi.org/10.1038/s41586-021-03695-w

Terlouw T, Treyer K, Bauer C, Mazzotti M (2021) Life cycle assessment of direct air carbon capture and storage with low-carbon energy sources. Environ Sci Technol 55:11397–11411. https://doi.org/10.1021/acs.est.1c03263

Titley MA, Butchart SHM, Jones VR, Whittingham MJ, Willis SG (2021) Global inequities and political borders challenge nature conservation under climate change. Proc Natl Acad Sci USA 118(7), article number e2011204118. https://doi.org/10.1073/pnas.2011204118

UN (1992) The United Nations framework convention on climate change. http://unfccc.int/resource/docs/convkp/conveng.pdf

van Andel TH, Heath GR, Moore TC Jr (1975) Cenozoic history and paleoceanography of the central equatorial Pacific Ocean. Geol Soc Am Mem 143:1–134. https://doi.org/10.1130/MEM143-p1

van Andel TH, Thiede J, Sclater JG, Hay WW (1977) Depositional history of the South Atlantic Ocean during the last 125 million years. J Geol 85:651–698. https://doi.org/10.1086/628357

WEF (2020) The global risks report 2020: an unsettled world. Cologny/Geneva: World Economic Forum. https://www3.weforum.org/docs/WEF_Global_Risk_Report_2020.pdf

Welden EA, Chausson A, Melanidis MS (2021) Leveraging nature-based solutions for transformation: reconnecting people and nature. People Nat (Brit Ecol Soc e-J). https://doi.org/10.1002/pan3.10212

Wuebbles DJ, Fahey DW, Hibbard KA, Dokken DJ, Stewart BC, Maycock TK (eds) (2017) Climate science special report: fourth national climate assessment, vol I. U.S. Global Change Research Program, Washington, DC, USA, 470 p. https://doi.org/10.7930/J0J964J6. https://science2017.globalchange.gov/

Yao B, Xiao T, Makgae OA, Jie X, Gonzalez-Cortes S, Guan S et al (2020) Transforming carbon dioxide into jet fuel using an organic combustion-synthesized Fe-Mn-K catalyst. Nat Commun 11, article number 6395. https://doi.org/10.1038/s41467-020-20214-z

Zahnle K, Schaefer L, Fegley B (2010) Earth's earliest atmospheres. Cold Spring Harbor Perspect Biol 2, article a004895. https://doi.org/10.1101/cshperspect.a004895

Cultivate Shellfish to Remediate the Atmosphere

2

David Moore, Matthias Heilweck, and Peter Petros

2.1 In this Chapter…

The very recent research that indicates that massive tree planting is not the panacea that many believe, is discussed. Photosynthetic carbon capture by trees and other green plants is widely thought to be our most effective strategy to limit the rise of CO_2 concentrations in the atmosphere by pulling carbon from the atmosphere into the sinks represented by the plant body and the soil. However, practical experience indicates that putting such plans into effect could do more harm than good to our environment. Planting trees can release more carbon from the soil sink than the plants sequester into their biomass. And, in all cases, the plant biomass sink is only ever a temporary sequestration because when the plant dies its biomass rots, and its sequestered carbon is returned to the atmosphere. Forests should be planted for the intrinsic values of forests; for clean, oxygenated air, natural biodiversity, and restorative conservation of terrestrial ecosystems, rather than tree planting as a means to sequester atmospheric CO_2. This chapter describes the basic message of the book, which is that shellfish cultivation as a carbon sequestration strategy is both more immediately rewarding and more helpful in the very long term. A considerable proportion of shellfish biomass is represented by the animals' shells, and shellfish shell is made by converting atmospheric CO_2 into crystalline calcium carbonate which is stable for geological periods of time. The essentials of habitat conservation, ecosystem balance and carbon sequestration for carbon-offsetting programmes are also introduced; topics developed in chapters which follow.

2.2 Plant Trees for the Intrinsic Value of Forests

Photosynthetic carbon capture by trees is widely considered to be possibly our most effective strategy to limit the rise of CO_2 concentrations in the atmosphere, and there are several ambitious targets to promote forest conservation, afforestation and atmosphere restoration on a global scale.

The *Intergovernmental Panel on Climate Change* Special Report of 2018 (Masson-Delmotte et al. 2019) suggested that an increase of 1 billion hectares of forest will be necessary to limit global warming to 1.5 °C by 2050. In the same publication year, Bastin et al. (2019) mapped the global potential tree coverage and estimated that the world's ecosystems could support an additional 0.9 billion (0.9×10^9) hectares of continuous forest (corresponding to more than a 25% increase in presently forested area) and that such a change has the potential to cut the atmospheric carbon pool by about 25%. We all like trees and we are all in favour of planting more of them, but as any mycologist would point out, there is a negative side to these estimations that seems to be escaping notice.

This is that forests don't only contain trees that can store gigatonnes of carbon in the wood they

D. Moore et al., *Aquaculture: Ocean Blue Carbon Meets UN-SDGS*,
Sustainable Development Goals Series, https://doi.org/10.1007/978-3-030-94846-7_2

make; forests also contain ***wood-decaying fungi*** that can (and do) digest that wood, releasing greenhouse gases, including CO_2, in the process.

Chlorinated hydrocarbons also make a normal every-day contribution to the degradation of timber by forest fungi. The fungal chloromethane contribution to the atmosphere has been estimated at around 150,000 tonnes per annum (Watling and Harper 1998), which, in the year of that publication, was about 60% ***more*** than was released into the atmosphere by industrial coal burning furnaces worldwide.

Of course, the ultimate end-product of food digestion by all aerobic living things, including those wood-digesting fungi, is CO_2. On a global scale, completely natural decomposition of dead wood in the world's forests releases billions of tonnes of CO_2 to the atmosphere each year, a similar magnitude, in fact, to the annual CO_2 emissions from fossil fuel combustion (Rinne-Garmston et al. 2019).

In recent years, an increasing number of studies have warned against too great a reliance on tree planting. For example, Boysen et al. (2017) noted that using biomass plantations to sequester carbon would reduce biodiversity, because they are likely to be monocultures of fast-growing species quite different from the native species. Furthermore, such plantations are likely to occupy scarce agricultural land that might otherwise be used for primary food production. These authors concluded: "…that this strategy of sequestering carbon is not a viable alternative to aggressive emission reductions". In the rest of this section we will discuss some more recent research that also, but for different reasons, casts doubt on the viability of tree planting as a method of long-term sequestration of carbon from the atmosphere.

Despite the fact that photosynthetic carbon capture by trees is most often the first thought in the minds of those hoping to limit the rise of CO_2 concentrations in the atmosphere, the problem with carbon capture by green plants (trees, kelp forests and peat mosses alike) is that it is **temporary**. When the plants die the plant debris is subject to decay and digestion and the ultimate end-product of digestion is the release of CO_2 back to the atmosphere. On a global scale, the world's forests release billions of tonnes of CO_2 to the atmosphere each year. In the temperate zone, we can all observe for ourselves every year that the decomposition of seasonally shed leaves, petals, ripe fruit and dead wood releases CO_2 to the atmosphere in the same year it was fixed (Fig. 2.1).

And even when the tree trunk itself dies, there are all those wood decay fungi in every forest waiting to help things along (Fig. 2.2).

If you hope that terrestrial green plants can effectively sequester carbon from the atmosphere, and meet the ambitious targets to promote forest conservation, afforestation and restoration on a global scale, you are bound to be disappointed; because you are expecting too much of them. And this applies as much to moorland and peat bogs as to forests.

According to the very useful Wikipedia article [URL: https://en.wikipedia.org/wiki/Peat] "Peat, also known as turf, is an accumulation of partially decayed vegetation or organic matter. It is unique to natural areas called peatlands, bogs, mires, moors, or muskegs… The peatland ecosystem is the most efficient carbon sink on the planet… In natural peatlands, the annual rate of biomass production is greater than the rate of decomposition, but it takes thousands of years for peatlands to develop the deposits of 1.5–2.3 m, which is the average depth of the boreal [northern] peatlands" (like those in Britain). Overall, in the northern hemisphere, peatlands cover an area of about 3.7 million km^2; about half this being permanently frozen (permafrost).

These northern peatlands are estimated to store around ***415 billion metric tonnes of carbon***, which is equivalent to over 45 years of current global CO_2 emissions. It is projected that global warming will cause the northern peatlands to become a major source of greenhouse gas emissions into the atmosphere (methane, carbon dioxide and nitrous oxide) as the peatlands warm up (Hugelius et al. 2020).

Unfortunately, planting trees on peatland will not help. Friggens et al. (2020) recorded a 58% ***reduction*** in soil organic carbon stocks 12 years after birch trees (*Betula pubescens*) had been planted in heather (*Calluna vulgaris*) moorland.

Fig. 2.1 Photographs of the same tree in summer (top) and winter (bottom) emphasising how deciduous trees shed their leaves at the end of the year. So, by the time the snow comes all the leaves, flowers and fruit of the summer season have been digested and their carbon returned to the atmosphere. Open access images from https://pixabay.com/

Significantly, this decline was not compensated for by the gains in carbon contained in the growing trees. This was a continuation of a long-term study of the effects of planting two native tree species (*Betula pubescens* and *Pinus sylvestris*), which have a wide Eurasian distribution, in *Calluna vulgaris* moorland with podzol and peaty podzol soils in Scotland. The study demonstrated that ***39 years after planting***, the carbon sequestered into tree biomass did offset the carbon lost from the soil but, crucially, there was ***no overall increase in carbon sequestered by the ecosystem***. The authors state that:

> "The results are of direct relevance to current policies, which promote tree planting on the assumption that this will increase net ecosystem C storage and contribute to climate change mitigation. Ecosystem-level biogeochemistry and C fluxes must be better quantified and understood before we can be assured that large-scale tree planting in regions with considerable pre-existing [soil organic carbon] stocks will have the intended policy and climate change mitigation outcomes" (Friggens et al. 2020).

The mosses (typically species of *Sphagnum*) that thrive in peatlands retain rainwater, so in addition to carbon sequestration, an important function of peatlands is the stabilisation of water

Fig. 2.2 Felled logs colonised by mycelia of *Trametes versicolor* (Basidiomycota; commonly called Turkey Tail in the United States) (A, B, C) and *Hypholoma fasciculare*, D, commonly known as the Sulphur Tuft. Early in the season the mycelia reach the end of the log and the differentiating sporophores outline the separate decay columns in the timber (A), which are formed by mycelia belonging to different compatibility groups. Sporophores are formed on these surfaces later in the season (B, C and D). Photographs by David Moore of logs in the Lovell Tree Collection Arboretum at Jodrell Bank Discovery Centre, Cheshire (https://www.jodrellbank.net/)

flows from hills, which reduces the risk of flash flooding. Peat bogs also filter and clean catchments around lakes used as domestic water reservoirs. As a traditional source of domestic fuel, and more recently as a source of horticultural composts, peat bogs have been greatly damaged by peat mining and most are certainly in urgent need of conservation. But the mosses grow slowly and although one hectare of healthy peatland holds as much carbon as one hectare of tropical rainforest they offer only limited promise for carbon sequestration. The Wikipedia entry goes on to explain that the water table of *Sphagnum* moss bogs must be maintained close to the surface to maintain the deeper layers of peat as a stable carbon sink. If they are drained or disturbed (by erosion or peat mining) the deeper layers are oxidised, and historical CO_2 is returned to the atmosphere. It comes down to deciding how much of your land do you want to cover in permanently waterlogged, and preferably frozen, peat bog?

The United Kingdom's *Office For National Statistics* (ONS 2016) estimated that in 2007 UK soils contained approximately 4 million tonnes of carbon, of which 57% was the carbon stored in peat soils, but as the majority of UK peatlands are degraded (Natural England 2010), they are a highly significant source of greenhouse gas emissions. Consequently, the aim of peatland restoration is to reduce the extent of these emissions as a contribution to the 'net zero future' (Natural Capital Committee 2020). The authors of the Natural Capital Committee report refer to the huge publicity given to the United Kingdom's plans for planting 11 million trees to sequester carbon emissions, but they warn that conserving carbon in soils is equally or more important. The report states that:

> "The right tree in the right place for the right reason can bring a multitude of benefits…" but adds "the wrong trees in the wrong places can have adverse impacts on soil (including soil carbon), water flows, water quality, recreation, biodiversity and air quality".

In the United Kingdom, the countryside charity ***CPRE*** has warned that emissions from UK peatland could cancel out all carbon reduction achieved through new and existing forests, in their August 2020 report entitled '*Net-zero virtually impossible without more ambition on peatlands*' [https://www.cpre.org.uk/]. Indeed, similar concerns about adverse impacts on carbon sequestration being caused by "the wrong trees in the wrong places" have been expressed by studies of ecosystems as far apart as Chile (Heilmayr et al. 2020) and China (Hong et al. 2020).

The overall conclusion seems to be that mass tree planting will harm the environment if not planned properly. Forests are only effective CO_2 sinks when they grow biomass or extend their area ***and remain alive***. Seasonally shed leaves, petals, ripe fruit and dead wood are digested and respired to CO_2 in the same year the CO_2 was fixed from the atmosphere (Fig. 2.1). And when the tree dies there are legions of animals, bacteria and, especially, fungi (Fig. 2.2) just waiting for the chance to digest the forest's biomass and convert it back to atmospheric CO_2 as quickly as possible. ***That's life***.

Despite these gloomy observations regarding trees and other photosynthetic plants, there remains some hope that better management of forests and their carbon stocks can help improve overall terrestrial carbon cycle management (Soudzilovskaia et al. 2019; Domeignoz-Horta et al. 2020; Manrique and Franco 2020) although the fact remains that we cannot rely on terrestrial vegetation to mitigate the effects of climate change for the simple reason that such a prospect expects too much of them. The fundamental problem with carbon capture by green plants is that it is **temporary**. When the plants die the plant debris is subject to decay and digestion that releases CO_2 back to the atmosphere. Globally, completely natural decomposition of dead wood in the world's forests releases billions of tonnes of CO_2 to the atmosphere each year, similar in magnitude to the annual CO_2 emissions from fossil fuel combustion (Rinne-Garmston et al. 2019). When a tree dies its entire body is digested within a few decades, the carbon being released back into the atmosphere as respiratory CO_2 in a global total of 10.9 billion metric tonnes per year (Seibold et al. 2021). Even when a kelp forest dies, it is digested and respired by a host of animals and microbes. Whether terrestrial or aquatic, photosynthetic plants **only** sequester atmospheric carbon while they are **alive**. To find an alternative we need to look further into Earth's history and recognise what **natural** ways of controlling atmospheric greenhouse gasses the planet has employed in its past.

Of course, sustainably managed forests can be harvested to provide wood fuels as environmentally benign alternative to fossil fuels (but still returning their CO_2 to the atmosphere), or timber for buildings and furniture. There are about 60 or so indoor wood decay fungi from which you need to protect your timber buildings and furniture, including dry rot, wet rot, cellar rot and oak rot. The longevity of the carbon pools represented by wood products derived from harvested timber depends upon their use: lifetimes may range from less than one year for fuelwood, to several decades or centuries for lumber; but still, timber is only ever a temporary remedy for the atmosphere.

Indeed, it has been suggested that there is firm evidence that current projections of global forest carbon sink ***persistence*** are too optimistic because the increased growth rates of trees caused by increased levels of CO_2 in the atmosphere may shorten the lifespan of forest trees (Brienen et al. 2020):

> "… Faster growth has a direct and negative effect on tree lifespan, independent of the environmental mechanisms driving growth rate variation. Growth increases, as recently documented across high latitude and tropical forests, are thus expected to reduce tree lifespans…" and that "… recent increases in forest carbon stocks may be transient due to lagged increases in mortality …" (quoted from Brienen et al. 2020).

So, current plans for tree planting on a massive scale are not the panaceas that many believe. Putting such plans into effect could do more harm than good (Elgin 2020; Friggens et al. 2020; Goswami 2020; Heilmayr et al. 2020; Hong et al. 2020; Natural Capital Committee 2020; and listen to The Climate Question 2021 podcast).

Sadly, our present forests are currently suffering from the effects of the climate changes that have already occurred. Many forested areas are dying due to drought, often amplified by more devastating wildfires, and virulent, newly emerged and invasive pests and diseases (Demeude and Gadault 2020). The threat to forests is worldwide and, in many cases, can be traced to invasions of non-native bark and ambrosia beetles which carry symbiotic fungi to feed their larvae within galleries they bore into the tree. It is the sudden appearance of pathogenicity in the fungus that is the new and currently uncontrollable threat to forest ecosystems, and fruit and timber industries, around the globe. Triggered by climate change, some invasive bark and ambrosia beetle/fungus symbioses are shifting from non-pathogenic saprotrophy in their native ranges to a prolific tree-killing in invaded ranges (Moore et al. 2020). Duffy et al. (2021) project an even more dramatic future. They estimate that the terrestrial carbon sink currently mitigates about 30% of anthropogenic carbon emissions but as global warming progresses, respiration rates will continue to rise in contrast to sharply declining rates of photosynthesis; they expect the land carbon sink to be halved by as early as 2040 under business-as-usual emissions. We cannot rely on forests and other terrestrial vegetation to mitigate the effects of climate change; it's dying because of it!

Despite all these negative reports and seemingly pessimistic facts regarding trees and other terrestrial vegetation, there remains some hope that better management of forests and their carbon stocks can help improve overall terrestrial carbon cycle management providing knowledge of the role of fungi and soil microbes in carbon cycling is implemented into sustainable forest management practices (Soudzilovskaia et al. 2019; Domeignoz-Horta et al. 2020).

The authors like trees (and other plants) and we are in favour of planting more of them, ***but they should be planted for their intrinsic ecosystem value***, because there are negative aspects to relying on them so heavily as a way to sequester carbon from the atmosphere on the long-term basis required for full and lasting benefit. The ***Trillion Tree Initiative*** is a World Economic Forum initiative, designed to support the UN *Decade on Ecosystem Restoration 2021–2030*, led by the United Nations Environment Programme (UNEP) and the Food and Agriculture Organization of the United Nations (FAO) [view: https://www.1t.org/]. This, and the parallel programme ***Trillion Trees***, which is a joint venture between BirdLife International, Wildlife Conservation Society (WCS) and the World Wide Fund for Nature (WWF) [view: https://trilliontrees.org/] sometimes seem to be the only nature-centric solutions catching the attention of mainstream media. Unfortunately, all is not properly managed down in the forest. Martin et al. (2021) surveyed 174 tree planting organisations to determine the type of enterprises involved in tree planting, their geographic locations and tree planting approaches. The number of organisations had increased almost three-fold in the past 30 years, especially among ***for-profit*** businesses. The organisations surveyed reported planting nearly 1.4 billion trees across 74 countries since 1961. Plantings were most frequently made to establish agroforestry systems using mixed species and single species plantations, or by assisted natural regeneration. Overall, tree planting programmes were intended to support local communities as well as environmental objectives. The most frequently reported species were commercial or utilitarian, but only 18% of organisations reported monitoring their plantations, and only 5% measured survival rates of their plantings. The overall impression gained from the survey is stated in the title of this paper: '*People plant trees for utility more often than for biodiversity or carbon*' (Martin et al. 2021) and this survey of real-life tree planting programmes demonstrates that better coordination, better planning. better management and better governance are all required if tree planting is to reach its full beneficial potential.

China is currently the world's single largest emitter of CO_2, being responsible for approximately 27% of global fossil fuel emissions in 2017. Several Chinese provinces have established a pattern of rapid afforestation of progressively larger regions, with provincial forest areas increasing by between 0.04 million and 0.44 million hectares per year during the past 10–15 years (Wang et al. 2020). This large-scale expansion of fast-growing plantation forests is estimated to correspond to a Chinese land biosphere sink equivalent to about 45 per cent of annual anthropogenic emissions in China over that 10–15 year period. Though this sound extremely encouraging, Wang et al. (2020) also state that the afforestation effort "... contributes to timber exports and the domestic production of paper ...", which means that the carbon sequestration is only temporary because the longevity of this impressive carbon sink is entirely dependent on the effectiveness and efficiency of future paper and timber ***recycling*** programmes. If these products are rapidly discarded, burnt or composted, the sequestered carbon they represent will be returned to the atmosphere.

Brienen et al. (2020) suggest that the lack of persistence of sequestered forest carbon raises the necessity of curbing greenhouse gas emissions; we, of course, would prefer to offer **an alternative *bio*technology** for really long-term carbon sequestration, as well as curbing the emissions. So, what about engineering solutions for 'aggressive emission reductions' to limit the rise of CO_2 concentrations in our atmosphere?

Most current research on 'aggressive emission reductions' is focussed on the integration of new technologies to capture CO_2 from flue gasses in power plants, which are responsible for about 80% of the worldwide CO_2 emissions (Romano et al. 2013). Methods based on exposing flue gas to water under suitable conditions ('hydrate-based processing') is a promising and high efficiency technology for CO_2 capture, but the high cost of maintaining suitable conditions for hydrate formation is preventing wide industrial application of this technology (Li et al. 2019). In Chap. 7 we discuss biotechnological carbon capture and storage (CCS), comparing nature-friendly solutions, such as kelp forests, shellfish and other aquaculture, with the heavy-engineering fossil fuel industrial solutions.

So, if expanding the forests and capturing CO_2 from flue gases are unlikely to save us, are we doomed? Well, no, actually; we just need to change our focus; turn away from trees (but still plant them; forests are good for us in so many ways) and concentrate on ***shellfish*** (Moore 2020). If we plant a trillion trees, then in about 70 years' time the trees will die and their timber will be digested by wood-degrading bacteria, fungi and insects and their sequestered CO_2 will be returned to the atmosphere. If we cultivate one trillion ***mussels*** (each of which is capable of making 10 g of shell) then in about 5 years the animals will die (either in the kitchen or in nature) and 10 million tonnes of shell will be deposited on the seafloor where it will remain unchanged, potentially for millions of years. The calcifying animals leave us a legacy of 4.4 million tonnes of CO_2 **permanently** drawn down from the atmosphere in a form which is never digested or degraded, and never returned to the atmosphere. And in the next five years we could cultivate another trillion mussels.

2.3 Cultivate Shellfish: Save the Atmosphere

There are four interesting and readily available facts about shellfish that are relevant to this discussion.

1. There's a lot of shell in shellfish.
2. Shellfish shell is mineralised carbon dioxide from the atmosphere.
3. Shellfish shell is not digested and is chemically stable for geological periods of time.
4. The shellfish cultivation industry offers unique opportunities for limiting climate change and enhancing conservation.

Let's examine each of these points in turn.

2.4 There's a Lot of Shell in Shellfish

Since our childhood days we have always enjoyed shellfish foods, especially cockles (clams), mussels and crab; and all the other seaside treats, too. And we've always been aware of the amount of shell left over after the meal. Think of your average shellfish meal for two (Figs. 2.3 and 2.4):

And on a global scale? Data from FAO Fisheries and Aquaculture Information and Statistics Branch (as of 25 May 2019) show that over the years 2010–2017 aquaculture harvests across the globe totalled 53,512,850 metric tonnes of crustaceans and 122,527,372 metric tonnes of molluscs (a combined total of 176,040,222 metric tonnes in 8 years (Table 2.1).

It's a reasonable guestimate (Figs. 2.3 and 2.4) that the shell represents an average 50% of the animal's mass, although Alonso et al. (2021) use a value for the average contribution of shell to total bivalve body weight which varies from 70 to 95%. The more conservative 50% makes for easy arithmetic so, in our case the total **shellfish shell** produced was 88 million tonnes over 8 years. An average of ***11 million tonnes of shell per year***.

If we further assume that the shell is made entirely from $CaCO_3$; then, on a molar mass basis, carbon represents 12% of the mass of calcium carbonate. So, 11 million tonnes of shell per year is equivalent to 1.32 million tonnes of carbon per year being captured from the atmosphere by current aquaculture activities.

Global carbon emissions from fossil fuel use were 9.8 **billion** tonnes in 2014 (equivalent to 35.9 billion tonnes of carbon dioxide) [source: https://www.co2.earth/global-co2-emissions]. So, a thousand-fold increase in today's aquaculture would permanently remove about 14% of the global carbon emissions ***in each year***. Extrapolating the figures in Table 2.1 suggests that this year's global aquaculture farming will remove about 5.5 million tonnes of CO_2 from the

Fig. 2.3 Mussels: fresh 810 g, shell dry weight 296 g = 36.5% of the fresh mussel is shell. Photographs by David Moore

Fig. 2.4 **Cockles (= Clams)**: fresh 950 g. shell dry weight 557 g = 58.6% of the fresh cockle is shell. Photographs by David Moore

Table 2.1 Quantities of shellfish harvested across the globe in each specified year

Species	Millions of metric tonnes each year, 2010–2017								Totals (metric tonnes)
	2010	2011	2012	2013	2014	2015	2016	2017	
Crustaceans	5.5	5.8	6.0	6.2	6.7	7.1	7.7	8.4	53,512,850
Molluscs	13.7	13.8	14.4	14.9	15.7	15.8	16.8	17.4	122,527,372
						Grand total			176,040,222

Data taken from Food and Agriculture Organisation of the United Nations, Fisheries & Aquaculture Department, Fishery Statistical Collections, Global Aquaculture Production. In this context, aquaculture means farming of aquatic organisms including fish, molluscs, crustaceans and aquatic plants. Farming implies some form of intervention in the rearing process to enhance production, such as regular stocking, feeding, protection from predators, etc. Farming also implies individual or corporate ownership of the stock being cultivated. Aquatic organisms which are exploitable by the public as a common property resource, with or without appropriate licences, are defined as fisheries and are not included in these data. Refer to Costa-Pierce and Chopin (2021) for a refreshingly candid discussion of the way FAO statistics are used and misused in descriptions of aquaculture for food production. Here we use FAO statistics to evaluate our argument that for atmosphere amelioration shellfish should be cultivated for their shells, taking their meat as a by product.

atmosphere. Biotechnological research on aquaculture is well-established (e.g. Rasmussen and Morrissey 2007; Xiang 2015); as is practical knowhow advice (Lovatelli 1990; Utting and Spencer 1991; Helm et al. 2004). A more unusual suggestion would be to cultivate coccolithophore algae in terrestrial ponds or even giant illuminated fermenters (Chap. 6; Moore 2021); some species of these single-celled green algae coat their cells with plates of microcrystalline $CaCO_3$, which they shed during their growth. It has been demonstrated that each hectare (10,000 m^2) of their pond cultivation could remove 0.66 tonnes of carbon (permanently) from the atmosphere each year (Moheimani and Borowitzka 2006).

Could we increase shellfish production to a level that would achieve very significant sequestration of atmospheric CO_2? Possibly. If we ***doubled*** aquaculture production of crustaceans and molluscs ***each year*** then from the 14th year we could be removing 10.7 **billion** tonnes of carbon from the atmosphere ***annually***.

Sustained annual doubling may not be realistic; but this simple calculation indicates that with determined effort (and adequate finance) to massively increase aquaculture production we could be permanently extracting significant amounts of carbon every year from the atmosphere within the timescale that is currently envisaged for carbon capture by afforestation.

According to Spanner Film's *Pie Net Zero* video "...if we turn a quarter of UK [coastal] waters over to mussel farming the shellfish would drawdown about an eighth of our [the UK's] total emissions" [https://www.youtube.com/watch?v=rhA5LgXMBkw&t=24s and https://www.youtube.com/watch?v=o-YuoWaCfhI&t=30s].

2.5 Shellfish Shell is Mineralised Carbon Dioxide from the Atmosphere

Molluscan shell is a ***biomineral composed of*** $\boldsymbol{CaCO_3}$ with a small amount of matrix proteins included, which are responsible in the live animal for directing the species-specific crystal growth. Arthropod (crab, shrimp, lobster) exoskeletons are composed largely of chitin but this is hardened by deposition of calcium-magnesium carbonate nanocrystals. In either group of organisms, shellfish shell is about 95% ***crystalline calcium carbonate***.

The animals make this by absorbing calcium through specific transporters in their tissues and reacting it with bicarbonate (HCO_3^-), which is synthesised from CO_2 (see Fig. 20 in Chap. 4). Some of the bicarbonate is absorbed directly from the surrounding water (or gaseous atmosphere for terrestrial species), the rest derives from CO_2 generated from the animal's food. Of course, most of these animals are filter feeders; the CO_2 generated by their metabolic cycles comes from digestion of plankton and is ultimately derived from planktonic photosynthesis (Tassanakajon et al. 2008). But since all food chains start with a photosynthetic producer organism that makes its own food, whatever the shellfish animal eats depends ultimately on fixation of photosynthetic carbon from the atmosphere. This is true of all the carbon in the food of predators, scavengers, filter feeders and detritus feeders alike, aquatic and terrestrial. Globally, metabolic carbon is derived from photosynthetic fixation of atmospheric carbon; there is no other source.

It is often pointed out by professional marine biologists that the formation of calcium carbonate by calcifier organisms is a ***source*** of carbon dioxide emitted to the atmosphere, and for this reason they dismiss calcifiers as a means to sequester atmospheric carbon. Indeed, some go further and brand the calcification reaction ($2HCO_3^- + Ca^{2+} \rightarrow CaCO_3 + CO_2 + H_2O$) as "a major way by which CO_2 is returned to the atmosphere". Unfortunately, this is an incomplete and misguided interpretation of the facts. This is certainly the calcifying reaction scheme. It is valid as written here in freshwater (with a 1:1 ratio between the CO_2 and $CaCO_3$ products). In seawater about 30% of this CO_2 dissolves and forms carbonic acid (H_2CO_3, which, of course dissociates into H^+ and HCO_3^-). As a result, the molar ratio of $CO_2/CaCO_3$ in seawater is about 0.6 rather than 1. Ignoring this minor detail, the calcifying reaction removes TWO bicarbonate ions from the environment ONE is precipitated as $CaCO_3$ and ONE is returned to the environment as CO_2. So, while it is true that "precipitation of calcium carbonate is a source of carbon dioxide (CO_2)" it is illogical to claim that returning **one** out of **two** carbons to the environment is a "major way by which CO_2 is returned to the atmosphere".

BUT, if you go one step further and ADD the consideration that there are a great many calcifying organisms in the oceans, which are all cycling through this reaction 24/7, then you can rightly claim that this, in aggregate, is a major way by which CO_2 is returned to the atmosphere; ***providing*** you remember that the other **one** of those **two** carbons on the left of the reaction scheme above is precipitated as $CaCO_3$ and ***admit*** the matching claim that if this ***is*** a major way by which CO_2 is returned to the atmosphere then it is ***also*** a major way by which **carbon is REMOVED FROM the atmosphere**.

Bivalves constitute a substantial component of communities at and within the coastal sea floor.

Among them, blue mussels, classified in the *Mytilus* species complex, are important foundation species throughout the temperate and polar littoral zones of the northern and southern hemispheres.

> "… Foundation species define ecosystems, control the biological diversity of associated species, modulate critical ecosystem processes, and often have important cultural values …" (Ellison 2019).

The 'cultural value' of mussels, of course, is that they are a vital economic resource for the aquaculture industry (see references in Telesca et al. 2018).

The environment in which mussels (and other bivalves) grow affects the animals' growth patterns in a complex way, with several interacting factors resulting in a variety of shell shapes. Growing awareness of climate change and its consequences for biodiversity, especially of habitat-forming bivalve species, has prompted use of *Mytilus* spp. as model organisms for studying ecological and physiological responses to environmental conditions. Shell shape variability in bivalves is an important aspect of the animals' ability to adapt to environmental stresses, and understanding, and quantifying, environmental effects on shell shape will contribute to our understanding of how bivalve communities might react to ongoing climate change (Telesca et al. 2018; and references therein).

Since all environmental conditions vary in both time and space, wide variation in shell morphology is to be expected, even in animals from the same locality. In any one bivalve population, variation in shell form can be attributed to any or all of the following differences (Seed 1968, 1980; Gimin et al. 2004; Telesca et al. 2018; and references therein):

- Temperature and food supply were the main drivers of mussel shape heterogeneity.
- Salinity had the strongest effect on the latitudinal geographic distribution of *Mytilus* shapes: lower salinities producing more elongated and narrower shells.
- Age also influences the shell; old mussels having proportionately heavier shells, in which width often exceeds shell height, than young.
- Genotypic differences and hybridisation within the *Mytilus edulis* species complex also influence shell variation.
- Population density coupled with population growth rate, which probably exert their effects through physical compression. This being maximal in localities of fast growth and high population density and least in areas of slow growth and low density. Compression (crowding) of mussels leads to an elongated shell form, while shells experiencing low crowding are more triangular in shape. Even in the same habitat, however, growth rates and population densities are very variable.
- Predation pressure influences changes in shell proportions and structure in mussels. Various species of aquatic birds are the main predators of mussels, some birds diving under water to take buried mussels, other predators are starfish and marine gastropods, such as the dog whelk, *Nucella lapillus*. In the Baltic Sea, mussels constitute 80–90% of the coastal animal biomass, this dominance is attributed to an almost complete absence of predators. Conversely, Northern Atlantic and Arctic *Mytilus* populations face predation pressures varying with latitude and competition for space (Telesca et al. 2018; and references therein).

Variability of shell shape is an adaptive feature allowing the animals to respond to less favourable environmental conditions. Food supply, temperature and salinity have the greatest influence on this variability. In intertidal bivalves, shell weight is supported by a smaller biomass in animals located high on the shore than in those at the low water mark. A response determined by the need to provide space within the shell for filter feeding by the gills. Mussels respond to shorter feeding periods at higher shore levels by maximising shell growth at the expense of body growth (although both are less than optimal) because gill area can be increased only by increasing shell length (Franz 1993) and thicker and heavier shell valves than normal, that can be closed tightly to protect the body against adverse environmental conditions (and predators) have been reported in *Mytilus edulis* inhabiting periodically dry zones.

When the emersion is periodic, bivalves need to maintain a large volume of water inside the shell to create a sufficiently watery environment for survival of the body tissue. A low rate of tissue growth is a useful strategy for survival under prolonged exposure. If the soft tissue continued to grow steadily and come to occupy a large part of the space, then there would not be enough water inside the shell to support the metabolic needs of the increased tissue (Seed 1968, 1980; Gimin et al. 2004).

2.6 Shellfish Shell is Not Digested and is Chemically Stable for Geological Periods of Time

Ultimately, then, the CO_2 for the shell comes from the atmosphere ***and it stays out of the atmosphere, PERMANENTLY***.

- Intact shellfish shells are excavated regularly from the middens associated with coastal communities of early humans (from around **120,000 years ago**, see Chap. 3; Moore et al. 2021).
- Intact shellfish shells abound in deepwater cores of ancient coastal sediments of **hundreds of thousands of years ago**.
- And remember the fossils from deep time? Ammonites (**65–240 million years ago**), trilobites (**520 million years ago**), brachiopods (**550 million years ago**), shellfish all. Certainly, these fossil shells are changed considerably in chemistry by now (over extended time periods carbonates can recrystallise into calcite, or exchange with silica or iron sulphide in the surrounding rock matrix), but the shell carbonates survive over geological time. The carbonates in shells are neither digested nor degraded. High temperatures are required to release the CO_2 from carbonates (to produce quicklime)—ask the cement industry, which uses fossiliferous limestone as a feedstock for cement production (cement production accounts for about 8% of the fossil CO_2 emissions from industrial sources). The sedimentary limestone rocks derive their calcium carbonate from the biological activities of bryozoa, corals, crinoids, microscopic algae and shellfish shells. Even chemical precipitation, which is an important method by which limestones form, depends on solution of biologically produced $CaCO_3$ as water currents agitate grains of sand and shell fragments together. Warm, shallow waters can be saturated with $CaCO_3$, which crystallises as aragonite on nuclei formed from shell fragments, and builds up in concentric layers to form small multilayered spheres called ooids. In the natural world, the carbonates of shells are only likely to release their $CO_2$when/if they encounter volcanic conditions, or sink in the ocean deeps beneath the Carbonate Compensation Depth of about 5,000 m. How much more permanent, do we need permanent carbon sequestration to be?

2.7 Change the Paradigm

The principal contrast of shellfish shell with trees is that shellfish shell is ***mineralised and permanently solidified*** atmospheric CO_2. Indeed, shellfish cultivation is the ONLY industry on the planet that currently ***removes*** serious quantities of carbon dioxide permanently from the atmosphere.

At present we cultivate shellfish for the meat (the shell is food-waste) and the industry is scaled according to that market. We must ***change the paradigm***: cultivate the bivalves, and those other shellfish, to sequester permanently CO_2 from the atmosphere and accept the food as a saleable by-product. Changing the paradigm means placing the ***value*** of the exercise of shellfish cultivation onto the **production of shell**, taking the **food value** of the animal protein as one of the peripheral or ***additional benefits*** (for which see next section).

There is a recently published study of mussel cultivation economics that can be used to estimate the broad costs of the project(s) we are proposing. Avdelas et al. (2020) reported a decline in mussel aquaculture in the European Union (EU) in the first decades of the twenty-first century, contrasting with increasing aquaculture production of mussels worldwide. In attempting to analyse potential causes for this, these authors investigated the economics (across eight EU nations) of the different mussel production techniques (bouchot, on-bottom, raft and long line; Table 2.2) and it is this aspect of their study we wish to use. In doing this we recognise that mussel production costs are highly variable, and that working with data averaged between different cultivation methods, which have different cost structures, and between different countries with different labour costs is not entirely satisfactory. Nevertheless, we believe that we can arrive at order-of-magnitude estimates this way.

Avdelas et al. (2020) provide a production cost (and farm gate sale price) for mussels produced by four different methods, averaged across the EU and across the years 2010–2016 (Table 2.2).

Table 2.2 shows that the overall average production cost of mussels in the EU over the years 2010–2016 was 0.87 € kg^{-1} (for a farm gate price of 1.08 (€ kg^{-1}). Other useful data from the same source (Table 2.1 in Avdelas et al. 2020) are:

- the grand average value of assets per enterprise = approximately €700,000.
- the grand average turnover per enterprise = approximately €384,000.
- total number of enterprises listed in Table 2.1 = 2,720.

Avdelas et al. (2020) also report that the eight EU countries they analysed produced 480,000 t of mussels in 2016 valued at €328 million. From these data, we calculate that the average EU mussel farming enterprise in 2016 produced an average of 176 t of mussels valued at approximately €121,000 = €687.5 t^{-1} = 0.69 (€ kg^{-1}).

Broadly similar economic statistics were computed by Parappurathu et al. (2017) for the rack (= raft) method of green mussel farming in Padanna, Kerala, India. They estimated the net operating income at USD$ 840 t^{-1} mussels (value of the income in Indian rupees in 2016 recalculated to 2021 USD$ values allowing for currency fluctuations. We thank Dr Balamurali Sreedhar for bringing this publication to our attention).

Using the European data, we can make the conservative estimate that investment of €1 million could create a mussel production facility able to cultivate at least 250 t y^{-1} of mussels; the food value of which is approximately €173,000. But we are interested in ***shell production***, so we calculate *that* 250 t y^{-1} of fresh mussels is approximately equivalent to 125 t of shell, which is equivalent to 15 t of ***carbon removed permanently from the atmosphere each year*** in those

Table 2.2 Economics of mussel production in the EU over the years 2010–2016

Production technique*	Bouchot	On-bottom	Raft	Long line
Number of enterprises	338	96	2039	247
Production cost (€ kg^{-1})	1.65	0.90	0.31	0.62
Farm gate price (€ kg^{-1})	2.04	1.25	0.37	0.66

*Data sourced from Table 2.1 in Avdelas et al. (2020), who consider these four production techniques: **bouchot culture**, also called pole or stake culture, uses poles (traditionally 4–6 m-long trunks of oak trees) which are staked across the intertidal zone in rows, 0.7 m apart; mussel seeds are collected on ropes hung around the poles; **bottom culture** is growing mussels directly on the seabed; **raft culture** uses mussel seeds that settle on ('collection') ropes suspended from a raft; and **long line culture** has collection ropes suspended from a long line which is supported by floats joined together by a cable or chain that is anchored at each end. These techniques are described more fully in Chap. 4; Heilweck and Moore 2021) shells. Consequently, we can estimate that an investment of €1 ***billion*** could remove **15,000 t carbon** from the atmosphere **each year**. But permanently removing carbon from the atmosphere is **not the only benefit** received from this mussel cultivation exercise; because of the food value of the mussel, an investment of €1 ***billion*** would attract an ***annual return***, in terms of mussel meat value, of 17%. So, run the facility for 6 years and by the end of that time you will have removed **90,000 t carbon** from the atmosphere **and** sold enough mussel flesh to **build a second facility**. Six years after that your record will be a total (permanent) removal from the atmosphere of **270,000 t carbon** and enough cash in hand to build ***two more facilities***. And so on; towards the end of this century (Table 2.3).

Who would not want to invest in a programme like that, which, in 50 years' time, could be ***removing*** so much carbon from the atmosphere annually that we'd have to stop it? Aren't pipe-dreams wonderful? And, ***of course***, this ***is*** a pipe dream. Sustained annual doubling is undoubtedly unrealistic; but our plan would be to accelerate the implementation of a historically proven ***natural solution*** to the Earth's climate crisis by **MASSIVELY** expanding the global-scale cultivation of oceanic calcifiers. Our fundamental claim being that cultivation of calcifying organisms (coccolithophores, foraminifera, corals, crustacea and molluscs) on a gigantic scale would make a sustained ameliorative contribution to climate change on this planet; potentially achieving UN's Sustainable Development Goals 13 and 14.

The projections in Table 2.3 indicate that with ***determined effort*** (and adequate finance) to enormously increase aquaculture production we could be permanently extracting significant amounts of carbon every year from the atmosphere within the timescale that is currently envisaged for significant carbon capture by afforestation. And calcifiers remove carbon from the atmosphere; permanently.

This plan may well be difficult to achieve. But bivalve aquaculture is an easy, readily deployed biotechnology that can be instituted across the entire range of scales from indigenous subsistence-level farming to mega scale industrial enterprises. The practicalities of these ideas are indicated in later **chapters** of this book, being brought together into our action plan in Chap. 8.

If you think that even ***determined effort and copious finance*** will be insufficient for this programme to have any effect, then consider:

- How many wind turbines were in operation ten years ago?
- How many all-electric cars did we have on our roads ten years ago?
- How many trees have we chopped down in the last ten years?
- How many extreme weather events, searing droughts, scorching wildfires, devastating flash floods; how many of 'the hottest years yet' have we suffered in the last decade?

Table 2.3 Forward projection of an investment, made in 2020, of €1 billion in a mussel shell production facility

Year	No. of facilities financed	Annual carbon sequestration (Mt)	Total carbon removed (Mt)
2026	1	0.09	0.09
2032	2	0.18	0.27
2038	4	0.36	0.63
2044	8	0.72	1.35
2050	16	1.44	2.79
2056	32	2.88	5.67
2062	64	5.76	11.43
2068	128	11.52	22.95

And also keep in mind that the shellfish cultivation harvest is (on average) **50% meat protein** and **50% shell**. So, if we ***could*** cultivate a few billion tonnes **of shell** each year by 2050, we would have a few billion tonnes **of meat** each year simply as a by-product of the primary shell production activity (at the time of writing, the world's terrestrial farmers are producing about 340 million tonnes of meat each year [source: https://ourworldindata.org]). And this ***would not*** represent an 'oversupply' of a seafood delicacy; rather, it would be a massive source of highly nutritious animal protein for human foods of all sorts (as wide a variety as the food technologists can devise), animal feed supplements for terrestrial farms (to free-up agricultural land for ***human*** food production), and shellfish meal for fish farms (to save all those capture-fish species we are currently over-fishing and driving to extinction).

2.8 Additional Benefits

We want shellfish producers to greatly expand their production ***specifically*** to generate more shell to sequester atmospheric carbon. But, in addition, shellfish shells are a valuable, sustainable, biomaterial with many potential uses ranging from calcium supplementation in poultry farming to pH regulation in hobbyist aquarium systems (Morris 2019; Morris et al. 2019). The EU report (Morris 2019) is written from the conventional point of view that shellfish shells are a waste product that is a candidate for valorisation rather than disposal as a sustainability goal:

> "… It is clear that shells are a potentially valuable commodity, and do not require high-energy processing to give them value. Where shells are produced in a significant volume, it should be possible to find an appropriate valorisation strategy for them within a close-enough proximity to make it both sustainably and economically viable".

Nevertheless, although the report seems not to recognise shells as being solidified atmospheric CO_2, the statement is made that

> "… in many cases, shells can provide more inherent value being returned to the marine environment rather than being used in land-based applications…" (Morris et al. 2019).

They can indeed, and the reason is simply that most cultivated bivalves are capable of building reefs of sufficient size that they provide coastal protection through their wave-calming effects (Ysebaert et al. 2019).

If we ***can*** change the paradigm to cultivate shellfish for their shells, the need to harvest the animals for food is removed. Farms placed in remote waters or waters hazardous to shipping, or in contaminated or toxic waters can be left to their own devices. Provide the habitat and the animals will occupy it and thrive as they have for millions of years. When they die, they will leave their shells behind and be replaced by the next generation. Job done. You do not need to kill them in your kitchen, leave that to nature.

On the other hand, if we do amplify farming ***and*** harvesting greatly, we will start to produce shellfish meat in excess of that required for the 'shellfish-as-a-delicacy' market. Then we could start thinking about processed shellfish meat as an alternative to red-meat products (burgers, steaks, sausages, etc.), in the expectation that reduced husbandry of farm animals for meat-eaters will release pastures for afforestation and reduce further destruction of existing natural forests. We suspect that pseudo-beef-burgers made from shellfish meat would be more readily acceptable than those made from the insects that some food technologists are keen to promote.

Another positive characteristic of shellfish is that they present no conflict between using land to grow food crops and using land to grow trees. Or, for that matter, between growing trees for biofuel and growing native trees to repair and re-establish natural forest ecosystems. Biofuel alleviates fossil fuel usage but does release the CO_2 back to the atmosphere when the fuel is burned. There isn't enough agricultural land on the planet to accomplish all these things.

The situation with regard to agricultural land is summed up in this quotation from Moore et al. (2020):

"Only about 7.5% of the Earth's surface provides the agricultural soil on which we depend for the world's food supply, and this fragment competes, sometimes unsuccessfully, with all other needs: housing, cities, schools, hospitals, shopping centres, landfills, etc. Indeed, there may not be enough ***soil*** in the first place. A subsistence diet requires about 180 kg of grain per person per year, and this can be produced on 0.045 hectares of land. In contrast, an affluent high-meat diet requires at least four times more grain (and four times more land, 0.18 hectares) because the animals are fed on grain and conversion of grain to meat is very inefficient. The Earth has about 0.25 hectares of farmland per person, but only about 0.12 hectares per person of farmland is suitable for producing grain crops. As it stands, the Earth does not have enough land for all inhabitants to enjoy an affluent diet as that is presently defined..." (Moore et al. 2020b).

In contrast, farming shellfish uses the shoreline and continental shelf and there is enormous scope for the shellfish sector to grow in those regions, let alone in the open sea. The *Views of the World* website states the total length of coastlines in the world as between 1.16 million kilometres and 1.63 million kilometres [http://www.viewsoftheworld.net/?p=5340]. Continental shelves cover an area of about 32 million km^2, which, according to the *Blue Habitats* website [https://www.bluehabitats.org/?page_id=1660] is only about 9% of the surface area of the world's oceans.

If we are willing to contemplate planting 1 billion hectares of new forest knowing both that (i) trees cannot solve the excess CO_2 problem and (ii) that we do not have enough land to grow food, then surely, we should be willing to contemplate developing towards 1 billion hectares (= 10,000 km^2, which is only 0.03% of global continental shelf) of coastal waters to production of animals that will ***permanently*** remove CO_2 from our tortured atmosphere.

There is no need for irrigation, food or fertiliser. Farming shellfish for food can be combined with restoration and conservation of overfished fisheries and usually involves so little intervention (beyond provision of habitats and, where necessary, protection of larvae and juveniles from predation in 'nurseries') that there is no inevitable conflict with other activities. About 70 per cent of the Earth's surface is covered by water, we might as well use it to rescue the atmosphere.

With so many positives, it is remarkable to me that there is so much debate about planting trees and no debate at all about cultivating shellfish for carbon sequestering. Two paragraphs in a letter received in 2020 from the *Scottish Government's Directorate For Marine Scotland (Marine Planning and Policy)* give some rationale for the current situation:

"... You rightly say that achieving net zero will require the use of natural carbon sinks for emissions and we must consider all options available, not just forestry, which is why we are also investing in peatland restoration".

"The national greenhouse gas emissions for Scotland reported through annual Official Statistics are calculated using the UK Greenhouse Gas Emissions Inventory, which is compiled in line with international (IPCC and UNFCCC) scientific guidelines. The inventory currently includes peatland and forestry, but not blue carbon habitats such as oyster reefs, seagrass beds and saltmarsh. This does not mean that these habitats are not important in tackling climate change, indeed they could play a vital role in adaptation by slowing wave action, reducing coastal erosion and lowering flood risk. These habitats also fulfil many important ecosystem functions, such as supporting biodiversity and improving water quality" (Personal communication 30 January 2020, MPP Reference: 202000011306).

If the International Agencies put more value and devote more intellectual attention to ***terrestrial*** remediation processes using green plants (methods that research shows make no net contribution to carbon sequestering), does that mean there is something very wrong with the aquaculture solution? Are we missing something? Well, no, we do not think so because we are not alone in believing that blue carbon habitats hold considerable promise (Macreadie et al. 2019).

Indeed, those who deny the potential of calcifying organisms to sequester carbon for the reason that the $CaCO_3$ they produce represents only a small fraction of the CO_2 they cycle during their lives are creating a similar **myth** to the climate change deniers and global warming sceptics discussed in Chap. 1 (Moore et al.2021a). The myths that deny the facts that human activities are causing climate change are missing the point that burning fossil fuels emits long-fossilised CO_2 additively into today's atmosphere. The myth that shellfish shells represent only a small (even negligible) fraction of the CO_2 the organisms respire are ***also missing the point***. And **The Point** is:

- When ***a tree dies*** its entire body, root and branch, is digested and respired and all its carbon is eventually released back into the atmosphere as CO_2. Over the forest's lifetime there is no net carbon sequestration; indeed during the forests's lifetime the trees are likely to release previously stored soil carbon.
- When ***any terrestrial ecosystem*** (moorland, peat bog, grassland, scrubland) dies, even when ***a kelp forest ecosystem dies***, it is digested and respired and all its carbon is eventually released back into the atmosphere as CO_2. The International Agencies expect photosynthetic plants to sequester atmospheric carbon into the lignocellulose of their cell walls. And that they will indeed do, ***while they are alive***. But when a plant dies all of the lignocellulose will be digested (mostly by fungi) and all of its carbon will be returned to the atmosphere as respiratory CO_2. ***A forest that cultivates 10,000 tonnes of timber a year will have all of its carbon returned to the atmosphere within a few decades.***
- When shellfish die half the mass of the animal is left behind as permanently stable crystalline $CaCO_3$. Shellfish farms that cultivate 10,000 tonnes of mussels a year, remove 1,606 metric tonnes of CO_2 ***permanently*** from the atmosphere every year.

What's wrong with that? What the International Agencies see wrong with that is revealed in another quotation from that letter from the *Scottish Government's Directorate For Marine Scotland*:

> "There is evidence that the bivalve shells do trap CO_2, but the scientific opinion seems to be that the process of doing so releases CO_2 from the water back into the atmosphere. There is therefore a net sequestering gain, but smaller than it may initially appear. The current evidence suggests the carbon value in oyster reefs is in the sediment they trap in the reef structure, so bivalve aquaculture would not necessarily result in significant climate mitigation" (Personal communication 30 January 2020, MPP Reference: 202000011306).

Over the past decade or so several major studies have published careful measurements of the CO_2 fluxes associated with shell formation, they tend to come to the not very surprising conclusion that more CO_2 is released to the atmosphere by the animal's respiration than is sequestered in the shell. This usually gives rise to the conclusion that shell formation in cultivated bivalves cannot be part of a carbon sequestering system, and this has been said so often over the years that the claim has reached the status of a self-evident truth. But it is not true.

One specific example will suffice but before we go any further we want to make it clear that we make no criticism of the science or its analysis done for this study (or other related studies); we criticise only the final conclusion. Munari et al. (2013) studied shell formation in the cultivated bivalve *Mytilus galloprovincialis* (the farmed Mediterranean mussel) in the River Po Delta, Italy. They measured respiration and calcium carbonate. The ratio of CO_2 released to $CaCO_3$ crystallised into the shell was calculated. They estimated that *M. galloprovincialis* **sequestered** 136.6 mol CO_2 m^{-2} $year^{-1}$ (= **6 kg CO_2 m^{-2} $year^{-1}$**) for shell formation, but the CO_2 fluxes (by which they mean release to the environment) due to respiration and calcification were 187.8 and 86.8 mol CO_2 m^{-2} $year^{-1}$ (equivalent to **8.26 kg CO_2 m^{-2} $year^{-1}$** and **3.82 kg CO_2 m^{-2} $year^{-1}$**), respectively. From which the authors suggest that mussel farming ***adds*** CO_2 to seawater, which leads them to conclude (in line with many other authors, before and since) that shell formation in cultivated shellfish cannot be part of carbon trading systems (and this final phrase is in the title of their paper).

But that is a conclusion too far because the 6 kg CO_2 m^{-2} $year^{-1}$ formed into the animals' shells is **permanently insoluble**. The correct conclusion from this study is that farmed Mediterranean mussels require a flux of 12.08 (= 8.26 + 3.82) kg CO_2 m^{-2} $year^{-1}$ in order to crystallise 6 kg CO_2 m^{-2} $year^{-1}$ into permanently sequestered $CaCO_3$. In other words, ***each year half the CO_2 which is churned between animals, seawater and atmosphere is REMOVED from the atmosphere***.

The emerged River Po delta area is estimated at about 700 km^2 (Ninfo et al. 2018). Shellfish cultivation, mainly of mussel, clam and oyster, is well-established, and this area is one of the most important such sites in Europe, involving about 1700 operators and 83 companies (Tamburini et al. 2020). Production of mussels by these farms is estimated to be 10,000 tonnes per year (Tamburini et al. 2020). This study makes an assessment of the ***sustainability*** of mussel (*Mytilus galloprovincialis*) farming in the Italian Po Delta; it is a **life cycle assessment** of mussel farming in this location. Life cycle assessment, of any commercial product, process, or service (it is not limited to living things) quantifies ***the environmental impact*** over the lifetime of the product, process, or service. A life cycle analysis of the environmental sustainability of the inshore great scallop (*Pecten maximus*) fishery of Galicia (northwest Spain) identified as determining factors in the environmental profile of the scallop meat product (a) the ***diesel fuel consumption*** of diesel powered fishing vessels, this being the major contributor to the environmental burden; and (b) the ***electricity consumption of the onshore evisceration plant***, which separates scallop shells from scallop flesh and removes the viscera (and any toxins it may contain) from the adductor muscles of the scallop flesh (the desired product) (Cortés et al. 2021).

Concerns over the quantitative contribution of the respiratory flux of CO_2 to production of the shell is made irrelevant by the ***facts*** that diesel fuel consumption and electricity consumption of harvest processing contribute most to any adverse carbon balance of the shellfish industry. These are factors that apply to all aspects of marine commerce; they can be managed independently of the nature of the commercial activity or harvest. They should not be allowed to obscure the most important ***facts*** that (a) permanently crystallised $CaCO_3$ in shellfish shell amounts to 30–50% of the shellfish farmers' harvest and (b) on a molar mass basis, CO_2 represents 44% of the mass of calcium carbonate. So, it's an easy calculation that 10,000 tonnes of mussels (the estimated harvest of the River Po delta) = 3,650 tonnes of shell = 1,606 tonnes CO_2 removed from the atmosphere ***annually***. Surely, that amount of carbon sequestered by the farmed Mediterranean mussels of the River Po Delta alone would be a useful part of a carbon trading systems? And that consideration must apply around the world.

Increase those values by a million-fold across the globe, and we would be removing 1.6 ***billion*** tonnes of CO_2 from the atmosphere every year, just by farming mussels. It is important to appreciate that this scale of operation is not expecting too much; The Global Agriculture website [https://www.globalagriculture.org/] estimates there are 570 million ***terrestrial farms*** in the world—and we are blest with far more ocean than farmland!

Globally, we consume around 350 million tonnes of meat a year. Meat has a much higher energy footprint than any other food. It takes 75 times more energy to produce meat than corn. And it takes an area of cropland 7 times the size of the EU to produce food for the livestock animals of Europe (most of the additional cropland required to meet this demand is in China, the United States and Indonesia) [https://www.theworldcounts.com/challenges/]. If we put that amount of effort into producing 350 million tonnes of ***shellfish*** meat (rather than livestock meat) then the shells of that shellfish harvest would be ***removing*** 42 million tonnes of carbon from the atmosphere each year.

We describe the quantitative measurement of the respiratory flux of CO_2 needed to ***produce*** the shellfish shell as irrelevant because ALL aerobic organisms release respiratory CO_2 to the atmosphere through their metabolic activities. And that includes forests (and other green plants, aquatic and terrestrial), though green plants have the advantage of photosynthesis. It is photosynthesis which enables green plants to fix more CO_2, initially into carbohydrates, than they release through respiration, ***in the daylight***. When the lights go out, though, all the plants in the forest ***breathe out*** $\boldsymbol{CO_2}$. ***That's LIFE***. The clincher is at the end of the organism's life cycle.

Atmospheric CO_2 sequestered by shellfish is undigestible, crystalline and chemically stable calcium and calcium-magnesium carbonates; when the animal dies the shell remains for geological periods of time. Effectively, the CO_2 is permanently removed from the atmosphere. That's the animal's generous legacy and our inheritance.

Shellfish cultivation is the only industry in which a massive increase in productivity will benefit the atmosphere. I'm not convinced that forestry can do what's needed, and the heavy-industry approach is alarming. Here we refer you to Chap. 7 by **Peter Petros & David Moore** for their analysis, but we will also give you another quotation, this time from the Wikipedia article on *Carbon Capture and Storage* [CSS]:

> "The increased energy required for the carbon capturing process is also called an energy penalty. It has been estimated that about 60% of the energy penalty originates from the capture process itself, 30% comes from compression of CO_2, while the remaining 10% comes from electricity requirements for necessary pumps and fans. CCS technology is expected to use between 10% and 40% of the energy produced by a power station. CCS would increase the fuel requirement of a plant with CCS by about 15% for a gas-fired plant. The cost of this extra fuel, as well as storage and other system costs, are estimated to increase the costs of energy from a power plant with CCS by 30–60%, depending on the specific circumstances" [source: https://en.wikipedia.org/wiki/Carbon_capture_and_storage].

Then compare the humble shellfish; cultivate the shellfish in order to sell a tasty food and get the CSS for free! And do it in the ocean, leaving you free to save the forests in the way that the forests need to be saved.

Another criticism that has been made about our argument that we should use shellfish for sequestering CO_2 is that there may be a major problem with large-scale ***intensive*** bivalve culture because of profoundly damaging effects on the overall marine ecosystems. This is said to be because bivalves of most species filter out huge quantities of both plant and animal plankton (including many larval stages), and when grown intensively they can create single species monocultures. The worst expression of this is when monocultures cause reduced growth rates, increased bivalve diseases and an environment dominated by a single bivalve species (Earll 2018).

We do not accept this as a criticism because while there are, indeed, considerable dangers in monocultures, we don't believe for one moment that there's any danger of damaging the overall marine ecosystem, because the conservation of that ecosystem would be part of the design planning for any shellfish cultivation exercise. We speak most often about mussel farming because we are most immediately aware of that species. But there are hundreds, perhaps thousands, of shellfish candidates for cultivation, molluscan and crustacean (Utting and Spencer 1991; Helm et al. 2004; Moore et al. 2021b) (see Chap. 3). Consequently, a range of cultivated ecosystems can be designed to suit different habitats and locations (integrated multi-trophic aquaculture), avoiding monocultures by combining filter feeders with detritus feeders, and phytoplankton farming with zooplankton cultivation (see Chaps. 4 and 6; Heilweck and Moore 2021; Moore 2021).

Furthermore, there are many shellfish fisheries around the world that have been fished out, or are overfished so a useful start could be made simply as a conservation measure to repair the ecosystem damage that's already been done. And, of course, 70% of the planet is covered in ocean; how many intensive bivalve farms would you need to damage that?

Instead of thinking about filling the ocean with mussel farms, think about seeding the oceans with appropriately designed communities of organisms, molluscs and crustaceans alike (for which see **Matthias Heilweck** in Chap. 4; Heilweck and Moore 2021) that will support the basic need: which is to ***cultivate shellfish for their shells***, and take whatever else they offer, whether that be food (human and/or animal), reef-building, pollution-filtration, coral reef reconstruction (for which see Chap. 5), as a free by-product. We point out in Chap. 5 that there is no shortage of scientific and practical knowhow about what ***could*** be done to rescue our distressed marine life, but there is a divide between what is possible and what we are doing due to poor financial, legislative and political support. Bindoff et al. (2019) state that "… blue carbon can contribute to mitigation for many nations but its global scope is modest …". Blue carbon science is relatively young, but has so well revealed the importance of aquatic ecosystems in the carbon balance and ecosystem services of the whole planet that it deserves significantly increased attention (Macreadie et al. 2019) (Fig. 2.5).

Ecosystem services have been defined as "… the benefits provided by ecosystems that contribute to making human life both possible and worth living …" (quoted from the ***UK National Ecosystem Assessment*** website at this URL: http://uknea.unep-wcmc.org/). Costanza et al. (1997) listed ecosystem services as follows:

1. "… Gas regulation. Regulation of atmospheric chemical composition.
2. Climate regulation. Regulation of global temperature, precipitation, and other biologically mediated climatic processes at global or local levels.
3. Disturbance regulation. Capacitance, damping, and integrity of ecosystem response to environmental fluctuations.
4. [Flash-Flood] Water regulation. Regulation of hydrological flows.
5. Water supply. Storage and retention of water.
6. Erosion control and sediment retention. Retention of soil within an ecosystem.
7. Soil formation processes.

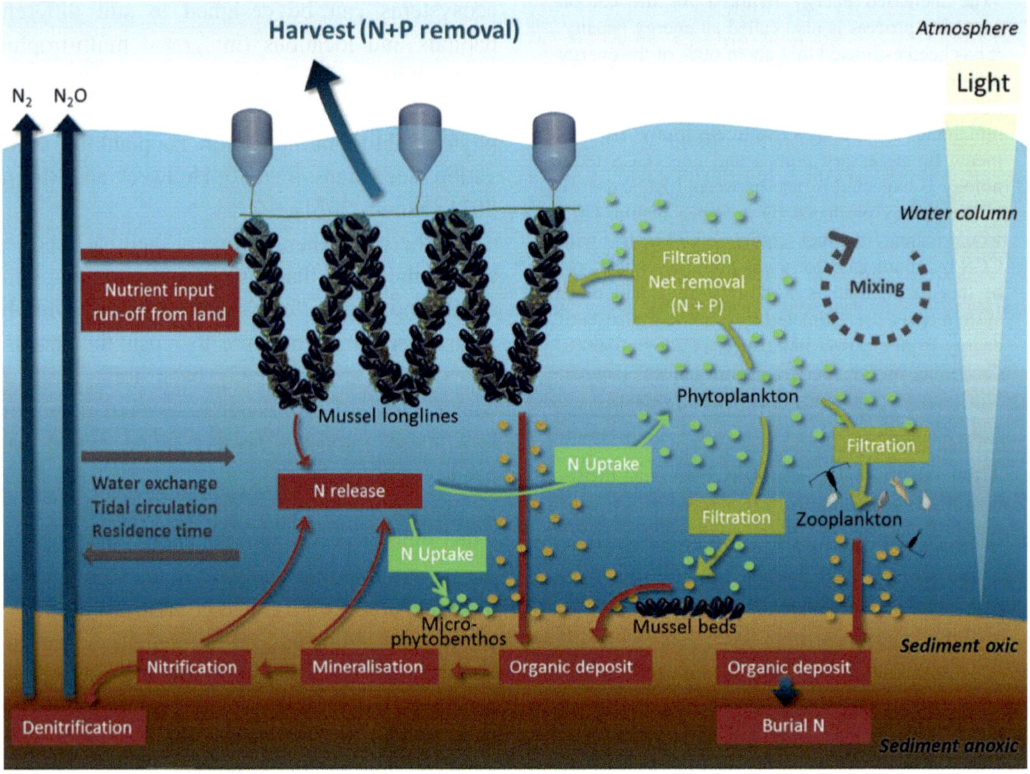

Fig. 2.5 Nutrient extraction services provided by bivalves. Blue mussels are shown as the example in this figure, but other bivalves like oysters and clams can also provide these nutrient extraction services. Figure from Petersen et al. (2019) under the Creative Commons Attribution 4.0 International License

8. Nutrient cycling. Storage, internal cycling, processing, and acquisition of nutrients.
9. Waste treatment. Recovery of mobile nutrients and removal or breakdown of excess or xenic nutrients and compounds.
10. Pollination. Movement of floral gametes.
11. Biological control. Trophic-dynamic regulations of populations.
12. Refugia Habitats for resident and transient populations.
13. Food production. That portion of gross primary production extractable as food.
14. Raw materials. That portion of gross primary production extractable as raw materials.
15. Genetic resources. Sources of unique biological materials and products.
16. Recreation. Providing opportunities for recreational activities.
17. Cultural. Providing opportunities for noncommercial uses". (quoted from Costanza et al. 1997; for further refinements of the concept, see also Daily 1997 and Costanza et al. 2014).

As excess loading in coastal zones of nitrogenous nutrients resulting from human activities is a major concern for the marine environment worldwide, nutrient extraction by bivalves is an extremely important ecosystem service offered by cultivation of these molluscs. It is explained in the following quotation:

> "Through their filtering of water, bivalves remove a proportion of the phytoplankton that in large concentrations otherwise is part of the negative effects of excess nutrient loading. By clearing the water column of particles, bivalves contribute to reductions in turbidity and concentrations of particulate organic nutrients, like nitrogen and phosphorous … The filtered material is either not ingested and ejected as pseudofaeces or is ingested and digested, then transformed into bivalve tissue or faecal material that settles in proximity of the bivalves. Nutrients in the ingested material that is transformed into bivalve tissue are immobilized, hence temporarily not accessible for primary production. If the bivalves are removed from the water column, e.g. through harvest, the nutrients are permanently made inaccessible. The material ejected as faeces or pseudofaeces can enter nutrient cycles that may result in either permanent burial in the sediment or removal through chemical processes; i.e. denitrification. Both processes will result in a nutrient extraction service provided by the bivalves that potentially can be used as a mitigation tool by managers seeking means of remediating effects of excess nutrient loading to coastal ecosystems. This can be realized as either bivalve aquaculture or by promoting or restoring natural bivalve populations…" (Petersen et al. 2019; and see Fig. 2.5).

The potential benefits of bivalve cultivation on a very large scale are being made clear but to reap those benefits requires huge effort, central governance, huge scale and huge finance. Gordon et al. (2020) state categorically:

> "… Marine restoration projects are undervalued …". In their final paragraph they conclude that "… Political agreements for global reductions in atmospheric carbon have been slow to emerge. Relying on their implementation as the only solution to the degradation of tropical habitats is a major gamble. In the meantime, restoration projects could help maintain species survival and ecosystem services, ultimately providing humanity with the breathing space to stabilize the climate …".

Duarte et al. (2020) conclude that:

> "… Rebuilding marine life represents a doable Grand Challenge for humanity, an ethical obligation and a smart economic objective to achieve a sustainable future …".

The political limitations of conventional ecosystem governance are discussed by Morrison et al. (2020), who conclude that securing a future for marine ecosystems (in their case, coral reefs) "… under climate change is a political challenge as much as an ecological or social one …". In terms of creating sustainable ecosystems globally the immense promise of blue carbon science is so strikingly evident that it must be taken more seriously. But more than anything else it requires the recognition that cultivation of coccolithophores, corals, crustacea and molluscs on a massive scale would have the effect of removing a massive amount of CO_2 directly from the atmosphere; ***here, now and permanently***, making a ***continued*** contribution to the health of the whole planetary ecosystem (Fig. 2.6). It would be a criminal dereliction of duty if humanity failed to grasp this last opportunity to carry out

Fig. 2.6 The Earthrise over the Moon, as seen from above Compton crater from the orbital vantage point of NASA's Lunar Reconnaissance Orbiter spacecraft. The large tan area in the upper right of the Earth's disc is the Sahara desert, and just beyond is Saudi Arabia. The Atlantic and Pacific coasts of South America are visible to the left [*image source* NASA/GSFC/Arizona State University]

this "doable Grand Challenge". And the sentence for such a criminal act is extinction.

2.9 The Shellfish Cultivation Industry Offers Unique Opportunities for Limiting Climate Change and Enhancing Conservation Strategies

Let's look at a few more numbers derived from the present situation to get an idea of the scale of the shellfish cultivation programme required in the future to make effective drawdown of CO_2 from the atmosphere and dramatically reduce climate change. In the section above entitled *ONE There's a lot of shell in shellfish*, we estimate that ***4.84 million tonnes*** of CO_2 per year is being captured, and mineralised, from the atmosphere by ***current*** aquaculture activities around the world. In carbon-offset terms, that's equivalent to one million business class return flights between London Heathrow and JFK New York (6 billion miles of flying per year, every year). Returning to my favoured mussels, we also estimate above that shellfish farms designed to produce 10,000 tonnes of mussels per year, such as the Po River delta farms (Tamburini et al. 2020), would permanently remove from the atmosphere an annual total of 1,606 metric tonnes of CO_2.

According to the *Carbon Offsetting* website [https://www.carbonfootprint.com/] one business class return flight LHR (London Heathrow) to JFK (New York) = 2.17 tonnes of CO_2 and would cost between US$8 and US$17 per tonne to offset, depending on the offsetting programme you wish to support (planting trees, renewable energy, community projects, etc.).

The same website calculates that the carbon footprint of ***my*** Ford Focus (EU 2015 FORD All New Focus, Model Year Post 2015 1/2 1.6 Duratorq TDCi (115PS) With Stop/Start - 5 Door) is 0.22 tonnes CO_2 per 1,000 miles (costing between US$3 and US$7 to offset, depending on the programme). So, the shells of 10,000 tonnes of mussels:

- offset 740 return business class tickets (equivalent to about US$19,000 in offsetting fees); ***OR***
- offset my driving 7,300,000 miles (equivalent to about US$33,500 in offsetting fees).

We have suggested above that a million bivalve shellfish farms like this might remove **1.606 billion (1.6 × 10^9) tonnes CO_2** permanently from the atmosphere each year. Global carbon emissions in 2014 from fossil fuel use were 35.9 billion tonnes of carbon dioxide [https://www.co2.earth/global-co2-emissions]. So, a million mussel farms would permanently remove about 4.5% of the global CO_2 emissions ***in each year***. The call for a million mussel farms is by no means an extreme or unrealistic proposition. Imagine a mussel farm on every offshore wind turbine, every oil and gas rig, every pier, wharf and jetty, every breakwater or harbour wall; imagine cultivating cockles (and other clams) in every shallow sandy/muddy bay.

Imagine restocking and extending every fished-out oyster fishery, every fished-out scallop fishery. We could start tomorrow.

We believe that a high priority should be given to having 'cultivation of blue carbon sequestration' included as projects on carbon-offset websites. There is a wide range of potential projects, ranging from support for developing/expanding local subsistence fisheries as a means to employ and feed communities in need, right through to the industrial scale of seamount installations (Chap. 4) and ocean-going factory ships (Chaps. 4 and 6; Heilweck and Moore 2021) that might be supported by high-energy industries needing to compensate their heavy carbon footprints, and have all the necessary skills and experience to take such large-scale efforts forward.

In 2011, the *Guardian* website published an article entitled *A complete guide to carbon offsetting* in which Duncan Clark summarised the offsetting debate in an edited extract from his book *The Rough Guide to Green Living* (Clark 2009) [https://www.theguardian.com/]. Sadly, the blue carbon industries do not feature in the debate.

We say again: the shellfish industry is the only industry on the planet that could save the atmosphere by a massive increase in its production.

So, why is carbon sequestration not being promoted more widely by the shellfish sector? I've asked myself that from the beginning. We can't understand why the guys who haul shellfish by the tonne into their boats in cold, driving rain haven't realised that half of what they are hauling aboard is solidified CO_2 from the atmosphere. And then used that fact to help sell the product. There's money in it; Alonso et al. (2021) estimate that the CO_2 sequestration potential of bivalve aquaculture, using the current value of 1 metric tonne of CO_2 in the carbon market is over 25 €, which would represent a value of around 125–175 million € y^{-1} to the European Union's bivalve aquaculture industry alone.

For the producers, restaurateurs and end users we imagine it's probably mind-set tinged with a lack of basic biological knowledge. The mind-set dominantly believes that shellfish is a food, and the shell is a waste that has to be disposed of. End of thinking: who's doing the washing-up?

For the man in the street this is understandable and excusable. You don't express concern that the cattle spent their lives farting methane ***into*** the atmosphere when you're tucking into your rib-eye steaks, so why be grateful that your 'moules marinière for two' has just taken about 20 g of CO_2 ***out of*** the atmosphere, and so will the next plate, and the next? What is unfortunate is that this unthinking attitude applies to the professionals as well.

For example, a research paper in the *Journal of Applied Poultry Research* in 2013 starts with this:

> "DESCRIPTION OF PROBLEM. One of the major industries in Atlantic Canada is harvesting and processing of shellfish. This activity generates approximately 140,000 t of raw material annually and accounts for almost 20,000 t of waste, with the majority coming from Atlantic snow crab, American lobster, and seashells such as whelk, abalone, blue mussel shell, surf-clam shell, soft-clam shell, scallop shell, quahog shell, and abalone shell ... The slow biodegradability of these waste products has raised concerns over disposal practices and their impact on the environment..." (quotation from Safamehr et al. 2013).

The paper goes on to consider shellfish shells as useful calcium sources for supplementing poultry feeds, which is a perfectly respectable use for them.

Another review of shell waste (Jović et al. 2019) points out that the aquaculture industry is a globally attractive source of cheap and healthy food for our growing population because of its need for relatively small investment and its low energy consumption, but as shellfish shells can account for up to 75% of total bivalve body weight, they claim that contamination of the ecosystem by waste shells can be a significant environmental problem if it accumulates at coastal sites. This publication aims to review recent trends in shell waste applications in the light of the European Union's *Circular Economy Action Plan* [https://ec.europa.eu/environment/strategy/circular-economy-action-plan_en] that stresses development of new technologies to exploit waste as a resource and contribute to sustainable development (Jović et al. 2019). But why do these publications and reports raise such concerns over shells as a waste disposal problem which will have a ***negative*** impact on the environment, when the ecological truth is quite ***the reverse***?

Bivalve molluscs are described as ***ecosystem engineers*** creating biogenic reef habitats of such significant size that they engineer entire ecosystems and can be used as a natural component of coastal protection schemes (Mann et al. 2009; Ysebaert et al. 2019). It is the accumulated shells of past generations of shellfish that is the foundation of the reef ecosystem.

> … If you have an oyster shell, the best thing to do is to put it back on the reef in order to start an ecological system. Once you've got the reef, then other fish come along and other flora and fauna can form around them. They purify the water and filter it, making it useful. But not everybody knows this … (Smith 2015).

And that's why, ***if you have a pile of shellfish 'food-waste', you should put it back where the animals came from***.

Murphy et al. (2020) suggest treating mussel shells "from aquaculture waste streams" with dilute acetic acid to make an absorbent sponge-like material, composed of recombined calcite crystals, that can absorb dyes and crude oil from polluted water. Waste shells from aquaculture have been used as a low cost $CaCO_3$ source to produce calcium oxide (CaO), which is widely used as a catalyst and chemical feedstock in industry. Heating $CaCO_3$ to over 800 °C (calcination) results in the conversion of solid $CaCO_3$ to solid CaO, liberating gaseous CO_2 in the process. Calcined oyster shells have been shown to be an effective catalyst in the conversion of soybean oil into biodiesel (Nakatani et al. 2009); calcined cockle shells to convert palm olein to biodiesel (Boey et al. 2011); and calcined freshwater mussel shells for conversion of Chinese tallow oil to biodiesel (Hu et al. 2011). Also, calcined oyster shell-derived calcium oxide was used to modify copper-based catalysts for synthesis of methanol (Wisaijorn et al. 2017). All of these studies demonstrate that calcined mollusc shells from the aquaculture industry can very effectively replace the more commonly used mined $CaCO_3$, such as limestone; thus replacing a calcination process that adds even more fossilised CO_2 to the atmosphere, with one that merely cycles present-day CO_2 from atmosphere-to-shell-to atmosphere, without making a net contribution to present-day atmospheric carbon.

Although bivalve shells are almost entirely composed of $CaCO_3$, with only a minor component of their structure being a protein component, the $CaCO_3$ crystals in the shells of crustacea are built into a significant scaffold of the polysaccharide chitin. Shell wastes from shrimp, crab, lobster and krill contain large amounts of chitin, that can be extracted by deproteinising and demineralising the exoskeletons. The chitin biopolymer, and its derivatives chitosan and related products, exhibit many biomedical activities, including as an antioxidant and as an immunomodulator (with potential for cancer treatments) and, aside from medical uses, can be exploited in various other applications, such as cosmetics, food processing, and textiles (Hamed et al. 2016), and even a film that is a biodegradeable alternative for single-use plastics (Srinivasa and Tharanathan 2007; Shamshina et al. 2019; and view https://www.theshellworks.com/).

The book ***Goods and Services of Marine Bivalves*** (Smaal et al. 2019) deals with a wide range of aquaculture topics including genomics-driven biotechnological innovations like new pharmaceuticals from molluscs, habitat and ecosystem-engineering modification in coastal protection by reef-building bivalves, water clarification services provided by their filter feeding and even shells as collector's items, but does not include a chapter dealing specifically with the potential service of extracting carbon from the atmosphere. Chapter 12 in this book comes closest to a revelation (Filgueira et al. 2019) in which these authors state clearly:

> … In valuing the ecosystem service of mussel farming in the carbon cycle a distinction has been made between the shell (waste) and the tissue (food). Following this rationale, the goods and services of mussel farming in deep fjords includes the valorization of the shells as a net sink of CO_2… (pp. 245–246).

Sadly, despite this promising start they apply this consideration only to harvested mussels, not to those that die in situ, and their final conclusion is that:

> … Under these considerations, bivalve shells can be considered net sinks of CO_2 and consequently provide additional ecosystem services besides the food provided by the tissue. A full life cycle analysis should be performed to account for the emissions required to properly dispose of the shells. The 0.45 g by the shell of each cultured mussel in Norway is hardly significant taking into account that a regular car produces more than 100 g CO_2 per km … (Filgueira et al. 2019; p. 246).

Now, we do not believe you can get much of a meal from ONE mussel. Rather, we know that a reasonable serving of *moules marinière* requires around 20 mussels (= 10 g CO_2 sequestered). So, a family of five, treating a similar family of friends to a meal out, could carbon-offset the drive to the restaurant. That consideration aside, the conclusions of Filgueira et al. (2019) miss the points that (a) disposal can mean no more than dropping the empty shells back into the fjord; and (b) that despite all its trees, Norway has no other industry that ***permanently*** removes carbon from the atmosphere.

The only publication we have found that recognises the true potential for marine calcification to remove CO_2 from the atmosphere is a newspaper article written by ***Steve Connor***, which was published in ***The Independent*** with the promising title "***Can seashells save the world?***" (Connor 2008). Although this article gives most prominence to coccolithophores (for which see Chap. 6; Moore 2021), which "… are microscopic marine plants that convert carbon dioxide into chalk … fighting global warming …" the article goes on to say:

> … Scientists have already estimated that some 118 billion tons of carbon released into the air as carbon dioxide between 1800 and 1994 have been taken up by the oceans worldwide. Indeed, about a third of the carbon dioxide produced by human activities since the start of the Industrial Revolution has been absorbed by the seas. So, without the capacity of the ocean to act as a natural carbon sink, the concentration of carbon dioxide in the air today—about 380 parts per million—would be significantly, and dangerously, higher (Connor 2008).

But, most importantly, this article recognises the central truth that professional marine scientists seem to be obscuring with irrelevant detail and caveats:

> … Marine calcification actually produces carbon dioxide in the short term, but in the long term it takes carbon out of the atmosphere, for example by the formation of limestone rock deposits on the seabed. Indeed, marine calcification is estimated to be the biggest carbon sink on earth over geological timescales by forming layers of calcium carbonate, the basic ingredient of chalk, limestone and marble (Connor 2008).

Of course, we would respond to Steve Connor's question "***Can seashells save the world?***" with a resounding "***Yes they CAN***".

References

Alonso AA, Álvarez-Salgado XA, Antelo LT (2021) Assessing the impact of bivalve aquaculture on the carbon circular economy. J Clean Prod 279. Article 123873. https://doi.org/10.1016/j.jclepro.2020.123873

Avdelas L, Avdic-Mravlje E, Borges Marques AC, Cano S, Capelle JJ, 17 others (2020) The decline of mussel aquaculture in the European Union: causes, economic impacts and opportunities. Rev Aquac. In press. online before inclusion in an issue. https://doi.org/10.1111/raq.12465

Bastin J-F, Finegold Y, Garcia C, Mollicone D, Rezende M, Routh D, Zohner CM, Crowthe TW (2019) The global tree restoration potential. Science 365:76–79. https://doi.org/10.1126/science.aax0848

Bindoff NL, Cheung WWL, Kairo JG, Arístegui J, Guinder VA, Hallberg R, et al (2019) Changing ocean, marine ecosystems, and dependent communities. Chap. 5 (pp 447–588) In: Pörtner HO, Roberts DC, Masson-Delmotte V, Zhai P, Tignor M, Poloczanska E, Mintenbeck K, Alegría A, Nicolai M, Okem A, Petzold J, Rama B, Weyer NM (eds) IPCC, 2019: IPCC Special Report on the Ocean and Cryosphere in a Changing Climate (Geneva: Intergovernmental Panel on Climate Change. Open access. Chap. 5 (142 pp). https://www.ipcc.ch/srocc/chapter/chapter-5/. Full Report (765 pp). https://www.ipcc.ch/site/assets/uploads/sites/3/2019/12/SROCC_FullReport_FINAL.pdf.

Boey P-LL, Maniam GP, Hamid SA, Ali DMH (2011) Utilization of waste cockle shell (*Anadara granosa*) in biodiesel production from palm olein: optimization using response surface methodology. Fuel 90:2353–2358. https://doi.org/10.1016/j.fuel.2011.03.002

Boysen LR, Lucht W, Gerten D, Heck V, Lenton TM, Schellnhuber HJ (2017) The limits to global-warming mitigation by terrestrial carbon removal. Earth's Future 5:463–474. https://doi.org/10.1002/2016EF000469

Brienen RJW, Caldwell L, Duchesne L, Voelker S, Barichivich J, Baliva M, Ceccantini G, Di Filippo A, Helama S, Locosselli GM, Lopez L, Piovesan G, Schöngart J, Villalba R, Gloor E (2020) Forest carbon sink neutralized by pervasive growth-lifespan trade-offs. Nature Commun 11. Article 4241. https://doi.org/10.1038/s41467-020-17966-z

Clark D (2009) The rough guide to green living. London: Rough Guides® (Apa Digital AG, Switzerland). pp 336. ISBN: 9781848361072

Connor SA (2008) Can seashells save the world? Science News in: The Independent. of Wednesday 23 April 2008. https://www.independent.co.uk/news/science/can-seashells-save-the-world-813915.html

Cortés A, González-García S, Franco-Uría A, Moreira MT, Feijoo G (2021) Evaluation of the environmental sustainability of the inshore great scallop (*Pecten maximus*) fishery in Galicia. J Ind Ecol 2021:1–14. https://doi.org/10.1111/jiec.13153

Costa-Pierce BA, Chopin T (2021) The hype, fantasies and realities of aquaculture development globally and in its new geographies. World Aquac 52(2):23–35. https://www.researchgate.net/profile/Thierry-Chopin-2/publication/352572613

Costanza R, d'Arge R, de Groot S, Farber M, Grasso B, Hannon S and seven others (1997) The value of the world's ecosystem services and natural capital. Nature 387:253–260. https://doi.org/10.1038/387253a0

Costanza R, Anderson S, Bohensky E, Butler J, Edyvane K, Howe S, six others (2014) Ecosystem services from healthy oceans and coasts. White Paper to support the [Australian] National Marine Science Plan. http://www.robertcostanza.com/wp-content/uploads/2017/02/2014_O_Costanza_etal._NMSP_ES_whitepaper.pdf

Daily GC (ed) (1997) Nature's Services, Societal Dependence on Natural Ecosystems. Washington, DC: Island Press. ISBN: 9781559634762. https://www.researchgate.net/publication/37717461_Nature's_Services_Societal_Dependence_On_Natural_Ecosystems

Demeude H, Gadault T (2020). Le cri de détresse des forêts françaises confrontées aux crises écologiques. Website of The GoodPlanet Info magazine. https://www.goodplanet.info/2020/11/26/le-cri-de-detresse-des-forets-francaises-confrontees-aux-crises-ecologiques/

Domeignoz-Horta LA, Pold G, Liu XJA, Frey SD, Melillo JM, DeAngelis KM (2020) Microbial diversity drives carbon use efficiency in a model soil. Nature Commun 11. Article number 3684. https://doi.org/10.1038/s41467-020-17502-z

Duarte CM, Agusti S, Barbier E, Britten GL, Castilla JC, Gattuso JP, Fulweiler RW, Hughes TP, Knowlton N, Lovelock CE, Lotze HK, Predragovic M, Poloczanska E, Roberts C, Worm B (2020) Rebuilding marine life. Nature 580:39–51. https://doi.org/10.1038/s41586-020-2146-7

Duffy KA, Schwalm CR, Arcus VL, Koch GW, Liang LL, Schipper LA (2021) How close are we to the temperature tipping point of the terrestrial biosphere? Sci Adv **7**. Article eaay1052. https://doi.org/10.1126/sciadv.aay1052

Earll R (2018) Marine conservation: people, ideas and action. Exeter, UK: Pelagic Publishing. 303 pp. ISBN: 9781784271763

Elgin B (2020) These trees are not what they seem. How the Nature Conservancy, the world's biggest environmental group, became a dealer of meaningless carbon offsets. Blog on the Bloomberg Green website. https://www.bloomberg.com

Ellison AM (2019) Foundation species, non-trophic interactions, and the value of being common. iScience 13:254–268. https://doi.org/10.1016/j.isci.2019.02.020

Filgueira R, Strohmeier T, Strand Ø (2019) Regulating services of bivalve molluscs in the context of the carbon cycle and implications for ecosystem valuation. In: Smaal A, Ferreira J, Grant J, Petersen J, Strand Ø (eds) Goods and services of marine bivalves. Springer Nature, Cham, Switzerland, pp 231–251. Open access download. https://link.springer.com/content/pdf/10.1007%2F978-3-319-96776-9_12.pdf

Franz DR (1993) Allometry of shell and body weight in relation to shore level in the intertidal bivalve *Geukensia demissa* (*Bivalvia: Mytilidae*). J Exp Mar Biol Ecol 174:193–207. https://doi.org/10.1016/0022-0981(93)90017-I

Friggens NL, Hester AJ, Mitchell RJ, Parker TC, Subke JA, Wookey PA (2020) Tree planting in organic soils does not result in net carbon sequestration on decadal timescales. Global Change Biology. Online Version of Record before inclusion in an issue. https://doi.org/10.1111/gcb.15229

Gimin R, Mohan R, Thinh LV, Griffiths AD (2004) The relationship of shell dimensions and shell volume to live weight and soft tissue weight in the mangrove clam, Polymesoda erosa (Solander, 1786) from northern Australia. NAGA, WorldFish Center Quarterly, 27 (3 & 4), Jul-Dec 2004. https://digitalarchive.worldfishcenter.org/bitstream/handle/20.500.12348/2026/na_2318.pdf?sequence1=

Gordon TAC, Radford AN, Simpson SD, Meekan MG (2020) Marine restoration projects are undervalued. Science 367:635–636. https://doi.org/10.1126/science.aba9141

Goswami A (2020) Relying on forests to achieve net zero targets not a good idea, and scientists agree. Blog on the *DownToEarth website* at this URL: https://tinyurl.com/2phdmjs4

Hamed I, Özogul F, Regenstein JM (2016) Industrial applications of crustacean by-products (chitin, chitosan, and chitooligosaccharides): a review. Trends Food Sci Technol 48:40–50. https://doi.org/10.1016/j.tifs.2015.11.007

Heilmayr R, Echeverría C, Lambin EF (2020) Impacts of Chilean forest subsidies on forest cover, carbon and biodiversity. Nature Sustain. https://doi.org/10.1038/s41893-020-0547-0.

Heilweck M, Moore D (2021) Saving the Planet with Appropriate Biotechnology: 3. The High Seas Solution/Salvando el planeta con biotecnología apropiada: 3. La solución de alta mar. Mexican J Biotechnol 6: 92–128. https://doi.org/10.29267/mxjb.2021.6.1.92

Helm MM, Bourne N, Lovatelli A (2004) Hatchery culture of bivalves. A practical manual. FAO Fisheries Technical Paper. No. 471, pp. 203. Rome: Food and Agriculture Organization of the United Nations. ISBN 9251052247. http://www.fao.org/3/y5720e/y5720e02.htm

Hong S, Yin G, Piao S, Dybzinski R, Cong N, Li X, Wang K, Peñuelas J, Zeng H, Chen A (2020) Divergent responses of soil organic carbon to afforestation. Nature Sustain. https://doi.org/10.1038/s41893-020-0557-y

Hu S, Wang Y, Han H (2011) Utilization of waste freshwater mussel shell as an economic catalyst for biodiesel production. Biomass Bioenerg 35:3627–3635. https://doi.org/10.1016/j.biombioe.2011.05.009

Hugelius G, Loisel J, Chadburn S, Jackson RB, Jones M, MacDonald G, Marushchak M, Olefeldt D, Packalen M, Siewert MB, Treat C, Turetsky M, Voigt C, Yu Z (2020) Large stocks of peatland carbon and nitrogen are vulnerable to permafrost thaw. Proceedings of the National Academy of Sciences of the United States of America, open access article 201916387. https://doi.org/10.1073/pnas.1916387117

Jović M, Mandić M, Šljivić-Ivanović M, Smičiklas I (2019) Recent trends in application of shell waste from mariculture. Studia Marina 32:47–62. https://doi.org/10.5281/zenodo.3274471

Li A, Wang J, Bao B (2019) High-efficiency CO_2 capture and separation based on hydrate technology: a review. Greenhouse Gases: Sci Technol 9:175–193. https://doi.org/10.1002/ghg.1861

Lovatelli A (1990) Artificial propagation of bivalves: techniques and methods. Technical paper prepared as part of the Mollusc-Shellfish Culture Course of the 7th NACA Training Course for Senior Aquaculturists in Asia and in the Pacific. FAO/UNDP Regional Seafarming Demonstration and Development Project (RAS/90/002) Bangkok, Thailand: National Inland Fisheries Institute, Kasetsart. http://www.fao.org/3/AB739E/AB739E00.htm

Macreadie PI, Anton A, Raven JA, Beaumont N, Connolly RM, Friess DA, Kelleway JJ, Kennedy H, Kuwae T, Lavery PS, Lovelock CE, Smale DA, Apostolaki ET, Atwood TB, Baldock J, Bianchi TS, Chmura GL, Eyre BD, Fourqurean JW, Hall-Spencer JM, Huxham M, Hendriks IE, Krause-Jensen D, Laffoley D, Luisetti T, Marbà N, Masque P, McGlathery KJ, Megonigal JP, Murdiyarso D, Russell BD, Santos R, Serrano O, Silliman BR, Watanabe K, Duarte CM (2019) The future of Blue Carbon science. Nat Commun 10:3998. https://doi.org/10.1038/s41467-019-11693-w

Mann R, Southworth MJ, Harding JM, Wesson JA (2009) Population studies of the native eastern oyster, Crassostrea virginica (Gmelin, 1791), in the James River, Virginia, USA. J Shellfish Res 28:193–220. https://doi.org/10.2983/035.028.0203

Manrique SM, Franco J (2020) Tree cover increase mitigation strategy: implications of the "replacement approach" in carbon storage of a subtropical ecosystem. Mitig Adapt Strat Glob Change 25(8):1481–1508. https://doi.org/10.1007/s11027-020-09930-5

Martin MP, Woodbury DJ, Doroski DA, Nagele E, Storace M, Cook-Patton SC, Pasternack R, Ashton MS (2021) People plant trees for utility more often than for biodiversity or carbon. Biol Conserv 261. Article number 109224. https://doi.org/10.1016/j.biocon.2021.109224

Masson-Delmotte V, Zhai P, Pörtner HO, Roberts D, Skea J, 14 others (eds) (2019) IPCC, 2018: Global Warming of 1.5°C. An IPCC Special Report on the impacts of global warming of 1.5°C above pre-industrial levels and related global greenhouse gas emission pathways, in the context of strengthening the global response to the threat of climate change, sustainable development, and efforts to eradicate poverty. Open access at this URL: https://www.ipcc.ch/sr15/ and for PDF download: https://www.ipcc.ch/sr15/download/

Moheimani NR, Borowitzka MA (2006) The long-term culture of the coccolithophore *Pleurochrysis carterae* (Haptophyta) in outdoor raceway ponds. J Appl Phycol 18:703–712. https://doi.org/10.1007/s10811-006-9075-1

Moore D (2020) A biotechnological expansion of shellfish cultivation could permanently remove carbon dioxide from the atmosphere/Una ampliación biotecnológica del cultivo de moluscos bivalvos podría eliminar permanentemente el dióxido de carbono de la atmósfera. Mexican J Biotechnol 5:1–10. https://doi.org/10.29267/mxjb.2020.5.1.1

Moore D (2021) Saving the Planet with Appropriate Biotechnology: 4. Coccolithophore cultivation and deployment/Salvando el planeta con biotecnología apropiada: 4. Cultivo de cocolitóforos e implementación. Mexican J Biotechnol 6:129–155. https://doi.org/10.29267/mxjb.2021.6.1.129

Moore D, Heilweck M, Petros P (2021a). Saving the Planet with Appropriate Biotechnology: 1. Diagnosing the Problems/Salvando el planeta con biotecnología apropiada: 1. Diagnóstico de los problemas. Mexican J Biotechnol 6:1–30. https://doi.org/10.29267/mxjb.2021.6.1.1

Moore D, Heilweck M. , Petros P (2021b) Saving the Planet with Appropriate Biotechnology: 2. Cultivate Shellfish to Remediate the Atmosphere/Salvando el planeta con biotecnología apropiada: 2. Cultivar mariscos para remediar la atmósfera. Mexican J Biotechnol 6:31–91. https://doi.org/10.29267/mxjb.2021.6.1.31

Moore D, Robson GD, Trinci APJ (2020) 21st Century Guidebook to Fungi. second edition. Cambridge University Press. London, UK. pp 600. ISBN: 9781108745680. https://www.cambridge.org/9781108745680

Morris JP (2019) Report synthesising the existing and potential uses of shells as by-products of the aquaculture industry. A Deliverable from the EU's FP7-funded Marie Curie ITN: *CACHE*: CAlcium in a CHanging Environment. Open access. https://doi.org/10.5281/zenodo.2662011

Morris JP, Backeljau T, Chapelle G (2019) Shells from aquaculture: a valuable biomaterial, not a nuisance waste product. Rev Aquac 11:42–57. https://doi.org/10.1111/raq.12225

Morrison TH, Adger N, Barnett J, Brown K, Possingham H, Hughes T (2020) Advancing coral reef governance into the Anthropocene. One Earth 2:64–74. https://doi.org/10.1016/j.oneear.2019.12.014

Munari C, Rossetti E, Mistri M (2013) Shell formation in cultivated bivalves cannot be part of carbon trading systems: a study case with *Mytilus galloprovincialis*. Mar Environ Res 92:264–267. https://doi.org/10.1016/j.marenvres.2013.10.006

Murphy JN, Schneider CM, Hawboldt K, Kerton FM (2020) Hard to soft: biogenic absorbent sponge-like material from waste mussel shells. Matter 3:2029–2041. https://doi.org/10.1016/j.matt.2020.09.022

Nakatani N, Takamori H, Takeda K, Sakugawa H (2009) Transesterification of soybean oil using combusted oyster shell waste as a catalyst. Biores Technol 100:1510–1513. https://doi.org/10.1016/j.biortech.2008.09.007

Natural Capital Committee (2020) Natural Capital Committee advice on reaching net zero by 2050: nature based interventions. Open access at this URL. https://www.gov.uk/government/publications/a-natural-capital-approach-to-attaining-net-zero-nature-based-interventions

Natural England (2010) *England's peatlands: carbon storage and greenhouse gases*. Report NE257. Open access and PDF download at http://publications.naturalengland.org.uk/publication/30021

Ninfo A, Ciavola P, Billi P (2018) The Po Delta is restarting progradation: geomorphological evolution based on a 47-years Earth Observation dataset. Scientific Reports **8**. Article number: 3457 (open access). https://doi.org/10.1038/s41598-018-21928-3

ONS (2016) UK Natural Capital: Experimental carbon stock accounts, preliminary estimates. Open access at this URL. https://www.ons.gov.uk/economy/environmentalaccounts/bulletins/uknaturalcapital/experimentalcarbonstockaccountspreliminaryestimates

Parappurathu S, Grinson G, Narayanakumar R, Aswathy N, Ramachandran C, Gopalakrishnan A (2017) Priorities and strategies to boost incomes of marine fisher folk in India. Agric Econ Res Rev 30:205–216. https://doi.org/10.5958/0974-0279.2017.00035.0

Petersen JK, Holmer M, Termansen M, Hasler B (2019) Nutrient extraction through bivalves. In: Smaal A, Ferreira J, Grant J, Petersen J, Ø. Strand (eds) Goods and services of marine bivalves. Springer Nature, Cham, Switzerland, pp 179–208. https://doi.org/10.1007/978-3-319-96776-9_10

Rasmussen RS, Morrissey MT (2007) Biotechnology in aquaculture: transgenics and polyploidy. Compr Food Sci Food Saf 6:2–16. https://doi.org/10.1111/j.1541-4337.2007.00013.x

Rinne-Garmston (Rinne) KT, Peltoniemi K, Chen J, Peltoniemi M, Fritze H, Peltoniemi M, Mäkipää R (2019) Carbon flux from decomposing wood and its dependency on temperature, wood N_2 fixation rate, moisture and fungal composition in a Norway spruce forest. Global Change Biol 25:1852–1867. https://doi.org/10.1111/gcb.14594

Romano MC, Anantharaman R, Arasto A, Ozcan DC, Ahn H, Dijkstra JW, Carbo M, Boavida D (2013) Application of advanced technologies for CO_2 capture from industrial sources. Energy Procedia 37:7176–7185. https://doi.org/10.1016/j.egypro.2013.06.655

Safamehr A, Langille ML, Anderson DM, MacIsaac JL (2013) Evaluation of composition and *in vitro* solubility rate of by-products of the Atlantic shellfish industry as alternative calcium sources. J Appl Poultry Res 22:529–538. https://doi.org/10.3382/japr.2013-00752

Seed R (1968) Factors influencing shell shape in the mussel *Mytilus edulis*. J Mar Biol Assoc UK 48:561–584. https://doi.org/10.1017/S0025315400019159

Seed R (1980) Shell growth and form in the *Bivalvia*. In: Skeletal growth of aquatic organisms: biological records of environmental change, Rhoads DC (ed) pp 23–68. New York: Springer Science+Business Media. 750 pp. ISBN 9780306402593

Seibold S, Rammer W, Hothorn T, Seidl R, Ulyshen MD, Lorz J, Cadotte MW, Lindenmayer DB and 65 others (2021) The contribution of insects to global forest deadwood decomposition. Nature 597:77–81. https://doi.org/10.1038/s41586-021-03740-8

Shamshina JL, Berton P, Rogers RD (2019) Chitin as a resource for eco-friendly bioplastics. In: Encyclopedia of Ionic Liquids, Zhang S (ed) Singapore: Springer Nature Switzerland AG. ISBN: 9789811067396. https://doi.org/10.1007/978-981-10-6739-6_43-1

Smaal A, Ferreira J, Grant J, Petersen J, Strand Ø (eds) (2019) Goods and services of marine bivalves. Springer Nature, Cham, Switzerland. 591 pp. Open access. https://link.springer.com/book/10.1007/978-3-319-96776-9

Smith D (2015) Oyster: A gastronomic history (with Recipes). London: Abrams & Chronicle Books. 256 pp. ISBN: 9781419719226

Soudzilovskaia NA, van Bodegom PM, Terrer C, Zelfde MV, McCallum I, McCormack ML, Fisher JB, Brundrett MC, de Sá NC, Tedersoo L (2019) Global mycorrhizal plant distribution linked to terrestrial carbon stocks. Nature Commun 10. Article number 5077. https://doi.org/10.1038/s41467-019-13019-2

Srinivasa PC, Tharanathan RN (2007) Chitin/Chitosan—safe, ecofriendly packaging materials with multiple potential uses. Food Rev Int 23:53–72. https://doi.org/10.1080/87559120600998163

Tamburini E, Turolla E, Fano EA, Castaldelli G (2020) Sustainability of Mussel (*Mytilus galloprovincialis*) farming in the Po River Delta, Northern Italy, based on a life cycle assessment approach. Sustainability 12. Article number 3814. https://doi.org/10.3390/su12093814

Tassanakajon A, Vatanavicharn T, Supungul P, Tang S, Amparyup P, Somboonwiwat K, Tharntada S, Takahashi J, Toyohara H (2008) Biotechnology of marine invertebrates—advances in shrimp and shellfish. In: Tsukamoto K, Kawamura T, Takeuchi T, Beard TD Jr, Kaiser MJ (eds) Fisheries for global welfare and environment, 5th World Fisheries Congress 2008, pp 221–239. TERRAPUB: Tokyo, Japan. https://www.researchgate.net/publication/228708594_Biotechnology_of_marine_invertebrates-recent_advances_in_shrimp_and_shellfish

Telesca L, Michalek K, Sanders T, Peck LS, Thyrring J, Harper EM (2018) Blue mussel shell shape plasticity and natural environments: a quantitative approach. Sci Rep 8. Article 2865. https://doi.org/10.1038/s41598-018-20122-9

The Climate Question (2021). Have we planted too much faith in trees? BBC World Service podcast [TheClimateQuestion-20210221-HaveWePlantedTooMuchFaithInTrees.mp3] broadcast 22 February 2021. https://www.bbc.co.uk/programmes/w3ct0xbf.

Utting SD, Spencer BE (1991) The hatchery culture of bivalve mollusc larvae and juveniles. Laboratory Leaflet Number 68. pp 32. Lowestoft, UK: Ministry of Agriculture, Fisheries and Food, Directorate of Fisheries Research. https://www.cefas.co.uk/publications/lableaflets/lableaflet68.pdf

Wang J, Feng L, Palmer PI, Liu Y, Fang S, Bösch H, O'Dell CW, Tang X, Yang D, Liu L, Xia C (2020) Large Chinese land carbon sink estimated from atmospheric carbon dioxide data. Nature 586:720–723. https://doi.org/10.1038/s41586-020-2849-9

Watling R, Harper DB (1998) Chloromethane production by wood-rotting fungi and an estimate of the global flux to the atmosphere. Mycol Res 102:769–787. https://doi.org/10.1017/S0953756298006157

Wisaijorn W, Poo-arporn Y, Marin P, Ordóñez S, Assabumrungrat S, Praserthdam P, Saebea D, Soisuwan S (2017) Reduction of carbon dioxide via catalytic hydrogenation over copper-based catalysts modified by oyster shell-derived calcium oxide. J Environ Chem Eng 5:3115–3121. https://doi.org/10.1016/j.jece.2017.06.018

Xiang J (2015) Recent major advances of biotechnology and sustainable aquaculture in China. Current Biotechnol 4:296–310. https://doi.org/10.2174/2211550105666151105190012

Ysebaert T, Walles B, Haner J, Hancock B (2019) Habitat modification and coastal protection by ecosystem-engineering reef-building bivalves. In: Smaal A, Ferreira J, Grant J, Petersen J, Strand Ø (eds) Goods and services of marine bivalves. Springer Nature, Cham, Switzerland. https://doi.org/10.1007/978-3-319-96776-9_13

Aquaculture: Prehistoric to Traditional to Modern

3

David Moore and Matthias Heilweck

3.1 In this Chapter...

It is pointed out that the human tradition of eating shellfish goes back to the time when *Homo sapiens* first started to migrate out of Africa, between 200,000 and 100,000 years ago. Archaeological finds of ancient meals of shellfish and ancient middens of shellfish shells track human migrations around the world. Middens do more than track migrations. They show that wooden artefacts and plant residues do not survive, but shells do. Illustrating the truth of our fundamental claim that shellfish shells sequester atmospheric carbon permanently. The coastal migrations of early humans continued across the Bering Strait to North America. Early humans along the northwest coast of North America, referred to as First Nations in Canada, actively, and sympathetically, managed the resources of their shoreline habitats, engineering intertidal rock-walled terraces as clam gardens, ancient sustainable mariculture technologies. When we reach recorded history, we enter a phase of increasing exploitation of marine resources for an ever-growing human population. By the end of the nineteenth century oysters had become a cheap staple food on both sides of the Atlantic. The working man could get a decent meal of oysters at any street corner for a few cents in New York or a penny or two in London. The real price we all paid for this was that oyster dredging on both sides of the Atlantic destroyed 85% of the world's oyster beds. New Yorkers in the 1800s ate about 600 oysters a year each; the average American today eats about 3 oysters each year. Farmed oysters account for 95% of the world's total present-day oyster consumption. The animal, which has been described as an ecosystem engineer for its reef-building abilities, is one of those that we have driven to the verge of extinction in the wild. In the twenty-first century, the oyster deserves to have the same vigour applied to its restoration and conservation as was applied to dredging it from the seabed during the nineteenth century.

3.2 Our Primeval Shellfish-Eating Tradition

The human tradition of eating shellfish goes back a long way; it might even rate as being a primeval behaviour. A variety of evidence indicates that *Homo sapiens* arose in Africa between 200,000 and 100,000 years ago. When their find of the remaims of a 164,000-year-old meal of shellfish in a cave on the South African coast was reported, the editorial summary of the published paper stated that "... the first thing *Homo sapiens* did once he and she had evolved was head for the beach ..." (Marean et al. 2007).

This was likely to have been more a matter of survival than a holiday trip because over this stretch of time in our history the world was in a changing but predominantly inhospitable glacial stage, causing cold and dry conditions that forced

D. Moore et al., *Aquaculture: Ocean Blue Carbon Meets UN-SDGS*,
Sustainable Development Goals Series, https://doi.org/10.1007/978-3-030-94846-7_3

early hominids to the coastline. It is thought that there was a gradual accumulation of a package of modern human behaviours in Africa, that included the use of bone tools, the making and use of flint tools, increasing geographic range (exploration), specialised hunting, gathering of aquatic resources, and eventually long distance trade and use of pigments for decorative arts. The generally accepted idea is that this 'package of behaviors' originated in Africa and were later exported to other regions of the world by migrations along the coastlines (Petit et al. 1999; McBrearty and Brooks 2000; Phillipson 2005; Marean and Assefa 2005).

The earliest fossils of *Homo sapiens* are located in Africa and dated to the late Middle Pleistocene (now called the Chibanian and estimated to span the time between 126- and 770-thousand years ago). At some times between 60 thousand and 200,000 years ago, modern humans dispersed, probably mainly along the coastlines initially, into Europe, Asia and more distant locations of Japan, the Philippines, Australia, and eventually the Americas.

This is the familiar 'Out of Africa' theory; that *Homo sapiens* developed first in Africa and then spread around the world supplanting all other hominid species present in Asia before the appearance of modern humans. A recent evaluation of issues raised by this theory, particularly single *versus* multiple waves of dispersal, and interactions with indigenous populations of other hominid species (interbreeding and/or replacement) concluded that there is growing evidence for ***multiple dispersals*** prior to 60,000 years ago in regions such as southern and eastern Asia, and that:

> … Modern humans moving into Asia met Neanderthals, Denisovans, mid-Pleistocene *Homo*, and possibly *H. floresiensis*, with some degree of interbreeding occurring. These early human dispersals, which left at least some genetic traces in modern populations, indicate that later replacements were not wholesale … (Bae et al. 2017).

Remains of the oldest known seafood dinner were those discovered by Marean et al. (2007) when excavating a sea cave near Pinnacle Point on the southern coast of South Africa (roughly midway between Cape Town and Port Elizabeth). Dated to 164,000 years ago, the find comprised about two dozen shells, mainly Brown Mussels (*Perna perna*) but also including at least one Whale Barnacle (a crustacean, subfamily *Coronulinae*), which indicates a diversity of taste for seafood. The next oldest known seafood dinner dates to 125,000 years ago, during the last interglacial, and was found on the Red Sea coast of Eritrea (Walter et al. 2000). Then, about 110,000 years ago, Neanderthals were cooking shellfish in caves on the coast of Italy (Erlandson 2001). This author maintains that:

> … aquatic and maritime adaptations (including seafaring) played a significantly greater role in the demographic and geographic expansion of anatomically modern humans after about 150,000 years ago. Another significant expansion occurred somewhat later in time, with the development of more sophisticated seafaring, fishing, and marine hunting technologies … (Erlandson 2001).

In fact, Erlandson (2001) gives a useful outline of what 'shellfish' means in an archaeological context:

> … The generic term shellfish is usually used to refer to a variety of aquatic invertebrates, dominated by molluscs (bivalves and univalves [for example, limpets]), but also including crabs, sea urchins, barnacles, shrimp, and other relatively common organisms … [which are usually] relatively small … What they lack in size, however, many shellfish make up for in quantity and accessibility — many types are found in large and sessile aggregations. While most shellfish provide nutritious sources of complete animal proteins and some vitamins or minerals, most are relatively low in fat, carbohydrates, and calories … (Erlandson 2001).

A few years earlier, Jones and Richman (1995) had studied the prehistoric resource value of the California mussel (*Mytilus californianus*). This is one of the most abundant bivalves found in the archaeological record of the west coast of North America and is still the most common among present-day intertidal populations. They found the mussels to be

> … high in protein, low in carbohydrates, and could contribute to complete diets among highly mobile foragers …" Significantly, they state that "… Mussels could not be overexploited to extinction, but resource value declines with frequent exploitation, rendering them of less dietary

significance in intensified economies … (Jones and Richman 1995).

A more recent study evaluates the usefulness of shellfish for human subsistence and the long-term mobility of human migration patterns, this time a continent and ocean away from the Californian coast, around the Red Sea (Hausmann et al. 2020). These authors studied shell remains found in a cluster of middens which date to approximately 7,000–5,000 years ago on the Farasan Islands of Saudi Arabia, which is an archipelago of coral reef islands located about 50 km offshore from the present-day city of Jizan. A midden is an old dump of domestic waste associated with past human occupation, and possibly, settlement. In this case, the shell remains were from *Conomurex fasciatus*, known as the lined conch, a species of sea snail common in the Red Sea. The purpose of the study was to determine the long-term sustainability of shellfish harvesting over the 2,000-year period covered by the middens, because the southern Red Sea is considered to be the southern gateway for migrations out of Africa and into Arabia. No indications of resource depletion during this occupation period were found.

> … These results have implications for the interpretation of shellfish harvesting during periods of terrestrial aridity and specifically the potential of ***shellfish as a reliable food source during Palaeolithic migrations out of Africa***… (Hausmann et al. 2020).

The variety and richness of marine resources that would have been available to the migrating humans are clear from a study of Mesolithic (15,000–5,000 years old) middens at an archaeological site at Sand on the Applecross Peninsula in Wester Ross, Scotland. Shellfish were an integral part of local life as part of a rich and varied diet that included other marine species such as crab and fish, as well as terrestrial animals and plants (Milner 2009).

The limpet (*Patella vulgata*) was the most abundant shell in these middens; the periwinkle (*Littorina littoralis*) was fairly common, as was the dog whelk (*Nucellus lapillus*), and the topshell (*Gibbula cineraria*) was also represented, but although this species can be consumed the shells are also decorative and may have been under-represented in the middens through being used for bead jewellery. Beads made from shells have been found at other sites up to 120,000 years old (Henshilwood et al. 2004; Vanhaeren et al. 2006).

Bivalves found on the site included mussels (*Mytilus edulis*), cockles (*Cardium edule*), scallops (*Pecten maximus*) although the latter is possibly also under-represented in the middens because the shells are useful as containers and for tools, and razor shells (*Ensis* sp.) though, in this case too, the empty shell is potentially useful as a spoon-like tool so may not have been discarded. One species absent from the Sand site middens is the oyster (*Ostrea edulis*). Milner (2009) suggested that oysters may not be available on this shore. But other archaeologists believe that because the early humans lacked tools that would open the shells of live oysters, they were placed over fire or heated stones and cooked until the oyster shells opened up. Oyster shells from the Mesolithic period do show evidence of fire scorching, and there is evidence for the use of fire by early humans back to about four hundred thousand years ago.

Possibly the most celebrated oyster shell middens in the United States were located on the upper Damariscotta River in Lincoln County, Maine, that empties into the Atlantic Ocean. It is famous for two enormous oyster shell heaps known as the Whaleback and Glidden middens created between 2,200 and 1,000 years ago by the Native Americans who once populated the area. When first found they were up to 30 feet (9 m) deep and covered several acres. So large, in fact, that they were mined in the late 1880s to supply a factory that processed the shells into chicken feed (Sanger and Elson-Sanger 1986).

Overall, then, it seems safe to say that whenever, wherever, and however often, early humans did come 'Out of Africa' they took the coastal route and carried with them a package of behaviours that included a ***taste for a wide variety of shellfish.***

3.3 Ancient Clam Gardening

So far, we have been describing the communities of early humans that migrated out of Africa as though they were exclusively hunter-gatherer peoples, beachcombing along the coastline collecting potential sources of food and other things of interest or use. That's not the whole story, because early humans had enough awareness and knowledge of their environment to undertake ***gardening*** activities to increase yields of various food sources.

The evidence that has so far emerged is that indigenous communities, referred to as First Nations in Canada, along the northwest coast of North America actively managed several resources of shoreline habitats. They pruned, and even fertilised shrubs and trees to increase berry and fruit production, and tended 'root gardens' for various plants and edible bulbs. In aquatic ecosystems, they made stone fish traps at the mouths of rivers and constructed wooden fish weirs in streams to harvest Pacific salmon, ***and*** managed their catches and catching methods to limit the impact of their fishing on the resource they were harvesting. Most remarkable of all in the context of our present discussion is that in ancient times indigenous human communities intervened in the management of their coastlines and engineered intertidal rock-walled terraces as **clam gardens**, ancient mariculture technologies, which have been documented from northwest Washington State, through British Columbia, and onto southeast Alaska (Jackley et al. 2016).

These terraces were made by constructing rock walls in the mid-intertidal zone, close to the lowest low water tide level. Then, through a combination of natural sedimentation as the tides ebbed and flowed and active addition of gravel and discarded shells by the people themselves, naturally sloping clam beaches were transformed into flattened terraces composed of rock and sediment on the landward side of the retaining wall of rocks, often with root gardens at the top of the beach (Fig. 3.1).

There does not seem to be any evidence for this clam garden technology along the southeast Asian coastlines of the Pacific, which might correspond to an 'Out of Africa' migration route around the coast of south east Asia and then North to cross the Bering Strait into North America:

"… we do not know of any examples of clam gardens elsewhere in the world - and I have really tried to find them … The terracing of the intertidal (with clam gardens and root gardens) seems to be unique to this region …" (Dana Lepofsky 2020, personal communication).

It is important to recognise that ***written*** details about the ecological function, use, and management of clam gardens is only now being accumulated by the present-day scientific community. The antiquity of this technology is truly prehistoric, meaning that there is no written tradition relating to it, but there ***is*** an aural tradition that is conserved by present-day First Nations peoples, some of which can be found in the YouTube videos listed in Table 3.1.

This coastal engineering activity, called clam gardens by Western science, has played a significant role for many millennia in First Nations communities of the Pacific Northwest; certainly

Fig. 3.1 Sectional schematic illustrating the basic architecture of a clam garden. The structure of the clam garden (at right) is compared with the profile of the original beach (at left). Clam gardens were often accompanied on the landward side by root gardens, which were tended for a variety of plants and edible bulbs and where shrubs and trees were pruned, and even nourished, to increase berry and fruit production. Redrawn and adapted from Groesbeck et al. (2014) and Wyatt (2015)

Table 3.1 YouTube videos describing how, for thousands of years, indigenous people all along the Pacific Northwest coast, from northwest Washington State, through British Columbia, and onto southeast Alaska, have cultivated clams in tidal clam gardens

Clam Gardens: Filling in the Gaps. Illustrating the research paper of Smith et al. 2019. https://www.youtube.com/watch?v=oJA3Erh81Oc
Simon Fraser University helicopter survey of clam gardens. Over Tracey Island, near the north end of Vancouver Island. https://www.youtube.com/watch?v=eWvkmcsXhtQ&t=12s
Mysteries of Ancient Clam Gardens. Native Watchman of the Mamalilikulla Qwe'Qwa'Sot'Em territory, Tom Sewid, takes us on a tour of the ancient clam gardens (lo'hewae) of coastal British Columbia https://www.youtube.com/watch?v=DIGn4yd15_I&t=183s
Prof. Anne Salomon describes Ancient Clam Gardens. https://www.youtube.com/watch?v=R4b1kVJrEPI&t=10s
Restoring a Coast Salish Clam Garden. On Gulf Islands National Park Reserve https://www.youtube.com/watch?v=cv247vHBllA&t=85s
A Wall Worth Building. Making Clam Habitat Great Again https://www.youtube.com/watch?v=22Nytmxw2Z8&t=141s
Clam Gardens—Learning Together. At Gulf Islands National Park Reserve https://www.youtube.com/watch?v=j2wPVx4sCN0&t=14s
Quadra Clam Gardens Time-Lapse. A very short video showing an archaeological excavation in the 3 h window in which a clam garden is exposed https://www.youtube.com/watch?v=viYc4u3NoDs

for up to 11,500 years (Smith et al. 2019; Toniello et al. 2019). However, most of the gardens have not been tended for many decades and are in need of restoration. Restoration efforts are currently focused on improving traditional and scientific knowledge of the gardens that remain through interviews with First Nations, researchers, and representatives of Tribal Park Reserves (McIntosh 2016).

Clam gardens were clearly developed by the First Nations communities to enhance the productivity of their favourite shellfish in a location convenient to them. Recent comparative research has demonstrated that clam gardens do, indeed, extend the optimal growing conditions for clams, increasing both densities and biomass of the native littleneck clams (*Leukoma staminea*) (Jackley et al. 2016). Overall, clam gardens contained 4 times as many butter clams (*Saxidomus gigantea*) and more than twice as many littleneck clams relative to nonwalled beaches. So, there is clear evidence that clam gardens increased clam biomass and density and were used together with other harvesting and processing behaviours to ensure the ***longevity*** of food security (Groesbeck et al. 2014). However, further than this, the local environments included a wide range of species that could also find the clam gardens to be an attractive habitat.

The dominant bivalves found on the beaches of this central coast of British Columbia include, as well as littleneck clams, butter clams (*Saxidomus gigantea*), macoma clams (*Macoma* spp.), horse clams (*Tresus* spp.), eastern soft shell clams (*Mya arenaria*), and heart cockles (*Clinocardium nuttallii*). The coastal regions also include eelgrass habitats, intertidal and subtidal kelp forest habitats, with the result that there is a diversity of reef-associated fish including lingcod (*Ophiodon elongatus*), rock fish (*Sebastes* spp.), and greenling (*Hexagrammidae*), and invertebrates, such as northern abalone (*Haliotis kamtschatkana*), sea urchin (*Strongylocentrotus* spp.), marine snails (*Astrea gibgerosa*, *Littorina* spp., *Nucella* spp.), mussels (*Mytilus* spp.), limpets (*Acmaea* spp., *Lottiidae*), chitons (*Polyplacophora*), and barnacles (*Semibalanus* spp., *Balanus* spp.); overall, more than enough to tickle the most jaded palate of a migrating traveller.

One more point of potential interest for the future is that quantification of the clam garden habitats constructed in northern Quadra Island, BC, Canada revealed that clam garden walls were built on 35% of the shoreline and an ***area*** of about 112,979 m^2 (=11.3 ha) of flat beach terrace was created by clam garden construction. Overall this increased the area of clam habitat by between

26 and 36%; and 35% of the final area of clam garden habitat had been constructed on bedrock shelves and rocky slopes where no clam habitat had existed previously and these features still have a significant impact on today's intertidal ecosystems. Judging from measurements of clams deposited in middens that had been collected in active clam gardens, the gardens enhanced clam production despite increased harvesting pressure. The decline of traditional management practices since European contact is associated with reductions of clam growth and clam size to levels similar to those of the early postglacial clams; that is, the advantages of the clam gardens are being lost in the contemporary world (Toniello et al. 2019).

Although the old clam gardens may not have been maintained properly in the more recent past, or even for many generations, they still maintain a legacy of increased shellfish productivity today, suggesting that clam gardens provided a reliable source of food for past populations through time; ***and if emulated could do the same again***.

Those who have so recently revealed this fascinating ancient technology remind us that it could be used today to provide sustainable food security with quotations like:

> … In many marine systems, current management approaches have demonstrably failed to halt or reverse fisheries declines, in part due to the inadequate recognition of the strong links between social and ecological processes. Ancient clam gardens and their governance by coastal communities are an example of an adaptive strategy that likely enhanced regional food security and thus conferred resilience to these coupled human-coastal ocean ecosystems. … The archaeological record is clear; abundant shellfish have supported large populations of people on the Northwest Coast through history. This new evidence helps emphasize the value of incorporating traditional management techniques into future strategies towards sustainable solutions, contributing to local food security efforts globally … (Groesbeck et al. 2014).
>
> … Clam habitat expansion facilitated by clam garden construction encouraged a sustainable and abundant food source in the past and could do so again in today's changing environmental conditions … Lepofsky et al. 2020).

It is intrinsically interesting that the First Nations Peoples of British Columbia have been harvesting shellfish from specially constructed clam gardens for several thousand years. But those of us who look to shellfish to sequester atmospheric carbon are mainly concerned to hold up this ancient activity as an example of a cheap, simple and locally managed procedure that could be used tomorrow on shorelines around the world. The expectation being that enhanced cultivation of molluscs and crustacea of all sorts in such gardens for the sake of their ability to sequester atmospheric CO_2 into crystalline $CaCO_3$ will also provide a sustainable source of food, as well as improved coastal defences.

3.4 What is Aquaculture?

As we have seen in the previous section, the origins of real aquaculture date back several thousand years with the First Nations' suite of activities including clam gardens, fish traps and fish weirs, coupled with their self-imposed management of harvesting to maintain resource sustainability. When we turn to recorded history there is not much mention of shellfish mariculture until the twentieth century; in earlier historical times attention was focussed on fish farming, and this was often of freshwater fish. Until very recently, marine resources remained as capture fisheries and foraged coastal resources; the responsibility of the fishermen and fishwives (in the original archaic meaning) of coastal communities.

Beveridge and Little (2002) discuss what aquaculture is and how it differs from fisheries (hunting) and agriculture. They offer a number of definitions that include what they consider to be key components: (a) some measure of care or cultivation to improve the yield of aquatic organisms by deliberate manipulation of their rates of growth, mortality and reproduction; and (b) the concept of ownership of access and exploitation rights. The Food and Agriculture Organization of the United Nations (FAO) introduced a definition of aquaculture which meets these criteria and reduces confusion with capture fisheries:

> … [aqualculture is] the farming of aquatic organisms including crocodiles, amphibians, finfish, molluscs, crustaceans and plants, where farming refers to their rearing to their juvenile and/or adult phase under captive conditions. Aquaculture also encompasses individual, corporate or state ownership of the organism being reared and harvested … (Rana 1997).

This definition does not cover all situations, of course. Many forms of aquaculture are based on the exploitation of multipurpose water bodies, for example, irrigation water storage ponds, rice paddies, etc., which might be 'common property'. So, Beveridge and Little (2002) suggest the key criteria distinguishing farming from hunting to be:

- Intervention(s) to increase yields.
- Defined ownership of stock or controls on access to and benefit from the activity.
- The end purpose is not specified; meaning that rearing shellfish for food, and rearing shellfish to sequester atmospheric CO_2 would both be regarded as aquaculture.

It remains the case, though, that differentiating between hunting and farming in the aquatic environment is difficult, largely because the range of methods used in exploitation of those environments has not been adequately documented. A suggestion intended to aid comparison of historical aquaculture with contemporary practices from an ecological standpoint, is the acceptance of a further set of three definitions that relate to resource use:

- **Intensive aquaculture**. Comparable to intensive terrestrial farming, here the aquatic animals are almost exclusively reliant on the farmer providing them with a supply of nutritionally balanced, high protein food, which is generally based on fishmeal, fish oils and mineral supplements (see Chap. 4; Heilweck and Moore 2021).
- **Semi-intensive aquaculture**. Which involves supplementing the level of natural food in the system in some way. The supplementary feed may be compounded from by-products or wastes from agricultural activities but could be cultivated in situ by arranging for high-nutrient waters to be mixed with low-nutrient water (see the discussion of perpetual salt fountains and Fig. 4.9 in Chap. 4; Heilweck and Moore 2021).
- **Extensive aquaculture**. In extensive aquaculture, the aquatic animals must rely solely on available natural food, such as the plankton (alive and dead), biological detritus and non-biological suspended matter (all together referred to as 'seston').

These definitions are useful enough to be applicable to the diversity of aquaculture practices to which reference is made in this book. They offer a general guide to the relative use of tangible environmental resources, which are known as '***ecosystem goods***' and include nutrients, water, light and energy. The intensity of aquaculture production methods has implications for use of '***ecosystem* services**', which are the activities provided by other components of the ecosystem, such as supplies of oxygen and recycling of wastes, and the consequential extent of intervention (by the farmer) that may be required. The more external food that is supplied per tonne production, the greater the wastes and the greater the demands on the environment to disperse and assimilate these wastes (balancing these services is discussed in Chap. 4; Heilweck and Moore 2021).

Beveridge and Little (2002) also discuss the origins of aquaculture practices, likening it to the origin of agriculture and, in particular, highlighting Colin Tudge's quotation:

> … People did not invent agriculture and shout for joy. They drifted or were forced into it, protesting all the way …

Tudge, (1998) argues against the traditional view that agriculture began in the Middle East around 10,000 years ago. He believes that this view under-plays the importance of 'proto'-farming activities, throughout much of our previous two-million-year history, to persuade more food out of the environment, especially from the late Paleolithic onwards (from about 40,000 years ago). The traditional view is that hunter-gathering is hard and that farming made life easier. Colin Tudge turns this notion on its head; asserting that farming is now, and always was, hard work and to be avoided unless absolutely necessary (Tudge 1998).

Tudge contends that when food supplies improved through upturns in abundance of game or more clement weather or death or emigration of people, the prehistoric communities returned to what they enjoyed best: hunting and gathering. Beveridge and Little (2002) point out that different peoples adopted farming activities at different times, and some cultures, such as the Chinese, developed agriculture independently, others (most of the continent of Europe) learned from neighbours or colonisers, while a few appear never to have acquired agriculture at all, such as the Indigenous Australians (humans first populated the Australian continent at least 65,000 years ago).

Beveridge and Little (2002) also claim that there is good evidence in ***aquaculture*** for Tudge's theory of people opting in and out of plant and animal cultivation according to their needs (Beveridge and Little 2002), being in many respects similar to agriculture, aquaculture too began in different ways in different societies, both agriculture-based and capture-fishing-based, and developed from there to the present day:

- Transplantation of fertilised eggs or juvenile animals, like any of the many contemporary shellfish hatcheries intended to replenish and/or expand harvested stocks.
- Entrapment of fish in areas where they could thrive and be harvested as required, like the stone fish traps and fish weirs of many indigenous communities.
- Environmental enhancements, such as development of spawning areas, enhancement of food, exclusion of competitors or predators, etc., like the clam gardens of British Columbia's First Nations Peoples.
- Holding fish and shellfish in systems like ponds, cages, and pens, until their biomass and/or food value improve, like any present-day salmon or whitefish farm.

On its own, each of these activities might be considered as no more than stock storage and/or stock enhancement and therefore fall under a 'managed fisheries' definition. But they can be woven into much more sustainable activities, and they were most likely implemented on a local scale by small communities that undoubtedly exhibited the concepts of community ownership and controlled access. These early 'proto-aquaculture' stages also feature low consumption of energy.

One other term used in aquacultures is **ranching**, which is used to describe the release of juveniles into the wild, only to be recaptured later as adults. The term has been used in relation to crustacean farming (Wickins and Lee 2002).

Although the details of ***fish farming*** are beyond the scope of this book, this activity is so bound up with the historical origins of aquaculture effort that we have to consider it, at least in outline, to help explain how aquaculture might have first developed. According to Beveridge and Little (2002):

> … aquaculture began in various parts of the world and at various points along the aquatic food supply line, between water and plate. The farming of fish and shellfish is generally considered to be an activity of settled societies, originating among both fishing and wetland farming cultures as well as at points of trade ….

We have indicated above that archaeological research demonstrates the importance of shellfish and fish in early (migrating) hunter-gatherer societies. It may therefore be assumed that when these communities settled and increased in size, foraging and fishing may have been insufficient to satisfy demands for shellfish and fish (usually called ***finfish*** to distinguish from shellfish), respectively, resulting in the development of the simpler proto-aquaculture stock management techniques.

> … Many proto-aquaculture activities relied on some sort of holding facility. The simplest to construct would have been earth ponds. In some parts of the world these would have been little more than mud walls constructed to temporarily hold water and fish following the seasonal flooding of a river. Such systems are still in use in some parts of the world today. The 'whedos' or fish holes of Benin are one such example … The

> practice of communal construction of weirs on small rivers and streams in Asia to store water outside of the monsoons principally to ensure adequate irrigation for wet rice cultivation … is also common. Attempts to increase fish yields would have been a logical next step …short-term storage of catches until there were sufficient fish or shellfish to make a journey to market worthwhile; the transport of live fish to market; the holding of catches until prices improved. These strategies are still seen among fisher folk today: modified traps, netted off shallow areas of lakes, cages of the sort still seen in parts of Indonesia, traditional floating cages used in the Great Lake area of Cambodia … (Beveridge and Little 2002).

The practice of harvesting the larger fish in a holding pond, leaving the smaller fish to remain as broodstock is a common procedure the world over. It is a natural step from there to primitive fish ***farming***.

3.5 Fish and Shellfish in Recorded History

Proto-aquaculture activities have been recorded in many parts of the ancient world, including Egypt, China and Mesoamerica. The earliest records seem to be from ancient **Egypt** where a 4,000-year-old bas relief in a tomb shows what appears to be a nobleman fishing, probably for tilapia, in an artificial, drainable pond (Bardach et al. 1972). This was the simplest proto-aquaculture; native tilapia being transferred from rivers to captivity in the ponds and involving little management. Rod and line fishing is believed to have been common among all classes in Egypt at that time, but the fishing activity of the nobility was limited to their ponds as it was more of a religious ritual, associated with death and rebirth, rather than a way of catching food. On the other hand, Roman writers of the day imply that fish was of great importance in the Egyptian commoners' diet; saying that:

> … the Nile supplies the native … with fish freshly caught [and] an unfailing multitude for salting … all Egyptians in the Nile Delta possess a net with which, during the day, they fish … (Beveridge and Little 2002).

This difference between nobles and commoners in the relevance of fish to the diet is something that emerges again in medieval Europe.

Modern aquaculture began in Egypt in the mid-1930s with the introduction of the common carp in two research finfish farms. According to FAO Fact Sheets (Salem and Saleh 2010) 14 different species of finfish (species of carp, catfish, meagre, mullet, seabass, seabream, and tilapia) and two species of crustacean are currently farmed in Egypt. The shellfish being ***penaeid shrimps*** (*Penaeus* spp.) (Kungvankij 1984; Briggs et al. 2005) and and the ***giant river freshwater prawn*** (*Macrobrachium rosenbergii*) (FAO 2020a, b).

Fish has almost always been important in the Asian diet and China's wetland-based agriculture was a strong incentive to develop inland fish farming as integrated polyculture aquaculture, and **China** is widely regarded as the cradle of aquaculture. Its history is carefully reviewed by Beveridge and Little (2002). A historical highlight is a document published 2,500 years ago (by the statesman Fan Li) that describes carp (*Cyprinus carpio*) farming in sufficient detail to show that aquaculture, as a 'semi-intensive' monoculture, was well-established by this time. There are other written records, dating from the period 2,200–1,720 years ago, telling of the integration of carp culture with that of aquatic plants and vegetables (Yang 1994). The most complex integrated aquaculture system was the fishpond-dyke-mulberry system (Ruddle and Zhong 1988), which supplied live fish, fruit, and leaves for silkworm cultivation.

With the possible exception of China, not many of today's aquaculture industries have a history extending back more than 30 or 40 years; they simply do not owe much to older traditions or technologies. The exception, China, has been the world's top (inland) fish producer for many years, and China's aquaculture is much more diverse than all other countries in terms of farmed species, of which there are over 200, and farming systems and methods. Nevertheless, until the founding of the **People's Republic of China** in 1949, the harvest of farmed fish was only a fraction of that obtained from capture fisheries. A 50-fold growth in production of cultivated freshwater fish occurred as a result of the political, economic, technical and demographic changes in 1949. Among those changes was the construction of over 82,000 man-made bodies of water as reservoirs for hydropower, flood control and irrigation purposes, amounting to an increase in the inland water surface area by over 2 million ha (De Silva et al. 1991). Most of these were used for farming carp, initially with wild-caught fry but techniques for spawning Chinese carp were developed in the late 1950s. Later progress of the mixed economy in the PRC incentivised greater increases in productivity of inland fish, although aquaculture was rare elsewhere in Asia at the time.

Today, over 700 species of freshwater fish and 60 species of marine freshwater migratory fish occur in the inland waters. Inland capture fisheries are still important, but the most commonly farmed species are several species of carp, bream, chub, and mandarin fish (*Synchiropus splendidus*; a small, brightly coloured saltwater aquarium fish), as well as soft-shelled turtle (a delicacy, particularly as turtle soup, in many parts of Asia). Farmed shellfish include the following (Bernal and Oliva 2016; FAO 2017):

- **Fresh-water shrimp**/prawn (*Macrobrachium rosenbergii*) (FAO 2020a), fresh-water mussel (*Margaritifera margaritifera*, reared to produce cultured pearls (FAO 1983; URL: https://en.wikipedia.org/wiki/Freshwater_pearl_mussel).
- **River-snail** (*Cipangopaludina chinensis*), eaten for its meat throughout China, its shells are abundant in Mid-Late Neolithic archaeological sites in the Guanzhong Basin of Northwestern China (Li et al. 2013).
- The **Chinese mitten-handed crab** (*Eriocheirn sinensis*). Cultivation begins from a larval stage that can adapt to freshwater, reared in a nursery until they become 'button-sized' juvenile crabs and then grown to market size in ponds, pens in lakes, or rice paddies. Paddy culture has increased rapidly in recent years as an environmentally friendly production approach that benefits both the crabs and the rice (Liu and Chen 2002).

Marine aquaculture has expanded recently, mostly in shallow waters, shoals and bays, and using (according to species) raft culture, net cage culture, vertical culture, seabed 'seeding', stone adhesion culture and pond culture. Farmed organisms include several fish species and various shellfish including shrimp (*Penaeus chinensis*), oyster, mussel, scallop, several clam species, abalone, crab, kelp and *Porphyra* sp., the edible red alga known as laver seaweed (Bernal and Oliva 2016; FAO 2017). Until recently, wild marine fish and shellfish resources in other Asian countries with extensive coastlines were sufficient to supply the needs of their populations; both freshwater and marine aquaculture are being developed extensively throughout Asia and Africa. Because of travel restrictions and bans on mass gatherings caused by the COVID-19 pandemic, ***World Aquaculture Society's Conferences*** planned for 2020 have been rescheduled to 2021; the Asian Conference in Singapore in June (https://www.was.org/meeting/code/WA2020), and *Aquaculture Africa* in Alexandria, Egypt, in December (https://www.was.org/meeting/code/AFRAQ20).

Pre-Hispanic **Mesoamerican** (Aztec) cultures that flourished from AD 1300 to AD 1521 developed several integrated wetland agriculture-with-aquaculture systems around lake margins in the Valley of Mexico. Although it is believed that these were once used to enhance fish storage and/or production this plays no part in the present-day systems that survive. (Micha and Chavez 1997). In the present day, **South American countries** are major producers and exporters of both capture-caught and aquaculture fish and shellfish; **Chile** and **Brazil** being two of the largest intensive fish producers in the world. However, while Chile relies primarily on marine fish, Brazil leads in continental production (Valladão et al. 2018; FAO 2020b). Production of native fish is beginning to overtake production of non-native species in some countries. In particular, the 'black pacu' (*Colossoma macropomum*), which is found in most rivers and streams in the Amazon and Orinoco river basins, is commonly farmed because farming black pacu competes well, economically, with tilapia production in South America.

There is no tradition of aquaculture in the **Caribbean region**, in contrast to capture fisheries, but developments are taking place. In 2014, the *State of World Fisheries and Aquaculture* (FAO 2014), published by the Food and Agriculture Organization of the United Nations, said that world food fish aquaculture production expanded at an average annual rate of 6.2% in the period 2000–2012 but over the same period in **Latin America** and the Caribbean the increase was 10% per annum. According to (Myvett et al. 2014): "… The practices mainly involve the use of ponds to culture such species as **penaeid shrimp** (*Penaeus* spp.), tilapia (*Oreochromis* spp.), carp (*Ctenopharyngodon idellus*, *Hypophthalmichthys nobilis*, *Hypophthalmichthys molitrix*) and cachama [the local name for black pacu] (*Colossoma macropomum*). Also, there is long line culture for algae (*Eucheuma* spp. and *Gracelaria* spp.) in **St. Lucia** and the **mangrove oyster** (*Crassostrea rhizophorae*) in **Jamaica**." (Myvett et al. 2014). A mangrove oyster farming system has been operating successfully in **Cuba** for many years (Nikolic et al. 1976).

The Caribbean Regional Fisheries Mechanism (CRFM) has identified the promotion and development of aquaculture as one of its priority programme areas and has established an Aquaculture Working Group tasked with identifying constraints to aquaculture development and making recommendations (Myvett et al. 2014). This CRFM Technical and Advisory Document considers some of the species that might be grown in the Caribbean culture systems (and elsewhere, also), they are:

- **Marine fish**. Several of the marine fish species occurring in the seas around the Caribbean islands could be cultured in floating net cages. Sea temperatures are ideal and the water is clean and good quality. Groupers are well suited to this this type of mariculture, which is well developed in Southeast Asia (Yang et al. 2014).
- **Turtle farming**. Sea turtle populations in the Caribbean region have been declining for a long time as taking turtles for their meat, shells and eggs has depleted stocks drastically. Turtle farming might be based on collection of eggs, or hatchlings, in the wild, though this could reduce the recruitment to the already dwindling stocks of wild sea turtles (unless the farms can be stocked with daytime hatchlings which would otherwise be taken by seabirds). Turtle farmers believe that the damage done through collection of eggs in the wild is compensated by their restocking efforts. There is a turtle farm located in the **Cayman Islands**, where the green turtle is cultured in large ponds, but it remains controversial (Bale 2017), although turtles are caught throughout the Caribbean and there is a market for the meat. Experience elsewhere is that sea turtle farms, whether for captive breeding or ranching, cannot be shown to be directly beneficial or proved to be detrimental to conservation of wild populations. What can be demonstrated is that they are very expensive, require advanced technical knowledge, and are of uncertain economic viability.
- **Seaweeds**. The technology for mariculture of seaweeds is so well known in most cases that it needs only adaptation to local conditions and is simple and well suited for unskilled coastal communities. The initial investment in materials and supplies is minimal and the installations required can be easily repaired and replaced. There is good potential for culture of *Gracilaria debilis*, which is used for food-grade agar production as a gelling agent, for local consumption and export (Veeragurunathan et al. 2019). It is also possible that culture of seaweeds producing carrageenan (also widely used for gelling, thickening and stabilising products in the food industry) such as *Eucheuma* sp. and *Hypnea* sp. could also be undertaken. These seaweeds are often found in the Caribbean, but the natural stocks are insufficient to sustain commercial

harvesting, though certainly sufficient as a source of 'seeds' for mariculture.

- **Cockle culture** technology is simple, cheap and well suited for use by unskilled local communities. Small seed cockles, collected from natural beds, are transferred to growing beds in shallow bays, being scattered as evenly as possible. Harvesting is done eight to nine months after sowing over a period of two or three months. Yields can be in the order of 35–40 t ha^{-1}.
- **Queen Conch (*Strombus gigas*) stocking**. Conches feed on a variety of plants including manatee grass and turtle grass which grow well on sandy bottoms in relatively silt-free water. The animals are usually found at depths between 3.5 and 16 m. Although they emphasise that improved management of the natural resource is essential, Myvett et al. (2014) suggest a hatchery could re-stock the many areas depleted by overfishing. Unfortunately, experience with this practice is not good; Stoner (2019) suggesting that "… every effort should be made to conserve wild populations. Hatchery production for stock restoration should be considered a last resort."
- **Oyster and mussel culture in the sea**. Farming oysters and mussels in the sea can be organised as smallholder activities that do not normally require high investments per unit area of production. A disincentive is that the smaller islands in the Caribbean do not have enough sites with sufficient planktonic food in the water for this to be a worthwhile mariculture development. It is also discouraging that only small quantities of these bivalves are consumed in the islands, so production volumes would have to be scaled to suit processing and transport capabilities. Nevertheless, the blue mussel (*Mytilus edulis*) aquaculture industry in **Chile** has solved all these problems of integration and could be a model worth following (Gonzalez-Poblete et al. 2018).
- **Pond-based aquaculture**. The advantage of constructing intensive mariculture systems on land is that they can be sited so far from the sea that the risk of storm damage is minimised. The trade winds, which blow northeast from the coast of Africa across most of the Caribbean at more or less constant speeds of 24–32 km h^{-1}, could be harnessed with new windmill technology to pump seawater into raceway ponds (see Chap. 6, Fig. 6.6; Moore 2021). **Penaeid shrimp**, marine reef food fish, like snappers and groupers, and ornamental fish are suitable for cultivation in land-based intensive mariculture facilities. Yields of shrimps in the order of 2 kg m^{-2} can be expected from such raceways, three times per year.

FAO's *Fishery and Aquaculture Country Profile* for the **United States of America** (FAO 2019) reports that the majority of the seafood consumed in the USA originates from imports, and in 2017, the USA was the world's leading importer of fish and fishery products. In the United States, capture fisheries and aquaculture occur in many of the country's coastal waters, rivers and lakes. Some fishing also takes place in the exclusive economic zones (EEZs) of other nations, and on the high seas, such as tuna caught in the Western Central Pacific. The two main capture species are Alaska pollock (*Gadus chalcogrammus*) caught in the Pacific Ocean and Bering Sea, and Gulf menhaden (*Brevoortia patronus*) caught in the Gulf of Mexico.

Inland commercial fisheries of the United States are limited to the Great Lakes and a few major rivers. Aquaculture production declined from a peak level of over 600,000 tonnes in 2004, due mainly to a decrease in the farming of channel catfish (*Ictalurus punctatus*), which is North America's most numerous and widely distributed catfish species. But despite this decline, channel catfish farming still dominates freshwater aquaculture, with rainbow trout (*Oncorhynchus mykiss*) representing only 8% of total production. US aquaculture produces food fish, ornamental fish, baitfish, molluscs, crustaceans, aquatic plants and algae, with some reptiles such as alligators and turtles.

Crayfish farming, of **red swamp crawfish** (*Procambarus clarkii*) comprises about a quarter of total freshwater aquaculture production.

Crawfish, as they are called locally, are cultivated and consumed for food in several southern states but Louisiana dominates the crawfish industry of North America, and includes the related species *Procambarus zonangulus* occurring in Louisiana that accounts for about 15% of production. The downside of crawfish farming is that in areas where it has been introduced *P. clarkii* is highly invasive and, more importantly, is a potential vector for the ***crayfish plague*** (caused by *Aphanomyces astaci*, a fungus-like‘water mould’ belonging to the Phylum *Oomycota*).

Crayfish plague was introduced into Europe in the 1960s along with American freshwater crayfish species and resulted in many indigenous freshwater crayfish populations being diminished or even eliminated (FAO 2016). US aquaculture also produces shrimp in brackish ponds in South Carolina, Texas, and Hawaii. The United States also farm-raises mollusc species such as **abalone**, **oysters**, **clams**, and **mussels**. Molluscs are grown in almost every coastal area of the United States and are produced using various systems (FAO 2019).

The US Department Of Agriculture (USDA) has several important programmes to assist the aquaculture industry. Details and reports are available from the USDA website at http://www.usda.gov by navigating to /Agriculture/Aquaculture.

Of particular interest are some programmes of the National Marine Fisheries Service (NMFS), which is the US federal agency responsible for the stewardship of national marine resources (its parent agency is the National Oceanic and Atmospheric Administration (NOAA; https://www.fisheries.noaa.gov/) of the US Department of Commerce. NMFS activities include: The National Sea Grant College Program (http://www.seagrant.noaa.gov/), that supports aquaculture in many topic areas; and the National Ocean Service (https://oceanservice.noaa.gov/), which administers Federal Coastal Zone Management Act funds for aquaculture facilities in the coastal zone.

The **common carp** (*Cyprinus carpio*) was undoubtedly vital to the development of aquaculture in continental **Europe** during most of the past two thousand years. The Romans are documented as being among the first to build coastal aquaculture ponds, most likely before the end of the second century BC. Excavating substantial areas of fish ponds (‘piscinae’) at their villas was common among members of the nobility, partly for the purpose of holding live food fish, but also as a demonstration of wealth and status, a practice that persisted for several centuries. But common people also built freshwater and saltwater ponds (‘dulces’) for food production and income generation, which were stocked with coarse fish and salmonids, eel, mullet, turbot and sea bass. Classical Roman literature gives the impression that the keeping of fish in artificial ponds was commonplace throughout the Mediterranean provinces of the Roman Empire (Beveridge and Little 2002).

During the first and second centuries AD the common carp was imported from the Danube and the practice of common carp aquaculture developed by the Romans gradually spread westwards across the continent of **Europe**, although it did not reach **England** until the late fourteenth century. In central and western Europe, fish farming was developed in the first instance by the monastic orders in order to ensure supplies of fish for days when eating meat was forbidden. Subsequent pond aquaculture developments in continental Europe were often practised in modified floodplains where soils were too poor to sustain agriculture because they were likely to flood.

In medieval England, pond fish culture was also used by the post-1066 AD Norman rulers to help consolidate their political power ***and*** the ecclesiastical power and influence of their church. The monasteries were the repository of knowledge throughout the Middle Ages, and that included knowledge of carp cultivation. Mill ponds, essential ‘header tanks’ for the waterwheels needed to drive the industry of the day were usually stocked with fish. Wealthy landowners also had such ponds constructed, so the privileged classes could exercise their preference for fresh rather than salted fish, and widespread development of carp farming ponds occurred throughout much of continental Europe during the fourteenth to sixteenth centuries (Hoffmann 1996; Beveridge and Little 2002).

With the dissolution of the monasteries by Henry VIII in the sixteenth century in England, many monastic fishponds were abandoned, though a few wealthy freemen began to create aquaculture ponds during the latter part of the medieval period. There is even evidence that in the Forest of Arden, Warwickshire (the forest in Shakespeare's *As You Like It*), wealthy peasants owned such ponds, and that part of their produce was sold. There is also evidence of fishmongers in the 1350s in London feeding fish they kept in ponds for sale, though the practice was never widespread. In post-medieval times, keeping common carp in ponds became popular for a while among land-owning nobility, the fish being used for both domestic consumption and sale (Currie 1991). However, there was a decline in eating freshwater fish such as carp, bream and perch during the seventeenth and eighteenth centuries. Indeed, the best-known cookbook in the English-speaking world, *Mrs Beeton's Book of Household Management*, first published in 1861, states that freshwater fish are seldom purchased (Chambers and Gray 1988; Beveridge and Little 2002).

Pond fish farming in Germany, France and Central Europe declined from the late sixteenth century onwards and did not recover until trout culture developed in the mid-nineteenth century. Fish was always more expensive than meat, and consumption of fish declined during the seventeenth century, first among the nobility and later among the middle classes. In eighteenth-century Europe fish farming withered and died (Beveridge and Little 2002).

For most of recorded history, therefore, aquaculture has meant the production of fish in freshwater ponds or coastal marine lagoons. Fish for religious purposes, fish for the nobility, very occasionally fish for the people of lower social standing, if they could afford it. Throughout our history, the hunt for animals on land has been a noble pursuit for the gentry. Fishing, the hunt for fish on the storm-tossed high seas, has been something that common people do, just to stay alive (Sahrhage and Lundbeck 1992). And for most of recorded history, with a few notable exceptions, aquaculture was rarely applied to shellfish. Despite the ancient example of the clam gardens of First Nations peoples, shellfish farming before about 1950 was insignificant.

Shellfish were delicacies that were collected from their natural habitat; foraged usually by the weaker, younger or older members of the families of the fisherfolk who were out on the high seas catching fish, all for sale to the gentry who could afford to load their tables with fine dishes filled with food (Fig. 3.2).

As populations increased and became concentrated in towns and cities, demand also grew, foraging became more intensive and shellfish stocks were overfished, in some cases close to extinction, like so many other targets of our capture fisheries (Clover 2005; Hilborn and Hilborn 2012). The prime shellfish example of this is the history of the oyster through the nineteenth century and into the early twentieth century.

3.6 Nineteenth Century Oyster and Scallop Dredging on Both Sides of the Atlantic

We have seen that oysters have been an important subsistence food for humans since the Neolithic period, although today they are considered by many to be a luxury delicacy. At the end of the seventeenth century and into the early 1800s, oysters were eaten mainly by the wealthy, except in local harvesting areas. However, by the mid-1800s, a surge in demand caused increase in production, and consumer prices for oysters dropped substantially; oysters became cheaper to purchase than meat, poultry, or finfish and oysters were eaten by people at all economic levels (MacKenzie 1996).

The issue of ***The Illustrated London News*** of 12th August, 1843 included an article entitled '***Oyster Day***'that reported:

> … August 4 was, in metropolitan parlance, '*Oyster Day*', i.e. the day on which oysters are first brought into the London market at Billingsgate … There were fifty sail of vessels at market from Rochester, Whitstable, Essex, and the Cheyney rock, near the Isle of Sheppy ….

A few years later, 1861, *The Illustrated London News* showed the excitement surrounding a

Fig. 3.2 Dishes with Oysters, Fruit, and Wine, by Osias Beert the Elder, probably the most important still-life painter in Antwerp during the early seventeenth century. This oil on panel still-life was painted about 1620–1625. The painting is in the National Gallery of Art, Washington DC, USA. Photo credit Daderot; image ID DSC09953.JPG. Creative Commons CC0 1.0 licence

street vendor in London on Oyster Day (Fig. 3.3). Those oysters that arrived at Billingsgate Market on Oyster Day (Fig. 3.3) were dredged from the river estuaries and seas off England's southern coastline by fishing smacks (Fig. 3.4).

One such river was the River Colne, a small river in England that runs through Essex. In fact, the Essex Colne (there are other rivers with that name in England) passes through the town of **Colchester**, once a Roman fort and settlement (called *Camulodunum*), which is the site of Britain's earliest known Christian church and is the earliest recorded town in England. Oysters are native to the river Colne and Colchester's oysters have been famous since before Roman times. Downstream of Colchester the Colne estuary joins the North Sea near Brightlingsea, just North of the Thames estuary. Dredging oysters is a long-established occupation of the fishermen of Essex, and by the end of the nineteenth century they were using iron-framed dredges with collection nets. The sailing smacks might tow 4 to 8 of these dredges across an oyster ground, gathering, besides oysters, much rubbish and gastropod pests/predators of the oyster from which the marketable oysters had to be sorted. The rubbish was thrown overboard, but the pests (mainly slipper limpets, *Crepidula fornicata*, but including tingle borers, *Ocenebra erinacea* or the American whelk tingle *Urosalpinx cinerea*, and dog whelks) were kept on deck to be dumped ashore when the day's work was done; an unceasing struggle to keep the oyster grounds clean (Fig. 3.4).

The slipper limpet is a non-native species in British waters; it was probably introduced from its natural habitat on the Atlantic coast of the

Fig. 3.3 The first day of oysters: a London street scene. Published in ***The Illustrated London News*** of 1861. Showing people gathered around the stall of an oyster seller, some opening oysters, others eating them. Taken from *Old Book Illustrations* (https://www.oldbookillustrations.com/illustrations/first-day-oysters/) under a Creative Commons 4.0 License

USA in about 1870 with consignments of American Bluepoint oysters imported for relaying in English waters (Cole 1942). *Crepidula* can starve and smother native shellfish, competing for food and space as well as drilling holes through bivalve shells to access the soft tissues within. Damage to shellfish by slipper limpets can make oysters, mussels and scallops impossible to sell. Slipper limpets are now established in the waters of South England and South Wales and have severely diminished oyster stocks. *Crepidula* eggs are deposited in capsules from which fully formed juveniles emerge. There is no free-swimming larval stage, so infestation of new areas requires transport by human activities. Today, it is an offence to use slipper limpets as bait for fishing or to release them to the sea in UK waters (https://www.gov.uk/government/news/slipper-limpets-not-permitted-to-be-used-as-bait-or-disposed-at-sea).

In the early days, all hauling of dredges was done by hand, and between 30 and 100 smacks of up to 15 tons displacement dredged the Colne estuary fishery. The nearby Blackwater and Crouch rivers had extensive oyster layings worked by other dredgermen. In the mid-nineteenth century the railways spread throughout Britain and brought fast transport for perishable goods like fish and shellfish, so all branches of trade in fresh foods expanded considerably as the steam locomotives of the railways brought the growing populations of major cities within reach. Expanding markets entailed expansion of the fishing fleets to meet the demand; more boats, bigger boats, and deeper dredging over an ever-widening area.

The Aldous shipyard of Brightlingsea built thirty-six big fishing smacks of 20 to 40 displacement tons between 1857 and 1867. Harris shipyard at Rowhedge and Harvey shipyard at Wivenhoe built a good number and more were launched on the River Blackwater. These smacks were the deep-sea trawlers of their day and they made up the most adventurous fleet of fishing vessels ever to sail from Essex, voyaging far out to sea for oysters and scallops.

Rich oyster beds were discovered off the island of Jersey in 1787 and within a few months over 300 smacks from Essex, Shoreham, Enisworth and Faversham were working there and continued through the Napoleonic wars. Later, a fleet of 60 Essex fishing smacks would sail there each spring and carry on dredging through all hazards. The Jersey fishery declined during the eighteen-forties and became exhausted by 1871 (Leather 1991). The Essex dredgermen went elsewhere.

Leather (1991) describes it like this:

> … Their quest for oysters and scallops led them at various times to work the Inner Dowsing and the Dudgeon banks [off the south Lincolnshire coast], landing catches at Grimsby or Blakeney in north Norfolk, the Ness grounds, stretching from Orfordness to Cromer in Norfolk, the Galloper and Kentish Knock areas of the North Sea, and the Terschelling and Hinder banks off the Dutch coast, landing catches at Brightlingsea. In the English Channel they dredged the Goodwin, Sandettie and the Varne grounds besides those on the French coast at Caen Bay, Dieppe, St Valéry-sur-Somme, Fécamp, Calais and Dunkirk, using Ramsgate, Dover, Shoreham or Newhaven to land catches.

Fig. 3.4 Dredging oysters in the River Colne. **Top**: Oyster dredging smacks at Brightlingsea going out for oysters. William Francis of Brightlingsea, foreman of the Colne Fishery Company, at the helm of the company smack NATIVE. Photo by Douglas Went, dated 1928, image ID BOXB5_017_031. **Bottom**: The Colne Fishery Company smack NATIVE hauls alongside an oyster skiff to dump thousands of slipper limpets dredged from the oyster grounds in the course of a tide's work.
The NATIVE probably intends to anchor when the despised limpets are unloaded. Photo by Douglas Went, dated 1936, image ID BOXB5_017_029. Both images from http://www.merseamuseum.org.uk/

Down Channel, towards the Atlantic, they dredged West Bay, off West Dorset's Jurassic Coast, and occasionally the Cornish Fal and Helford rivers were visited by the Essexmen.

> … Others sailed round Land's End to work on the south Pembrokeshire coast, based at Swansea and Bangor, and southern Ireland, north west Ireland and the Solway Firth [off the south west coast of Scotland] regularly saw the rakish Colne topmasts … (quoted from Leather 1991).

Oyster and scallop dredging at this intensity (truly 'over-fishing') caused widespread stock exhaustion of both oysters and scallops. The flourishing oyster fisheries they exploited at Swansea and Cardigan Bays in Wales, Largo Bay in Fife, Scotland, in the Solway Firth and off north Norfolk were rapidly worked out by fleets of fishing smacks from Colne. It's interesting to note that in the present day many of the locations

Table 3.2 Website URLs of a few of the conservation and restoration projects underway at present

Cardigan Bay—natural resources wales (https://cdn.cyfoethnaturiol.cymru/media/687993/eng-cardigan-bay-reg-37-report-2018.pdf)
Solway firth marine conservation zone (https://assets.publishing.service.gov.uk/government/uploads/system/uploads/attachment_data/file/915681/mcz-solway-firth-2019.pdf)
Offshore and English inshore marine conservation zones (https://jncc.gov.uk/our-work/marine-conservation-zones/)
Billion Oyster Project is restoring oyster reefs to New York Harbor in collaboration with New York City communities (https://www.billionoysterproject.org/)
Oyster restoration—Chesapeake Bay foundation (https://www.cbf.org/about-cbf/our-mission/restore/oyster-restoration/)
Galveston Bay foundation—habitat restoration (https://galvbay.org/work/habitat-restoration/)
Oyster restoration in the USA—The grand tour by the blue marine foundation (https://www.bluemarinefoundation.com/2017/06/27/oyster-restoration-in-the-usa-the-grand-tour/)
The Nature Conservancy oyster restoration in the US (https://www.nature.org/en-us/about-us/where-we-work/united-states/oyster-restoration/)
The European native oyster habitat restoration alliance (https://www.noraeurope.eu/)
… there are many more … use your favourite search engine to search for restoration+place+organism

mentioned in the quotations above are protected conservation areas and/or restoration areas (Table 3.2).

After the turn of the century, 1900, a series of poisoning scares raised fears regarding the safety of oysters and killed the demand for sea oysters. At least some of these fears were well-founded and seemed likely to be due to the zeal of Victorian civil engineers who were keen to take advantage of the ability of filter-feeding bivalves to purify foul water. When they built the town sewer networks they were creating to collect sewage from the growing urban populations they laid their outfalls into the estuary and coastal oyster beds in the hope that they might clean up the effluent.

We can't be too critical of the civil engineers of the 1830–1870s because it wasn't an entirely illogical plan. Final proof of the germ theory of disease (by Louis Pasteur and Robert Koch) only came in the 1880s and viruses were first discovered in the 1890s. At the time the engineers were designing and building their sewer systems, diseases were thought to be caused by a ***miasma***, a noxious form of 'bad air'; so if that's what the doctors tell you, go ahead, shift the sewage downstream and get the oysters to clean it up. What's miasma got to do with oysters?

The majority of the shellfish-associated infections that have been reported over the last century have been linked to oysters, followed by clams and mussels. We now know that they have been caused by viruses, particularly Hepatitis A virus, which causes a potentially serious liver infection, and caliciviruses, which include noroviruses, and cause acute nonbacterial gastroenteritis. All these viruses are spread in human faeces; so there's a good reason to be more careful with your sewage. *Vibrio* species, which thrive in warm seawaters, head the list of bacterial pathogens associated with shellfish, several species of which can cause foodborne infection, usually linked to eating undercooked seafood as the bacterium occurs naturally in the gut of oysters and other shellfish, and in the intestines of fish that inhabit oyster reefs. The vast majority of people who develop sepsis from *Vibrio vulnificus* infection became ill after eating ***raw*** oysters (Berg et al. 2000; Potasman et al. 2002).

While the dredgermen of the Essex fishing fleets were busily fishing out the estuaries and waters around Britain, something very similar was happening across the Atlantic in US waters. It is an exactly parallel story to the English one. Oysters were still only eaten by the wealthy in the early 1800s United States, but a rising production through the mid-1800s brought oyster prices below that of other protein foods. In 1885, oysters cost $0.03 each (equivalent to $0.80 today) and by 1889 dropped to $0.01 each (equivalent to $0.27 today).

> … The low prices meant anyone could eat them, and oysters quickly became popular with the working class as it was a rich, cheap source of protein, with major oyster markets such as New York City, Philadelphia, Baltimore, and New Orleans supplying the high demand for oysters in the US … (MacKenzie 1996; Foodworthwritingfor.com 2018).

And, another parallel with Britain, the growing network of railways transported the oyster harvest throughout the United States.

According to MacKenzie (1996) the eight greatest oyster dredging estuaries in the continental United States and eastern Canada, starting during the eighteenth century, were:

- **Bedeque Bay**, Prince Edward Island, Canada.
- **New Haven Harbor**, an inlet on the north side of Long Island Sound in the state of Connecticut in the US.
- **Delaware Bay**, the estuary outlet of the Delaware River on the northeast seaboard of the United States.
- **Upper Chesapeake Bay**, the largest estuary in the United States, separated from the Atlantic Ocean by the Delmarva Peninsula.
- **James River**, the longest river in Virginia, US; it runs into Hampton Roads where the Chesapeake Bay flows into the Atlantic Ocean.
- **Apalachicola Bay**, an estuary and lagoon located on the northwest coast of the US state of Florida.
- **Louisiana Estuaries**, which are the various estuaries through Louisiana's wetlands of the Mississippi River delta confluence with the Gulf of Mexico.
- **Puget Sound** and **Willapa Bay**; Puget Sound is an inlet of the Pacific Ocean, on the northwestern coast of the United States, Washington State, that leads to Seattle. **Willapa Bay** is a bay located on the Pacific coast of Washington state south of Seattle. The Long Beach Peninsula separates Willapa Bay from the Pacific Ocean.

Production from these oyster beds was so prolific through the nineteenth century that, because of the size of its population, New York City and its restaurants became the centre of the industry (Fig. 3.5).

> … In fact, oysters were such a key part of New York City's economy that [oyster] shells were used for roads, cement, fertilizer, and many other items … (MacKenzie 1996).

Furthermore, oyster houses, or oyster saloons, restaurants that specialised in serving oysters, raw or cooked, arose throughout major cities in the United States in the nineteenth century to meet the popular demand for oysters (Fig. 3.5).

It was very similar in Europe, especially in the United Kingdom, where English towns such as Whitstable and Colchester had held oyster markets for the Romans. So, on both sides of the Atlantic ocean oysters became a standard part of the diet in big cities in the nineteenth century because of their abundance, accessibility, and low cost. Indeed, thanks to the landmark in transport history of the inaugural sailing on 4 July 1840 of the first regular Atlantic steam ferry operated by the *British and North American Royal Mail Steam Packet Company* (later known as ***Cunard Steamship Company***) on the Liverpool–Halifax–Boston route, New York oysters were supplied to markets in the United Kingdom. And, as we have mentioned above, were supplied (together with their pests) for relaying in UK waters.

The creation of railroad transportation during the nineteenth century allowed oysters to be shipped all over the United States and Europe. Canning and refrigeration were two more groundbreaking developments that expanded the oyster trade (MacKenzie 1996).

Oysters were eaten raw or steamed, fried, grilled, roasted, or stewed with other meats and ingredients. In London, oysters were sold at

Fig. 3.5 Oyster saloons in the Fulton Market of New York City towards the end of the nineteenth century. **Left**: print of an oyster saloon in the Fulton Market in 1877. **Right**: A print from 1870s of oyster stands in Fulton Market, New York City at a time when oysters were extremely abundant and cheap (1 cent per oyster). Both images in the Public Domain

every street corner (Fig. 3.3) during the nineteenth century, and street vendors were just as common alongside the oyster saloons in New York City as well; the street oyster vendors being supplanted by the hot dog stands of the New York City of today. Oysters were also very popular with the bars, gin palaces and pubs of the two cities, being so cheap they could be used as loss-leaders to stimulate sales of the more profitable beer, gin and other liquors (MacKenzie 1996).

And another parallel between the United States and Europe was overharvesting of oysters. By the time the twentieth century ticked around it was becoming difficult to meet market demands. Decline in oyster stocks in the United States and Europe was obvious. In the United States, the oyster beds around New York and New Jersey were the first to fail. Harvesting moved to the Chesapeake Bay and other areas, only to be eventually shut down as well due to overharvesting. Add to overharvesting the effects of pests, like slipper limpets and the American whelk tingle that brought several oysters to the brink of extinction, and, further, the human infections that became associated with fresh shellfish; almost inevitably, during the first decade of the twentieth century both the demand for, and the supply of, oysters massively declined. And then, of course, the First World War killed and changed so many millions of lives for ever on both sides of the Atlantic.

3.7 Bringing the Oyster Back to the Table

Consumption of staple foods such as beef or pork has increased over time, whereas the consumption of oysters, once a staple food for many, has decreased over time, and quite considerably decreased in recent history. Since the 1950s, consumer demand for oysters has grown, and though natural stocks are still depleted to the extent that natural oyster beds no longer supply much of the demand, attention has shifted to oyster farming and cultivation to produce enough oysters to meet the demands of consumers in a sustainable fashion. According to the *Monterey Bay Aquarium Seafood Watch*:

> … Farmed oysters account for 95% of the world's total oyster consumption. Most oyster farming operations are very well managed and produce a sustainable product … (https://www.seafoodwatch.org/).

But in the process of developing farming operations for the oyster, the animal has lost its position as a cheap, staple food in the life and nutrition of what might be deemed the working classes.

Oysters have once again become a relatively expensive and exclusive food item. It has been stated that the average American today eats about 3 oysters each year, while **New Yorkers** in the 1800s ate about 600 oysters a year each (MacKenzie 1996); and in 1864 in **London**, 700 million European flat oysters (*Ostrea edulis*) were consumed, and nearly 120,000 workers were employed in the various tasks involved in oyster dredging in Britain (Beck et al. 2011). We must all applaud the efforts of those who have developed, and continue to develop, farming operations for oysters and other shellfish to meet today's market demand for these nutritious primary foods. But this appreciation of oyster farming efforts does not go far enough; greater prizes are being won by this activity than merely meeting the economic market demand for a delicacy.

First, it is worth remembering that the oyster has been described as one of Nature's most perfect foods, and this applies to other filter-feeding shellfish, too. As well as being a good source of easily digested protein, oysters are low in cholesterol and fat, high in omega-3 fatty acids, and an excellent source of vitamins (A, B1, B2, B3, C, and D), and minerals (iron, magnesium, calcium, selenium, and zinc).

A 100 g serving of uncooked oyster meat contains about 70 cal. Even oyster farming itself is climate friendly. Oyster farmers don't feed or add chemicals to their crop. Instead, they keep the oysters in an area where they can grow and be fed naturally. Far less energy input is required to produce an ounce of oyster protein than most other protein sources, from beans to beef (McMurray 2018). But the benefits go much further even than this.

Second, it is well known that production of animal protein through livestock production by terrestrial farming is associated with high greenhouse gas (GHG) emissions, which have three main sources: anaerobic fermentation in the animal gut (most farm animals being ruminants), manure management, and fodder production. But oyster aquaculture is different (Ray et al. 2019).

Oysters ***release no methane*** (CH_4) and only negligible amounts of nitrous oxide (N_2O) and carbon dioxide (CO_2). Even the ocean sediment, which might be considered the equivalent of the livestock farmer's slurry, showed broadly unchanged fluxes of N_2O and CH_4 during oyster aquaculture.

> … There is no GHG-release from oyster fodder production. Considering the main drivers of GHG-release in terrestrial livestock systems, oyster aquaculture has less than 0.5% of the GHG-cost of beef, small ruminants, pork, and poultry in terms of CO_2-equivalents per kg protein, ***suggesting that shellfish aquaculture may provide a low GHG alternative for future animal protein production compared to land based sources***. We estimate that if 10% of the protein from beef consumption in the United States was replaced with protein from oysters, the GHG savings would be equivalent to 10.8 million fewer cars on the road … (Ray et al. 2019. the emphasis is mine).

Another estimate is that ***85% of the world's oyster beds*** have essentially disappeared (though in many bays more than 99% of oyster reefs have been lost), making those oyster beds the marine habitat that has been most severely impacted by human activities on the planet (Beck et al. 2011). Any response to the question 'what can be done now to restore exhausted fisheries?' must include the establishment of regimes that protect and nurture all of our invaluable marine resources, so that our generation uses them sustainably and leaves them in good order for all the generations that follow ours.

This ambition must be applied to all marine resources, not just shellfish, but here we concentrate on the ***native oyster reefs*** that once dominated so many estuaries, bays and coastlines. They dominated economically, of course, by providing for centuries of resource extraction, but this, worsened by coastal degradation is what pushed oyster reefs to the brink of functional extinction worldwide.

Oysters are unusual in that they create their own habitat; they have been described as ***ecosystem engineers***, a status defined as the one or a few species that produce a reef habitat for entire ecosystems. Meaning that they create ***biogenic reef habitats*** of such significant size that they become important to general estuarine biodiversity. By providing habitats at different depths they enhance benthic–pelagic coupling, and the result is an overall improvement in fishery production (Lenihan and Peterson 1998; Beck et al. 2011). On the other hand, it is a major loss to the entire ecosystems if the abundance and biomass of native oyster populations are decimated by the combined impacts of exploitation, disease and habitat loss. Lenihan and Peterson (1998) showed that the height reduction of the oyster reef habitat caused by intensive dredging lowered the height of the reef leaving habitats that other organisisms might use in deeper, less oxygenated, waters and, consequently, adversely affecting the abundance and distribution of fish and invertebrate species that utilise this temperate reef habitat.

In their survey of restored oyster reefs in the Neuse River estuary, North Carolina, USA, Lenihan and Peterson (1998) found that ***a single season's oyster dredging*** reduced the height of restored oyster reefs by about 30%. On deep experimental reefs they found that when the water column stratified in summer, oxygen depletion near the seafloor at 6 m depth caused mass mortality of oysters, other invertebrates, and fishes, though oysters and other organisms raised closer to the surface by sufficient reef height survived. Also, they found that the highly mobile blue crabs (*Callinectes sapidus*) abandoned burrows located in hypoxic/anoxic bottom waters in favour of those in shallow water.

Thus, to rebuild oyster populations it is essential to appreciate the dynamics of both the oyster population and the entire ecosystem the oysters engineer (Mann et al. 2009). It is the accumulated shells of past generations of shellfish that is the foundation of the reef ecosystem and that's why we should recycle the oyster shells and put them back where they came from:

> …***If you have an oyster shell, the best thing to do is to put it back on the reef*** in order to start an ecological system. Once you've got the reef, then other fish come along and other flora and fauna can form around them. They purify the water and filter it, making it useful. But not everybody knows this … (Smith 2015; the emphasis is mine).

Clover (2005) makes another point that not everybody knows, which is that, over in Europe, parts of the North Sea owe their modern-day turbidity to the removal of oyster beds that, more than a century ago, were producing 100 times more oysters than they do today. Maps made in the nineteenth century show oyster beds 200 km in length off the eastern shores of the North Sea (the continental Europe shore), the last of which were fished out before the Second World War.

On this scale the filter-feeding oysters would have cleared the nutrients in suspension, clarifying the water of the day. Furthermore, such extensive bivalve reefs forming a hard substrate across so much of the seabed just offshore would have resulted in far less sediment being stirred up by wave action. If previous generations had been able to put into effect the conservation actions we know about today, present-day Europeans would not only have a much larger resource of oysters in the North Sea but also clearer waters for swimmers and divers to enjoy.

Few of the research papers to which we have referred so far even mention this next point: specifically, that oyster shells are made from atmospheric CO_2 which is permanently solidified in the form of crystalline calcium carbonate (as detailed in Chaps. 2 and 6; Moore 2021; Moore et al. 2021).

Lee et al. (2020) tabulated annual carbon deposition estimates for a variety of ecosystems to show that European flat oyster beds (at a density of 75 oysters m^{-2}) in the Northern Hemisphere have the potential to deposit more carbon per square metre than terrestrial forests in the Northern Hemisphere, through biodeposition to the seabed alone, and that oyster beds compare favourably with other shellfish habitats (Table 3.3).

Table 3.3 Annual values of carbon deposition defined as sedimentary, carbonate, or sedimentary+carbonate per ecosystem

Ecosystem	Carbon store type	Carbon deposition per annum (g m^{-2})
Seagrass	Sedimentary	83
Saltmarsh	Sedimentary	210
Mangroves	Sedimentary	174
Maerl (coralline red algae)	Carbonate	74
Horse mussel (density 40 m^{-2})	Carbonate (+?sedimentary)	40 (+about 360 organic matter deposition[a])
Oyster (density 75 m^{-2})	Sedimentary (+?carbonate)	50
Terrestrial forests[b]	Net sink	29

Notes +? indicates data deficiency
[a]Data are available on organic content of sediment deposits rather than carbon deposition
[b]Net global sink/global forest cover
Data from Lee et al. (2020)

In the United States, oysters are usually sold by the bushel, which is a volume measure used for dry goods equal to 64 US pints (=35.2 L). One bushel of oysters contains between 100 and 150 oysters and weighs approximately 53 pounds (=24 kg) and yields approximately 7 pounds (=3.2 kg) of meat. So, a bushel of oysters contains about 20 kg of shell, which will remain intact for thousands, even hundreds of thousands of years if simply discarded. Now, 20 kg of shell is equivalent to 2.4 kg of carbon captured, and permanently removed, from the atmosphere (see Chap. 2; Moore et al. 2021). FAO statistics (Helm 2005) show total production of oyster ***farming*** of *Crassostrea gigas* (Pacific cupped oyster) over the 5 years 2010–2015 to be 3 million tonnes, so in those five years farming of this species alone ***removed 360,000 tonnes of carbon*** from the atmosphere. We already know (see above) that today's annual consumption of oysters is only about 1/200th of that reached at the end of the nineteenth century. If we could turn back the clock and restore oyster ***production*** to the level of oyster ***dredging*** in the year 1900, we could remove ***14,400,000 tonnes of carbon*** from the atmosphere every year. Just with oyster cultivation.

Potentially, therefore, if we could expand present-day oyster farms to a sufficient extent with animals collected from the wild to provide new recruits to existing self-sustaining oyster populations, we would achieve more than just restoration of an important part of the marine ecosystem. We would be making a serious contribution towards restoring the Earth's wider ecosystem by returning our atmosphere to its natural, pre-industrial, condition. It will not be easy. Most of the Pacific oysters (*Crassostrea gigas*) farmed today are of a triploid strain. Unfortunately, the market success of the strain is due to its partial reproductive sterility (Allen and Downing 1986).

Natural (diploid) oysters tend to be unsavoury during the warm summer months of their spawning season, because their body consists mostly of gonads before spawning and is left thin and watery after spawning. The triploids are fat and marketable throughout the year because they produce so few eggs and sperm (Hollier 2014). Even when triploids can be made to breed successfully, the survival of fertilised eggs to metamorphosis and settlement was only about 0.0085% (Guo and Allen 1994). Established farming methods could easily be applied to natural diploid oysters collected from the wild locally, with the aim of rearing them in farm-protected conditions through their first year before transfer to the locations of old (exhausted/extinct) oyster reefs with the specific aim of re-establishing those oyster beds to the scale that existed at the start of the nineteenth century. This is doable if we apply the same zeal to restoration as was applied to dredging operations during the 100-year exploitation of oysters described and illustrated above, which destroyed so many of the native oyster beds in European waters. Farming the more robust Pacific species could deliver many of the important ecosystem functions that have been lost with the virtual extinction of the native oyster. Experimental trials in Dutch and English waters show that this can even aid restoration of the native species which will settle amongst Pacific Oysters. The ***Shellfish Association of Great Britain*** has published several reports that effectively advocate farming (or at least encouraging) the Pacific Oyster in UK waters (Herbert et al. 2012; Syvret et al. 2021). Unfortunately, in England and Wales, the Pacific Oyster is currently classified as an invasive, non-native species under the Wildlife and Countryside Act 1981, and there are concerns that the species may have negative impacts on native ecosystems. However:

> … These well voiced concerns, combined with a uniquely stringent application of EU marine conservation legislation and lack of settled national policy, have meant that quite different approaches have been adopted towards Pacific Oyster aquaculture in different areas around the country. If the UK is to rise to meet ambitions to produce more sustainable healthy food, including those targets set out in the English Aquaculture Strategy (Huntington and Cappell 2020) and the Blue New Deal action plan (New Economics Foundation 2016), then clarification of the legality and status of farming Pacific oysters must be provided to enable investment and provide security for existing businesses. (Syvret et al. 2021).

3.8 Present-Day Aquaculture

Aquaculture should not be thought of solely in terms of "… a broader food landscape of wild aquatic and terrestrial food sources…" (Little et al. 2016) but in the even broader ecological context of the promise it holds for the ***restoration of the ecosystems*** we have used in the past to the level of quality that existed a few centuries ago when they first came to the attention as a food source to the civilisations such as the native Americans, the Romans, and then the great cities of the world in the 1800s: London, Bristol, Liverpool, New York, New Orleans and San Francisco.

Commercially grown bivalves are the only sustainable form of human food that has ***no negative impact on the environment*** (https://www.eco-business.com/opinion/sustainable-shellfish-aquaculture/). This is because bivalve molluscs offer several ecosystem services that add value to their environment beyond their food value. These additional bivalve ecosystem services in the habitat restoration context have been listed (National Research Council, 2010) as:

- Turbidity reduction by filtration (and see Swift 2021).
- Biodeposition of organics containing plant nutrients.
- Induction of denitrification associated with organic deposition.
- Sequestration of carbon
- Provision of structural habitats (Reef structures) that promote diversity of fish, crustacea and and other organisms.
- Habitat and shoreline stabilisation.

While Jacquet et al. (2017) add these advantages of ***bivalve farming*** to the above list:

- Bivalves don't require feeding.
- Bivalves build food security.
- Bivalve welfare is not as serious a concern as it is for terrestrial farm animals (i.e. bivalve cultivation is more ethical).

Carranza and zu Ermgassen (2020) call bivalve cultivation ***restorative shellfish mariculture*** (RSM), which they define:

> …as the farming of marine shellfish, implying some form of intervention during the species life cycle, in order to address negative socio-ecological issues arising from the unsustainable use of marine ecosystems …

To take full advantage of the services that bivalve cultivation can provide to the marine ecosystem, we must also **change the paradigm**;

Table 3.4 A few of the printed papers on aquaculture methods

References	Title
Ansa and Bashir (2007)	Fishery and culture potentials of the mangrove oyster (*Crassostrea gasar*) in Nigeria. URL: https://agris.fao.org/agris-search/search.do?recordID=AV20120138411
Baker and Baker (2019)	Carbon mineralization associated with aquaculture of the Northern Quahog *Mercenaria mercenaria*. https://doi.org/10.2983/035.038.0302
Bernal and Oliva (2016)	Aquaculture. Chapter 12 in *The First Global Integrated Marine Assessment: World Ocean Assessment I*, (Division for Ocean Affairs and the Law of the Sea, Office of Legal Affairs, United Nations). https://doi.org/10.1017/9781108186148.015
Costa-Pierce (2002)	Ecological Aquaculture: The Evolution of the Blue Revolution. https://doi.org/10.1002/9780470995051
Cragg (2016)	Biology and Ecology of Scallop Larvae. Chapter 2 in: *Scallops—Biology, Ecology, Aquaculture, and Fisheries*. https://doi.org/10.1016/B978-0-444-62710-0.00002-X
Pogoda (2019)	Current status of European Oyster decline and restoration in Germany. https://doi.org/10.3390/h8010009
Pogoda et al. (2019)	The native oyster restoration alliance (NORA) and the Berlin Oyster Recommendation: bringing back a key ecosystem engineer by developing and supporting best practice in Europe. https://doi.org/10.1051/alr/2019012
Sarkis and Lovatelli (2007)	Installation and operation of a modular bivalve hatchery. FAO Fisheries Technical Paper. No. 492. PDF download: http://www.vliz.be/imisdocs/publications/ocrd/121106.pdf
Shumway and Parsons (2016)	Scallops—Biology, Ecology, Aquaculture, and Fisheries. ISBN: 9,780,444,627,100
Walne (1979)	Culture of bivalve molluscs: 50 years' experience at Conwy. ISBN: 9,780,852,380,635
Wickins and Lee (2002)	*Crustacean Farming: Ranching and Culture*. URL: https://www.epdf.pub/crustacean-farming-ranching-and-culture.html

from shellfish farming for food to shellfish farming for whole-planet ecosystem repair and restoration. Take the food represented by shellfish meat as a byproduct from the production of shell, and leave or return the shell to the seabed from which it was harvested.

Shellfish farming has been a part of our history for over 100 years and the systems in use to farm shellfish have evolved from simple transfer of eggs or juveniles from their natural (inconvenient?) location to a more convenient location, to what are now technology-based systems that are designed for specific species and farming sites. There is an enormous amount of information available about aquaculture methods, both online and in print, a few of which are referenced in Table 3.4 (in addition to those referenced in this text). Google and Bing will find you any more you need, in an instant.

The British Columbia Shellfish Growers' Association (BCSGA) website has a useful brief summary of the overall process at this URL: https://www.seawestnews.com/what-is-shellfish-aquaculture/, which we have used as a framework for my discussion. By definition, shellfish aquaculture is the farming (i.e. cultivation and harvest) of aquatic invertebrates, such as oysters, clams and mussels, but also the crustacea, crabs, lobsters, shrimp and prawns (some illustrated in Table 3.5). Cultivation implies involvement in the rearing process to enhance production, such as regular stocking and protection from predators.

Most shellfish reproduce by releasing eggs and sperm into the water, generally in summer when the water is warm and planktonic food is abundant. After fertilisation of an egg, cell division produces swimming larvae that "… feed and feel an urgent need to grow more like their mothers …" (Garstang 1951). These eventually metamorphose into tiny bivalves, crabs or lobsters that settle to the seabed. This settling phase in oysters is the stage considered suitable for transplanting and may be called ***oyster spat*** or

Table 3.5 Aquaculture and foraging for other shellfish videos on YouTube

Clam farming: A forgotten industry in South Carolina, USA at https://www.youtube.com/watch?v=3FN8NVC4v1Y&t=292s
Coastal Foraging Wild Clams! Catch and Cook Homemade Clam Chowder in New England at https://www.youtube.com/watch?v=O9Kzd3RcvDg&t=314s
Onshore Abalone: The risks and rewards of onshore abalone farming in Australia at https://www.youtube.com/watch?v=Pwmb8u6Z44g&t=165s
Abalone Farm in California at https://www.youtube.com/watch?v=uDYbK0zUvFE&t=123s
Crawfish Agriculture in the Southern Regional Aquaculture Center with details of the crawfish farming methods presently used in Louisiana and Southern Texas at https://www.youtube.com/watch?v=WAufhGUvCjA&t=156s
Noal Farm Lobster Farming and processing at https://www.youtube.com/watch?v=-NcAH2nbqMI
Studying lobsters and crabs on Canada's east coast at https://www.youtube.com/watch?v=li_61ldtL4o&t=163s
Feeding ground for rock lobsters on South Africa's western coast at https://www.youtube.com/watch?v=QhLqN2oIDaA&t=120s
The National Lobster Hatchery, Padstow (Cornwall, England). Filmed by Manchester Museum at https://www.youtube.com/watch?v=Z6rbwuLmMw8

oyster seed. The farming cycle begins with the collection of larvae, which may be gathered in the wild or produced in ***farm hatcheries*** (depending on the species and location) (see Robert and Gérard 1999; Helm et al. 2004; for practical guides to bivalve hatcheries).

- Clam larvae are kept in hatchery tanks where they transform into seed.
- Mussel larvae transform to juvenile animals.
- Scallop larvae settle and become juvenile animals.
- Oyster larvae are kept suspended in tanks by circulating water until they transform into seed.

Farmers acquire clam and oyster seed at various stages of its development, depending on the requirements of their operation. The seed is put into a nursery environment where it is nurtured into juvenile animals. Generally speaking, the juvenile animals then graduate to the ***growout phase*** of their development during which they mature to marketable size.

- Clams are spread on subtidal locations, which are the licenced aquaculture farms (called tenures in British Columbia), where they burrow and mature to marketable size over a period of two to four years.
- Mussels are relocated to deepwater tenures where they are suspended in mesh socks to mature to marketable size over a period of 18–36 months.
- Scallops are transferred to deepwater tenures where they are suspended in a mesh bag or tray (suspension culture) or are seeded on the ocean floor (bottom culture). Maturation to marketable size takes 6–36 months in suspension culture and an additional 24–36 months in bottom culture. Scallops are examples of the very few bivalves that do not attach to the seafloor. They are 'free-living' in the sense that many species are capable of swimming rapidly over short distances and even of migrating greater distances across the ocean floor.
- Oysters are frequently moved to a floating upwelling system called a flupsy (view the Long Island Shellfish Restoration Project YouTube video in Table 3.6). Ocean water is circulated through the flupsy and juvenile animals, kept in trays, are able to grow to a larger size. When they are large enough, the young oysters are moved to be reared in a growout system. The most common growout techniques are raft, longline and intertidal (edited text from http://www.bcsga.ca/shellfish-farming-101/shellfish-aquaculture/).

Table 3.6 Oyster aquaculture websites and videos on YouTube

Explanation of the **flupsy** (used to grow oysters) by the *Long Island Shellfish Restoration Project* at https://www.youtube.com/watch?v=FjSNOIJCG8s
FlipFarm is an innovative oyster growing system designed and operated by *Marlborough Oysters Ltd* in New Zealand. FlipFarm provides an ideal environment for oyster growth, conditioning and hardening along with the ability to have complete control over fouling, pests and predators (https://www.youtube.com/watch?v=CGe3wSV3B8w&t=8s)
Video showing the sea conditions at some FlipFarm sites. The baskets ride this out and oysters are rumbling so this site is good for finishing stock ready for harvest (https://www.youtube.com/watch?v=RbXH20P-ADk&t=10s)
FlipFarm harvesting intermediate oysters for sale at https://www.youtube.com/watch?v=ye-a-vAaL5k
The latest addition to the FlipFarm gear is a helix flipper, used to rotate the baskets into the drying position, which kills biofouling, creates more even and better shaped shell growth, hardens oysters and improves meat condition (https://www.youtube.com/watch?v=ZX54JdEccXc&t=27s)
OysterGro® Aquafarming Systems (Bouctouche Bay, Saint-Édouard-de-Kent, New Brunswick, Canada) at https://www.youtube.com/watch?v=e33ZII2DMN4&t=29s
Oyster farming and harvesting in Japan—Big Oyster Cultivation at https://www.youtube.com/watch?v=cdKBZpOMDfk&t=122s
Oyster Farming in the Southern United States Using the OysterGro System funded by the USDA at https://www.youtube.com/watch?v=S0OpUeovaLQ&t=153s
Gregg Morris owner of 2 Rock Oyster Farm (Duxbury, MA, US) shows how oyster farming is done in Duxbury Bay at https://www.youtube.com/watch?v=bbSRv8QDyTc&t=256s
Farming the Sea; aquaculture in Florida. A full 26 min episode from Changing Seas TV (https://www.changingseas.tv/) at https://www.youtube.com/watch?v=EAN-VRvD8_k&t=512s
Oyster Farming at Tio Point (District of Marlborough on the South Island of New Zealand) at https://www.youtube.com/watch?v=m5VdJ0Ed03c&t=324s
The Solar Oysters website hosts a video showing how to automate the world of shellfish aquaculture at https://www.solaroysters.com/
The Wild Oysters project is aiming to restore Britain's seas to health through the restoration of the native oyster URL: https://wild-oysters.org/

References

Allen SK, Downing SL (1986) Performance of triploid Pacific oysters, *Crassostreagigas* (Thunberg). I. Survival, growth, glycogen content, and sexual maturation in yearlings. J Exp Mar Biol Ecol 102:197–208. https://doi.org/10.1016/0022-0981(86)90176-0

Ansa EJ, Bashir RM (2007) Fishery and culture potentials of the mangrove oyster (*Crassostrea gasar*) in Nigeria. *AGRIS* (bibliographic database maintained by the Food and Agriculture Organization of the United Nations), article AV20120138242. https://agris.fao.org/agris-search/search.do?recordID=AV20120133115

Bae CJ, Douka K, Petraglia MD (2017) On the origin of modern humans: Asian perspectives. Science 358, article eaai9067. https://doi.org/10.1126/science.aai9067

Baker P, Baker SM (2019) Carbon mineralization associated with aquaculture of the Northern Quahog *Mercenariamercenaria*. J Shellfish Res 38:519–527. https://doi.org/10.2983/035.038.0302

Bale R (2017) This turtle tourist center also raises endangered turtles for meat. Natl Geogr Mag. https://www.nationalgeographic.com/animals/article/wildlife-watch-cayman-turtle-farm-welfare-controversy

Bardach JE, Ryther JH, McLarney WO (1972) Aquaculture—the farming and husbandry of freshwater and marine organisms. John Wiley & Sons, Inc., New Jersey, USA, 351 p. ISBN 13: 9780471048268

Beck MW, Brumbaugh RD, Airoldi L, Carranza A, Coen LD, Crawford C, Defeo O, Edgar GJ, Hancock B, Kay MC, Lenihan HS, Luckenbach MW, Toropova CL, Zhang G, Guo X (2011) Oyster reefs at risk and recommendations for conservation, restoration, and management. Bioscience 61:107–116. https://doi.org/10.1525/bio.2011.61.2.5

Berg DE, Kohn MA, Farley TA, McFarland LM (2000) Multi-state outbreaks of acute gastroenteritis traced to fecal-contaminated oysters harvested in Louisiana. J Infect Dis 181(Supplement 2):S381–S386. https://doi.org/10.1086/315581

Bernal P, Oliva D (2016) Aquaculture. In: Inniss L, Simcock A et al (eds) Chapter 12 in The first global integrated marine assessment: world ocean assessment

I. For the Division for Ocean Affairs and the Law of the Sea, Office of Legal Affairs, United Nations, pp 203–212. Cambridge University Press, Cambridge, UK. https://doi.org/10.1017/9781108186148.015

Beveridge MCM, Little DC (2002) The history of aquaculture in traditional societies. In: Costa-Pierce BA (ed) Chapter 1, Ecological aquaculture. The evolution of the blue revolution, pp. 1–29. Blackwells, Oxford, 382 p. ISBN: 9780632049615. https://doi.org/10.1002/9780470995051. https://doi.org/10.1002/9780470995051.ch1. https://www.researchgate.net/publication/257655025_The_History_of_Aquaculture_in_Traditional_Societies

Briggs M, Funge-Smith S, Subasinghe RP, Phillips M (2005) Introductions and movement of two Penaeid shrimp species in Asia and the Pacific. FAO Fisheries technical paper 476. Food and Agriculture Organization of the United Nations, Rome. ISBN: 9251053626. https://www.fao.org/documents/card/en/c/769f22e1-ccba-5e3d-a8ad-fd1380541789/

Carranza A, zu Ermgassen PSE (2020) A global overview of Restorative Shellfish Mariculture. Front Mar Sci 7, article 722. https://doi.org/10.3389/fmars.2020.00722

Chambers RA, Gray M (1988) The excavation of fishponds. In: Aston M (ed) Medieval fish, fisheries and fishponds in England, vol 182, pp 113–128. BAR Publishing (British Archaeological Reports Publishing) British Series, Oxford, UK. 484 p. ISBN: 978 0860545095. https://www.barpublishing.com/ and https://www.worldcat.org/title/medieval-fish-fisheries-and-fishponds-in-england/oclc/19722923

Clover C (2005) The end of the line: how overfishing is changing the world and what we eat. Ebury Press, London. The non-fiction imprint of Penguin Random House, 320 p. ISBN: 9780091897819

Cole H (1942) The American whelk tingle, *Urosalpinxcinerea* (Say), on British oyster beds. J Mar Biol Assoc UK 25:477–508. https://doi.org/10.1017/S0025315400055119

Costa-Pierce BA (ed) (2002) Ecological aquaculture: the evolution of the blue revolution. Wiley-Blackwell, New Jersey, US, 320 p. ISBN: 9780632049615. https://doi.org/10.1002/9780470995051

Cragg SM (2016) Biology and ecology of Scallop Larvae. In: Shumway SE, Parsons GJ (eds) Chapter 2, Scallops—Biology, ecology, aquaculture, and fisheries. Developments in aquaculture and fisheries science, vol 40, pp 31–83. https://doi.org/10.1016/B978-0-444-62710-0.00002-X

Currie CK (1991) The early history of the carp and its economic significance in England. Agricult Hist Rev 11:97–107. https://www.bahs.org.uk/AGHR/ARTICLES/39n2a1.pdf

De Silva SS, Zhitang Y, Lin-Hu X (1991) A brief review of the status and practices of the reservoir fishery in mainland China. Aquac Res 22:73–84. https://doi.org/10.1111/j.1365-2109.1991.tb00496.x

Erlandson JM (2001) The archaeology of aquatic adaptations: paradigms for a new millennium. J Archaeol Res 9:287–350. https://doi.org/10.1023/A:1013062712695

FAO (1983) Freshwater aquaculture development in China. Report of the FAO/UNDP study tour organized for French-speaking African countries. 22 April–20 May 1980. FAO Fisheries Technical Paper 215. 125 p. Rome: Food and Agriculture Organization of the United Nations. ISBN: 9251011133. http://www.fao.org/3/AD016E/AD016E00.htm#TOC

FAO (2014) The state of world fisheries and aquaculture 2014. Opportunities and challenges. FAO, Rome. http://www.fao.org/3/a-i3720e.pdf

FAO (2016) Species fact sheets: Procambarus clarkii (Girard, 1852). FAO Fisheries Division. http://www.fao.org/fishery/species/3454/en

FAO (2017) Fishery and aquaculture country profiles: The People's Republic of China (2017). Country Profile Fact Sheets. In: FAO Fisheries Division [online]. FAO, Rome. http://www.fao.org/fishery/facp/chn/en

FAO (2019) Fishery and aquaculture country profiles: The United States of America (2019). Country Profile Fact Sheets. In: FAO Fisheries Division [online]. FAO, Rome. http://www.fao.org/fishery/facp/USA/en

FAO (2020a) Species fact sheets: Macrobrachium rosenbergii (De Man, 1879). FAO Fisheries Division. http://www.fao.org/fishery/species/2608/en

FAO (2020b) The state of world fisheries and aquaculture 2020. Sustainability in action. FAO, Rome. https://doi.org/10.4060/ca9229en

Foodworthwritingfor.com (2018) The history of oysters: its rise as a delicacy and a staple food beloved by many. Foodworthwritingfor blog, written by C.P., July 31, 2018. https://foodworthwritingfor.com/2018/07/31/the-history-of-oysters-its-rise-as-a-delicacy-and-a-staple-food-beloved-by-many/

Garstang W (1951) Larval forms, and other zoological verses. Basil Blackwell, Oxford, 85 p. ISBN: 9780631070900. URL: http://www.booktryst.com/2012/09/poetry-in-larvae-love-songs-of-walter.html

Gonzalez-Poblete E, Hurtado F, C.F., (2018) Blue mussel aquaculture in Chile: small or large scale industry? Aquaculture 493:113–122. https://doi.org/10.1016/j.aquaculture.2018.04.026

Groesbeck AS, Rowell K, Lepofsky D, Salomon AK (2014) Ancient clam gardens increased shellfish production: adaptive strategies from the past can inform food security today. PLoS ONE 9, article e91235. https://doi.org/10.1371/journal.pone.0091235

Guo X, Allen SK (1994) Reproductive potential and genetics of triploid Pacific oysters, *Crassostreagigas* (Thunberg). Biol Bull 187:309–318. https://doi.org/10.2307/1542288

Hausmann N, Meredith-Williams M, Laurie E (2020) Shellfish resilience to prehistoric human consumption in the southern Red Sea: variability in *Conomurex fasciatus* across time and space. Q Int. https://doi.org/10.1016/j.quaint.2020.04.034

Heilweck M, Moore D (2021) Saving the planet with appropriate biotechnology: 3. The high seas solution/Salvando el planeta con biotecnología apropiada: 3. La solución de alta mar. Mex J Biotechnol 6:92–128. https://doi.org/10.29267/mxjb.2021.6.1.92

Helm MM (2005) Cultured aquatic species information programme: *Crassostrea gigas*. In: FAO Fisheries Division [online]. Food and Agriculture Organization of the United Nations, Rome. http://www.fao.org/fishery/culturedspecies/Crassostrea_gigas/en

Helm MM, Bourne N, Lovatelli A (2004) Hatchery culture of bivalves. A practical manual. FAO Fisheries Technical Paper, No. 471. 177 p. Information Division, Food and Agriculture Organization of the United Nations, Rome. ISBN: 9251052247. https://www.fao.org/documents/pub_dett.asp?pub_id=166536

Henshilwood C, d'Errico F, Vanhaeren M, van Niekerk K, Jacobs Z (2004) Middle Stone Age shell beads from South Africa. Science 304:404. https://doi.org/10.1126/science.1095905

Herbert RJH, Roberts C, Humphreys J, Fletcher S (2012) The Pacific oyster in the UK: Economic, legal and environmental issues associated with its cultivation, wild establishment and exploitation. Published for: Shellfish Association of Great Britain, London. http://www.shellfish.org.uk/files/PDF/73434Pacific%20Oysters%20Issue%20Paper_final_241012.pdf

Hilborn R, Hilborn U (2012) Overfishing: what everyone needs to know. Oxford University Press Inc., New York, USA, 168 pp. ISBN: 9780199798148

Hoffman RC (1996) Economic development and aquatic ecosystems in Medieval Europe. Am Hist Rev 101:631–669. https://doi.org/10.1086/ahr/101.3.631

Hollier D (2014) Tasty mutants: the invention of the modern oyster. Posting on *The Atlantic magazine*'s website (https://www.theatlantic.com/world/). https://www.theatlantic.com/technology/archive/2014/09/todays-oysters-are-mutants/380858/

Huntington T, Cappell R (2020) English aquaculture strategy. Final report. Produced by Poseidon Aquatic Resources Management Ltd. Published for: The Sea Fish Industry Authority. 80p + appendices. https://www.seafish.org/document/?id=9efe670c-847b-4a4f-b8ec-72f2e5396df6

Jackley J, Gardner L, Djunaedi AF, Salomon AK (2016) Ancient clam gardens, traditional management portfolios, and the resilience of coupled human-ocean systems. Ecol Soc 21, article 20. https://doi.org/10.5751/ES-08747-210420

Jacquet J, Sebo J, Elder M (2017) Seafood in the future: bivalves are better. Sol J 8:27–32. https://www.thesolutionsjournal.com/article/seafood-future-bivalves-better

Jones TL, Richman JR (1995) On mussels: *Mytiluscalifornianus* as a prehistoric resource. North Am Archaeol 16:33–58. https://doi.org/10.2190/G5TT-YFHP-JE6A-P2TX

Kungvankij P (1984) Overview of penaeid shrimp culture in Asia. In: First International conference on shrimp and prawn culture in December 1984, at the SEAFDEC Aquaculture Department, Iloilo, Philippines. NACA/WP/84/11. http://www.fao.org/3/ac242e/ac242e00.htm

Leather J (1991) Smacks and Bawleys. Terence Dalton Ltd., Sudbury, Suffolk, UK, 160 pp. ISBN: 9780861380794

Lee HZL, Davies IM, Baxter JM, Diele K, Sanderson WG (2020) Missing the full story: first estimates of carbon deposition rates for the European flat oyster, *Ostrea edulis*. Aquat Conserv Mar Freshwat Ecosyst 30:2076–2086. https://doi.org/10.1002/aqc.3402

Lenihan HS, Peterson CH (1998) How habitat degradation through fishery disturbance enhances impacts of hypoxia on oyster reefs. Ecol Appl 8:128–140. https://doi.org/10.1890/1051-0761(1998)008[0128:HHDTFD]2.0.CO;2

Lepofsky D, Toniello G, Earnshaw J, Roberts C, Wilson L, Rowell K, Holmes K (2020) Ancient anthropogenic clam gardens of the northwest coast expand clam habitat. Ecosystems. https://doi.org/10.1007/s10021-020-00515-6

Li F, Wu N, Lu H, Zhang J, Wang W, Ma M, Zhang X, Yang X (2013) Mid-Neolithic exploitation of mollusks in the Guanzhong Basin of Northwestern China: preliminary results. PLoS ONE 8, article e58999. https://doi.org/10.1371/journal.pone.0058999

Little D, Newton R, Beveridge M (2016) Aquaculture: a rapidly growing and significant source of sustainable food? Status, transitions and potential. Proc Nutr Soc 75:274–286. https://doi.org/10.1017/S0029665116000665

Liu Q, Chen L (2002) Paddy culture: Chinese mitten-handed crabs. Glob Aquacult Advoc. https://www.aquaculturealliance.org/advocate/paddy-culture-chinese-mitten-handed-crabs/

MacKenzie Jr CL (1996) History of oystering in the United States and Canada, featuring the eight greatest oyster estuaries. Mar Fisher Rev 58:1–78. http://www.aquaticcommons.org/9826/1/mfr5841.pdf

Mann R, Southworth MJ, Harding JM, Wesson JA (2009) Population studies of the native eastern oyster, *Crassostreavirginica* (Gmelin, 1791), in the James River, Virginia, USA. J Shellfish Res 28:193–220. https://doi.org/10.2983/035.028.0203

Marean CW, Assefa Z (2005) The middle and upper Pleistocene African record for the biological and behavioral origins of modern humans. In: Stahl AB (ed) African archaeology: a critical introduction, pp 93–129. Blackwell Studies in Global Archaeology Series New York, USA and Oxford, UK, p 490. ISBN: 9781405101561

Marean CW, Bar-Matthews M, Bernatchez J, Fisher E, Goldberg P, Herries AIR, Jacobs Z, Jerardino A, Karkanas P, Minichillo T, Nilssen PJ, Thompson E, Watts I, Williams HM (2007) Early human use of marine resources and pigment in South Africa during the Middle Pleistocene. Nature 449:905–908. https://doi.org/10.1038/nature06204

McBrearty S, Brooks AS (2000) The revolution that wasn't: a new interpretation of the origin of modern

human behavior. J Hum Evol 39:453–563. https://doi.org/10.1006/jhev.2000.0435
McIntosh C (2016) Clam garden governance in the Gulf Islands National Park Reserve. MA (Environment and Management) Thesis, Faculty of Social and Applied Sciences, Royal Roads University, Victoria, BC, Canada. https://viurrspace.ca/handle/10170/960
McMurray P (2018) The oyster companion: a field guide. Firefly Books Ltd., Ontario, Canada, 256 p. ISBN: 9780228101581
Micha J-C, Chavez M (1997) Development of agro-piscicultual ecosystems in tropical marshlands. In: Mathias JA, Charles AT, Baotong H (eds) Integrated fish farming, pp 347–358. CRC Press, Boca Raton. https://www.researchgate.net/publication/335252739_Development_of_Agro-Piscicultural_Ecosystems_in_Tropical_Marshland
Milner N (2009) 3.12 Mesolithic middens and marine molluscs, procurement and consumption of shellfish at the site of Sand. In: Section 3 of Scotland's first settlers. Scot Archaeol Internet Rep 31:384–400. http://journals.socantscot.org/index.php/sair/article/view/2067
Moore D (2021) Saving the planet with appropriate biotechnology: 4. Coccolithophore cultivation and deployment/Salvando el planeta con biotecnología apropiada: 4. Cultivo de cocolitóforos e implementación. Mex J Biotechnol 6:129–155. https://doi.org/10.29267/mxjb.2021.6.1.129
Moore D, Heilweck M, Petros P (2021b) Saving the planet with appropriate biotechnology: 2. Cultivate Shellfish to Remediate the Atmosphere/Salvando el planeta con biotecnología apropiada: 2. Cultivar mariscos para remediar la atmósfera. Mex J Biotechnol 6:31–91. https://doi.org/10.29267/mxjb.2021.6.1.31
Myvett G, Haughton M, Murray PA (2014) Study on the potential of fish farming in the Caribbean. CRFM Technical & Advisory Document No 2014/2. Caribbean Regional Fisheries Mechanism Secretariat, Belize, 78 p. ISSN: 19951132. https://brusselsbriefings.files.wordpress.com/2013/07/study-on-the-potential-of-aquaculture-in-the-caribbean.pdf
National Research Council (2010) Ecosystem services of bivalves: implications for restoration. In: Ecosystem concepts for sustainable bivalve mariculture, chap 7, pp 123–132. Report of the Committee on Best Practices for Shellfish Mariculture. The National Academies Press, Washington, DC, 190 p. ISBN: 9780309146951. https://doi.org/10.17226/12802
New Economics Foundation (2016) Turning back to the sea: a blue new deal to revitalise coastal communities, 116 p. https://neweconomics.org/uploads/files/NEF-Blue-New-Deal-AP-HighRes.pdf
Nikolić M, Bosch A, Alfonso S (1976) A system for farming the mangrove oyster (*Crassostrearhizophorae* Guilding, 1828). Aquaculture 9:1–18. https://doi.org/10.1016/0044-8486(76)90044-2
Petit JR, Jouzel J, Raynaud D, Barkov NI, Barnola J-M et al (1999) Climate and atmospheric history of the past 420,000 years from the Vostok ice core Antarctica. Nature 399:429–436. https://doi.org/10.1038/20859
Phillipson D (2005) The emergence of humankind in Africa. In: African archaeology, 3rd edn, chap 3, pp 15–51. Cambridge University Press, Cambridge, UK. ISBN: 9780511800313. https://doi.org/10.1017/CBO9780511800313. https://doi.org/10.1017/CBO9780511800313.003
Pogoda B (2019) Current status of European Oyster decline and restoration in Germany. Humanities 8, article 9. https://doi.org/10.3390/h8010009
Pogoda B, Brown J, Hancock B, Preston J, Pouvreau S, Kamermans P, Sanderson W, von Nordheim H (2019) The native oyster restoration alliance (NORA) and the Berlin Oyster Recommendation: bringing back a key ecosystem engineer by developing and supporting best practice in Europe. Aquat Living Resourc 32, article 13. https://doi.org/10.1051/alr/2019012
Potasman I, Paz A, Odeh M (2002) Infectious outbreaks associated with bivalve shellfish consumption: a worldwide perspective. Clin Infect Dis 35:921–928. https://doi.org/10.1086/342330
Rana K (1997) Recent developments in aquaculture statistics. In: Fishery and aquaculture statistics in Asia: Proceedings of the FAO/SEAFDEC regional workshop on fishery statistics, Bangkok, Thailand, 19–21 August 1997, vol II. Fishery and aquaculture statistical systems and programs, pp 242–254. Secretariat, Southeast Asian Fisheries Development Center, Bangkok, Thailand. FAO, Rome. https://aquadocs.org/handle/1834/41101
Ray NE, Maguire TJ, Al-Haj AN, Henning MC, Fulweiler RW (2019) Low greenhouse gas emissions from oyster aquaculture. Environ Sci Technol 53:9118–9127. https://doi.org/10.1021/acs.est.9b02965
Robert R, Gérard A (1999) Bivalve hatchery technology: the current situation for the Pacific oyster *Crassostreagigas* and the scallop *Pecten maximus* in France. Aquat Living Resour 12:121–130. https://doi.org/10.1016/S0990-7440(99)80021-7
Ruddle K, Zhong G (1988) Integrated agriculture-aquaculture in South China. The dike-pond system of the Zhujiang Delta. Cambridge University Press, Cambridge, UK, 192 p. ISBN: 9780521341936. https://www.researchgate.net/publication/258212339_Integrated_Agriculture-Aquaculture_in_South_China_The_Dike-Pond_System_of_the_Zhujiang_Delta
Sahrhage D, Lundbeck J (1992) A history of fishing. Springer-Verlag, Berlin, Heidelberg, 348 p. ISBN: 9783642774133. https://doi.org/10.1007/978-3-642-77411-9
Salem AM, Saleh MA (2010) National aquaculture sector overview. Egypt. In: National aquaculture sector overview fact sheets (2003–2020). FAO Fisheries Division, FAO, Rome. http://www.fao.org/fishery/countrysector/naso_egypt/en
Sanger D, Elson-Sanger MJ (1986) Boom and bust on the river: the story of the Damariscotta oyster shell heaps.

Archaeol Eastern North Am 14:65–78. http://www.jstor.org/stable/40914267

Sarkis S, Lovatelli A (2007) Installation and operation of a modular bivalve hatchery. FAO Fisheries Technical Paper, No. 492, 173 p. Information Division, Food and Agriculture Organization of the United Nations, Rome. ISBN: 9789251055953. http://www.vliz.be/imisdocs/publications/ocrd/121106.pdf

Shumway SE, Parsons GJ (eds) (2016) Scallops—biology, ecology, aquaculture, and fisheries. Dev Aquacult Fisher Sci 40, 1162 p. Elsevier Science Ltd., Amsterdam. ISBN: 9780444627100

Smith D (2015) Oyster: A gastronomic history (with recipes). Abrams & Chronicle Books, London, 256 p. ISBN: 9781419719226

Smith NF, Lepofsky D, Toniello G, Holmes K, Wilson L, Neudorf CM, Roberts C (2019) 3500 years of shellfish mariculture on the Northwest Coast of North America. PLoS ONE 14, article e0211194. https://doi.org/10.1371/journal.pone.0211194

Stoner AW (2019) Forty years of field and laboratory experiments with hatchery-reared Queen Conch: the case for conserving natural stocks. Rev Fisher Sci Aquacult 27:490–516. https://doi.org/10.1080/23308249.2019.1628705

Swift R (2021) Oysters, obviously: Restoring the hidden heroes of ocean health. Article on the Asia-Pacific Boating.com website, posted 3 March 2021. https://asia-pacificboating.com/oysters-obviously-restoring-the-hidden-heroes-of-ocean-health/

Syvret M, Horsfall S, Humphreys J, Williams C, Woolmer A, Adamson E (2021) The Pacific Oyster: why we should love them. Shellfish Association of Great Britain. https://www.fishmongers.org.uk/wp-content/uploads/2021/06/Pacific-Oyster-Report-for-Policy-June-2021.pdf

Toniello G, Lepofsky D, Lertzman-Lepofsky G, Salomon AK, Rowell K (2019) 11,500 y of human–clam relationships provide long-term context for intertidal management in the Salish Sea, British Columbia. Proc Natl Acad Sci USA 116:22106–22114. https://doi.org/10.1073/pnas.1905921116

Tudge C (1998) Neanderthals, bandits and farmers. How agriculture really began (darwinism today series). Weidenfield & Nicolson, an imprint of the Orion Publishing Group, London, p. 64. ISBN: 9780297842583

Valladão GMR, Gallani SU, Pilarski F (2018) South American fish for continental aquaculture. Rev Aquac 10:351–369. https://doi.org/10.1111/raq.12164

Vanhaeren M, d'Errico F, Stringer C, James SL, Todd JA, Mienis HK (2006) Middle Paleolithic shell beads in Israel and Algeria. Science 312:1785–1788. https://doi.org/10.1126/science.1128139

Veeragurunathan V, Prasad K, Malar VJ, Singh N, Meena R, Mantri VA (2019) *Gracilariadebilis* cultivation, agar characterization and economics: bringing new species in the ambit of commercial farming in India. J Appl Phycol 31:2609–2621. https://doi.org/10.1007/s10811-019-01775-z

Walne PR (1979) Culture of bivalve molluscs: 50 years' experience at Conwy, 190 p. Fishing News Books Ltd., Farnham, Surrey, UK. ISBN: 9780852380635

Walter RC, Buffler RT, Bruggemann JH, Guillaume MM, Berhe SM, Negassi B, Libsekal Y, Cheng H, Edwards RL, von Cosel R, Néraudeau D, Gagnon M (2000) Early human occupation of the Red Sea coast of Eritrea during the last interglacial. Nature 405:65–69. https://doi.org/10.1038/35011048

Wickins JF, Lee DO'C (2002) Crustacean farming: ranching and culture, 2nd edn, p 464. Blackwell Science Ltd., Oxford, UK. ISBN: 9780632054640. https://epdf.tips/crustacean-farming-ranching-and-culture.html

Wyatt GJ (2015) Sea level change in the Gulf Islands National Park Reserve, southern British Columbia: implications for the interpretation of nearshore archaeological features. MSc (Environment and Management) Thesis. Faculty of Social and Applied Sciences, Royal Roads University, Victoria, BC, Canada. https://viurrspace.ca/handle/10170/862

Yang H (1994) Integrated fish farming. In: Li S, Mathias J (eds) Freshwater fish culture in China: principles and practice, pp 219–270. (Developments in aquaculture and fisheries science, vol 28), 450 p. Elsevier, Amsterdam. ISBN: 9780444888822

Yang H-L, Lin H-Y, Lin C-C (2014) The search for sustainable grouper farming. Glob Aquacult Adv. https://www.aquaculturealliance.org/advocate/search-for-sustainable-grouper-farming/

4 The High Seas Solution

Matthias Heilweck

4.1 In this Chapter...

The case is made for greater use of the High Seas to replace forage fish with mussels in the diet of farmed fish and produce the increasing amounts of food that will be required by the growing human population, while at the same time pulling down carbon from the atmosphere with bivalve cultivation. The vision is to preserve the oceans as a healthy and sustainable food source for mankind by emphasising conservation and ecosystem balance beyond coastal waters. The plans are for huge (centralised) bivalve mollusc farming facilities on the high seas, using factory ships and offshore factory rigs (re-purposed disused oil rigs?) located on seamounts outside Exclusive Economic Zones and employing Perpetual Salt Fountains on the flanks of the seamount to bring nutrients to the farms. If properly designed (and the design and building capabilities exist throughout the offshore industries around the world), this will immediately provide (i) feed for animals and food for humans, (ii) sustainable marine ecosystems, and (iii) permanent atmospheric carbon sequestration in the form of reefs of bivalve shells.

4.2 The Context

Since ancient times, the seas have always been an abundant source of healthy food; this is about to change dramatically! Since World War II, the industrialisation of fishing has led to a drastic decline of the biomass of large, high trophic level fish; they are the carnivorous ones we prefer to eat, like tuna, swordfish, grouper, salmon or cod.

Since the 1990s or even before, the global wild fishing production has stagnated, despite greatly increased effort devoted to fishing with sonar-assisted super trawlers, fishing ever further from their home port and ever deeper in the oceans searching for new stocks to replace the already overfished traditional ones.

With traditional fisheries in danger of being fished out, aquaculture was considered as the best alternative to wild fishing and became the fastest growing food-producing sector, contributing efficiently to the food security of our expanding human population.

The other side of this coin, though, is that marine animal proteins are needed, in quantity, to farm aquaculture fish species. Consequently, forage fish, which means low trophic level fish, like sardines, anchovy, herring, sprats, capelin, and other organisms, like krill or even copepods, are now harvested from all possible ocean locations to feed this aquaculture industry.

This in itself is environmentally destructive, because these low trophic species are already the subsistence food for about one billion people who live in coastal communities as well as all the other higher trophic animals that depend on a healthy marine environment like other fish, seabirds and sea mammals. Indeed, forage fish stocks are diminishing more and more, yet it is essential that the aquaculture industry remains able to produce

D. Moore et al., *Aquaculture: Ocean Blue Carbon Meets UN-SDGS*,
Sustainable Development Goals Series, https://doi.org/10.1007/978-3-030-94846-7_4

Fig. 4.1 Fishing down the food web, a North Sea perspective. Image © Hans Hillewaert inspired by the work of Daniel Pauly (Pauly et al. 1998) who describe the concept like this: "… Fishing down food webs (that is, at lower trophic levels) leads at first to increasing catches, then to a phase transition associated with stagnating or declining catches … [making] present exploitation patterns … unsustainable …" Image taken from https://en.wikipedia.org/wiki/Fishing_down_the_food_web under the Creative Commons Attribution-Share Alike 4.0 International license

healthy fish both for the human food markets, and for the open ocean to maintain healthy food chains and continue to support all the wild species from upper trophic levels, including, of course, all those other wild fish species humans also want to eat. We cannot use forage fish twice; their uncontrolled harvesting as fish food is not sustainable. Sea mammals, seabirds and wild fish populations are drastically declining, while forage fish are progressively replaced by jellyfish in the habitats these fish previously occupied, and artisan fisherfolk have to work longer and sail further to catch less and earn less. The growth of world aquaculture is also unsustainable if forage fish stocks continue their decline because less wild fish means less natural fishmeal and fish oil can be included in farmed fish feed and the aquaculture farms are likely to produce a much poorer food for humans. Currently, we are duped twice!

Even without considering the effects of other marine threats like toxins, nutrient runoff eutrophication and petroleum or plastic pollution, this questionable practice of fishing down the food web (Fig. 4.1), if unchecked, could ultimately drive so many fish species to extinction that only coelenterates will be left to dominate the diminished ecosystems. Do you want to eat polluted jellyfish in the near future? I don't.

4.3 A Novel Idea

I'm Matthias Heilweck, from Alsace, France. For over 30 years, alongside my bread-and-butter job, oceanography has been the focus of my spare time. Although I live about 1000 km away from the coast, I have always loved to read all the press articles and scientific publications I could find

Fig. 4.2 My message in a bottle [https://www.commonseagood.com]

about this topic. Being a pragmatic ecologist with a Cartesian preference for logical analysis I believe I have found a good way to help preserve the oceans as a healthy and sustainable food source for mankind. Between global pollution, overfishing and artificialisation of aquaculture, this may sound utopian, even pretentious (Fig. 4.2). However, I decided to publish here the synthesis of my constructive, I hope, reflections about the topic and thereby confront critical analysis of them. My ideas are also explained on my website [view: https://www.commonseagood.com], which is, in a sense, my message in a bottle, optimistically thrown into the ocean of the World Wide Web, where it has bobbed around since June 8th, 2014, a ***World Oceans Day***.

It all comes down to the development of a huge and healthy marine protein source, able to take over from the wild forage fish, those small pelagic fish of low trophic levels, which are harvested today in the open ocean solely to feed fish farms.

To achieve this my project still needs many relevant contributions from scientists, engineers, lawyers and other specialists. Obviously, I will also need huge funds to finance the project. This is unavoidable, even simple operations on the high seas are tremendously expensive. But the goal is a sustainable, ecologically-friendly, healthy and scalable production of aquafeed on the high seas, which is created from almost nothing and is not taken from anyone. That, you must believe, is worth paying for.

In the following sections of this **Chapter**, I will explain in detail:

- WHY it should be done.
- HOW it can be achieved.
- WHERE it can take place.
- WHAT sort of marine organisms can be cultivated.

What I consider to be the facts on which these plans are based need to be audited, considered creatively and complemented by experts in each domain. If you have the patience to follow my initial brainstorming step by step through to the end, I am confident that you will admit that it makes sense.

4.4 Why Should It Be Done?

There are two good reasons to proceed. First, because we have to find new food resources to feed a growing human population; and second, because we need to keep the ocean resources (and, consequentially, ourselves) fit and healthy.

We have to find new food resources simply because there will soon be almost 10 billion humans living on this planet. As of August 2020, the current world population is 7.8 billion, and that population is projected to reach 9.8 billion in 2050 [source: https://twitter.com/UN/status/877551686537027585]. This corresponds to a 25% increase over the next 30 years, and, obviously, our current food production must also be increased by 25% over its current level if those new members of the human population are to be fed. It is clearly evident that a 25% increase in agricultural production would require the adoption of broad open areas of currently unused land, as well as quality irrigation water in quantity. Whereas, in contrast, what we see around us is increasing urbanisation and desertification promoted by global warming. As these factors progress, available arable lands shrink dramatically. Our remaining forests and wilderness reserves are either highly coveted for agriculture or already acquired. Availability of fresh

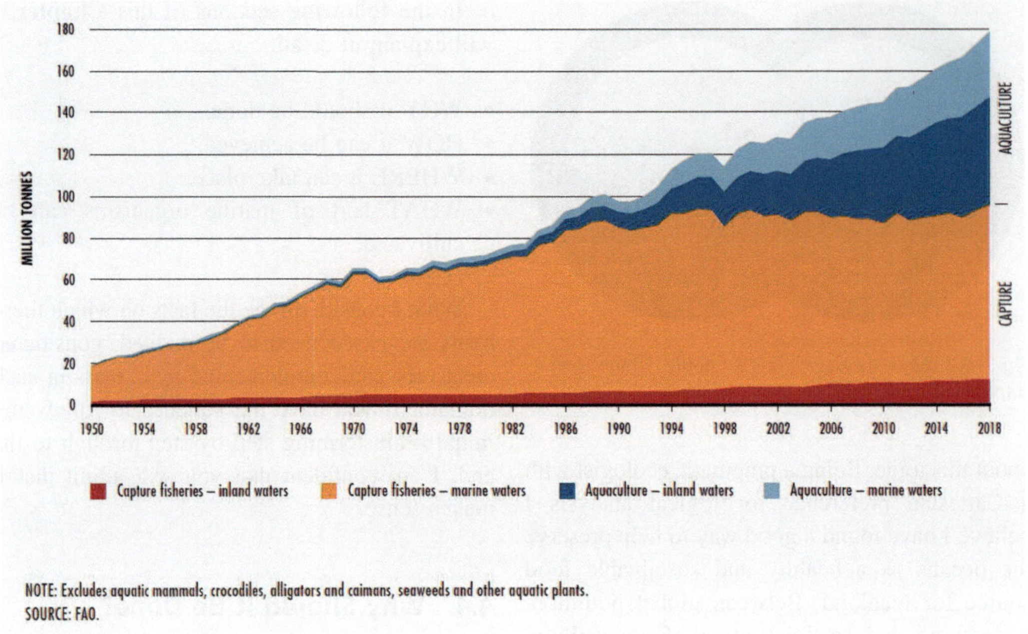

Fig. 4.3 World capture fisheries and aquaculture production 1950 to 2018, revealing the trend to stagnation of production through marine capture being compensated by an increase in aquaculture production. *Source* FAO report *The State of World Fisheries and Aquaculture 2020* (FAO 2020)

potable water, let alone irrigation water, is lessened by scarcity or pollution, and even our coastal seas are no longer free for additional exploitation.

In **Chapter 2** of this book, David Moore points out that the fundamental problem is actually that there isn't enough agricultural land on the planet to feed generously its entire current population, he states:

> '… Only about 7.5% of the Earth's surface provides the agricultural soil on which we depend for the world's food supply …' (Moore et al. 2020).

So, where, then, is there still enough space and water that we can exploit? Well, I would say: "On the High Seas!", but we need to be careful to avoid the errors of the past (and present) if we really intend to keep the ocean resources (and ourselves) healthy.

Ocean fish resources are currently overexploited. For several years now, despite the technical efforts of the fishing industry which has driven capture fisheries ever further and deeper, production by wild capture has stagnated. At the same time, the overall production of seafood has increased thanks to aquaculture. This activity is growing so rapidly that it is projected to overtake capture production in the foreseeable future (Fig. 4.3).

The inland farming of herbivorous fish (carp, *Tilapia*) is still able to increase as long as appropriate locations can be found on land (though such locations are becoming rare); recycling water reduces the impact on fresh water supplies and the farming activity can even be combined with recreational activities. View Mark Driscoll's *Tasting the Future* website at this URL: https://tastingthefuture.com.

The marine farming of carnivorous fish (salmon, bass, cobia [Fig. 4.4, and view https://pacificreef.com.au]) or shrimp is following a course that is more difficult to justify, despite its current success (view Columbia University's *Earth Institute* website: https://blogs.ei.columbia.edu/2016/04/13/making-fish-farming-more-sustainable/.

I believe that in the context of its global expansion, this food system bites its own tail,

Fig. 4.4 Farmed cobia (*Rachycentron canadum*) being fed in a netpen. [View: https://thefishsite.com/articles/cultured-aquatic-species-cobia] Image source: Philip Chou/SeaWeb/Marine Photobank

because as this sector expands increasing quantities of wild captured forage fish are necessary to feed the farms and, as I have mentioned before, these catches are declining. Consequently, the fish farming industry searches for substitutes for the fishmeal and the fish oil of their feed. Current attempts to feed farmed carnivorous marine fish with materials from terrestrial agricultural resources, instead of fishmeal and fish oil, may work technically, but have adverse effects on the wellbeing of the farmed fish, and thus on the health of our food through loss of the benefits derived from these well-balanced natural compounds.

Once caught, forage fish does not preserve well in its raw state. The whole fish has to be converted into fishmeal and fish oil (by cooking, screw-pressing and drying) if it is to be stored at room temperature for some time before use in aquafeed formulations. Fishmeal can be packed in bags. Fish oil for its part is more fragile and must be hermetically sealed to avoid becoming rancid (view https://www.iffo.com/production).

Apart from the good amino acid profile of their constituent proteins (concentrated in fishmeal after the fish is processed) and their content of vitamins, minerals and trace elements, marine foods are healthy nutritional sources because, they contain high levels of long chain omega-3 polyunsaturated fatty acids (LC-PUFA), which is concentrated in fish oil after the fish is processed. Many research findings have demonstrated the value of omega-3 fatty acids in human nutrition (Bernasconi et al. 2020). LC-PUFA are known to lower blood pressure, slow the development of arterial plaque, reduce the chance of abnormal heart rhythm and the likelihood of heart attack and stroke. They are responsible for many functions in animal cells, such as signaling, cell membrane fluidity, and structural maintenance through which they also influence the nervous system (Gammone et al. 2019).

Fishes, indeed all aquatic organisms, need large amounts of these omega-3 fatty acids in their diets, as they are the fundamental component of all their cell membranes. These essential LC-PUFA cannot be synthesised by the fish themselves, but are nearly exclusively produced by phytoplankton, the first and basic marine trophic level of the food chain in which the phytoplankton is assimilated by zooplankton, which is assimilated by the forage fish, and so on.

Without an appropriate amount of LC-PUFA, fish are much more sensitive to stress, prone to diseases, and have to suffer greater loads of parasites like sea lice. In most aquaculture farms, in order to grow fish properly despite those issues, the feed has to be supplemented with antibiotics and the water treated with pesticides; both of which are ultimately ingested by the humans who eat this fish. After some significant scandals, the fish farming industry has sought to improve its methods using vaccines instead of antibiotics and lumpfish (sea lice eating fish) or mechanical treatments, instead of pesticides. Other ways of ongoing improvements are to go further offshore, where the stronger currents can sweep the pollution away, or to filter the water in recirculating aquaculture systems (RAS) on land. But the main problem of the low level of LC-PUFA in aquafeed remains. With a long chain omega-3 rich diet, farmed fishes are much healthier; which is desirable for the fish themselves, of course, but also for us, the eventual diners.

For our part, as with all terrestrial organisms, our nerve cell membranes require LC-PUFAs for efficient synaptic vesicle recycling (Marza et al. 2008). DHA (docosahexaenoic acid) is an omega-3 fatty acid that is a primary structural component of membranes in nerves in the human brain, cerebral cortex, skin, and retina. Further, EPA (eicosapentaenoic acid) is another important LC-PUFA that is essential for cardiovascular health. Consequently, the benefit to the human heart, eyes and brain is important enough for us to need to obtain LC-PUFA in sufficient quantities from our nutrition for healthy growth and development. It thus becomes quite critical to us to preserve farmed fish as a good LC-PUFAomega-3 source, particularly since a recent study has shown that the more LC-PUFA you ingest, the better your body feels (Bernasconi et al. 2020).

I am not alone in coming to this conclusion. Two kinds of attempts are in progress today to replace wild captured fish oil and supply more LC-PUFAomega-3 to the expanding aquaculture industry.

- Aerobic fermentation of heterotrophic microalgae discovered in muddy marine waters and fed with cane molasses in fermenters. *AlgaPrime DHA* [http://algaprime.com/] is produced by Corbion for BioMar. *Veramaris*, a joint venture of DSM and Evonik [https://www.veramaris.com/], produces a similar omega-3 algal oil using sugar derived from corn. As a supplement obtained by feeding a simple compound (cane molasses or corn sugar) to a single strain of one algal species (*Schizochytrium* sp.), it cannot provide a well-balanced diet profile for the farmed fish, even if it contains the important long chain omega-3 s. Compared to forage fish reduction, it is also a costlier process reserved for high-end markets and unable to fill the global gap, especially in developing countries.
- Growth trials with a genetically modified canola (rapeseed, *Brassica napus*) with an added gene from a microalga (Opsahl-Ferstad et al. 2003), started in Australia in 2019 (Nuseeds' Aquaterra™ [https://aquaterraomega3.com/]) and in the USA (Cargill Inc's Latitude™ [https://www.cargill.com/page/latitude]). This omega-3 canola technology promises great things. I quote from the Nuseeds website: "… Grown on just a fraction of the world's existing, converted canola farmland, Aquaterra reduces pressure on the oceanic environment and delivers many nutritional, environmental and economic benefits. One hectare of omega-3 canola provides the DHA yield from 10,000 kg of fish …" Use of omega-3 canola in fish farming has been well researched (Ruyter et al. 2019; and see https://nofima.com/publication/1732999/) and shows that the material is a good alternative to fish oil from capture fishing but two aspects continue to worry me. One is the fact that this feed supplement alone does not produce the natural well-balanced diet profile for the farmed fish; because it is deficient in the very important EPA, and vitamins and minerals are also lacking. The other is the ethical consideration of using genetic manipulation to feed an animal that will become a primary human food. Acceptance, and even the definition, of 'GM foods' varies between administrations around the world. To quote from Opsahl-Ferstad et al. (2003): "… Combinations of basic understanding of gene function, transgene integration and expression, gene interactions, fatty acid metabolism in plants and animals and finally public acceptance have to be gained …" I am not sure that the public's acceptance has been, or even will be, secured as this is a highly politicised and emotive topic.
- Insect meal is often presented as a third alternative, but the breeding of Black Soldier Fly (BSF), which is the most heavily industrialised, produces only proteins and saturated oils of value, but not the scarce LC-PUFA. The latter could eventually be produced by mealworms given an appropriate diet, but their rate of development is probably too slow for this to be a profitable enterprise.

In any case, all these single compounds are only suitable for "modern"agri-food industries which are fond of cracking technologies down to their components. Thanks to mechanical denaturation processes (heat, freezing, pressure) and chemical denaturation processes (organic solvent, surfactant, acid or alkali), each constituent of a natural food is separated into several low-cost nutrifunctional compounds, which in combination generate greater profits than the original (natural) food could have brought. Listed in catalogues, anyone can select different components to combine in any chosen food, with the required biochemical composition, aspect, flavour and palatability. This is the origin of junk food; "la malbouffe" as it's called in France, or ready to eat meals. A diet strongly suspected to contribute to the dangers of diabetes, cardiovascular failures and hormonal dysfunctions and their related chronic diseases.

The same occurs in the animal feed industry's pellet manufacturing. We are not only what we eat, we are also what we feed. Biochemists and nutritionists should use their knowledge to enhance food production in accordance with nature's laws rather than going deliberately against them. This plea comes from the heart. Let it be said, I believe in organic farming with some biodynamic precepts, agroforestry and multi-trophic aquaculture and the principles of permaculture to feed the world, rather than GMOs or industrialised farming and animal husbandry. To work with nature rather than against nature because if we replace intensive agriculture using the principles of permaculture, they will have positive effects on biodiversity, food production and carbon storage (Berners-Lee2019; Holmgren 2011; Shepard 2013).

My ambition is to develop a natural and healthy animal protein source in the present desert of the high seas. Enough to supply some long chain omega-3 rich feed to fish farms around the world (first to family ponds and small-scale farms in developing countries). Hopefully, this would enable fish farming to provide healthy food for humans, while leaving forage fish unmolested in the open sea to nourish penguins, seals, dolphins, whales, sharks, seabirds … and wild fish from the upper trophic levels to support artisan fisheries. I have tried to re-think the whole process of aquaculture management in a pragmatic way to the benefit of all: fish farms, small-scale fisheries and natural marine ecosystems. In any case, industrial fisheries are rapidly approaching the time by which they must be finished. The main thread of my reflections is based upon an implacable logic: the better our aquaculture, the less we fish; and we let the oceans live.

4.5 How Are We Going to Achieve It?

To begin, let us take a look at the issues we may encounter. The first issue encountered when operating any infrastructure on the high seas is likely to be the effects of **local weather conditions**. So, let us first choose the only ocean which seems not to be prone to major hurricanes: the South Atlantic (Fig. 4.5). It is also the ocean in which there are fewer fisheries and maritime traffic, so we should be able to find an unoccupied and safe place there.

The lack of mineral nutrients for phytoplankton growth is the second issue on the high seas in general, and in the South Atlantic in particular. Except in some eastern areas, where constant wind from the land causes upwelling, algal bloom and consequential important fish production, it is a real desert out there. Anyhow, we have to content ourselves with these desert areas because the naturally rich zones along the African coast are already occupied and exploited. Further west, the nutrients present in the photic zone (above 100 m depth where light is available; Fig. 4.6) are rapidly completely consumed by the photosynthetic activity of phytoplankton. In these conditions, the phytoplankton cannot multiply sufficiently to feed many zooplankton. That is why the high seas are biological deserts. No wind-driven currents are strong enough here to mix the layers, and the minerals are not massively renewed from the deeper layers, where they remain present in quantity.

The challenge now is to bring together the existing nutrients from the depths and the light from above. Logically, there are only two possibilities, take the light down or bring nutrients up. Let us consider both.

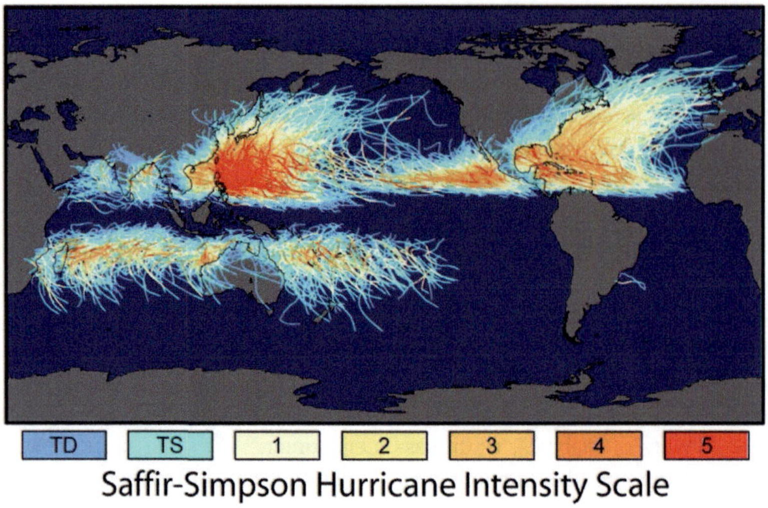

Fig. 4.5 Global distribution of the tracks and intensities of all tropical storms (hurricanes/cyclones) 1851 to 2006. Tropical cyclones do not form close to the equator and there is only one recorded tropical cyclone recorded along the Atlantic and Pacific coasts of South America (hurricane Catarina in 2004, rare and perhaps unique). Source: Historic Tropical Cyclone Tracks, NASA Earth Observatory [https://earthobservatory.nasa.gov/images/7079/historic-tropical-cyclone-tracks]

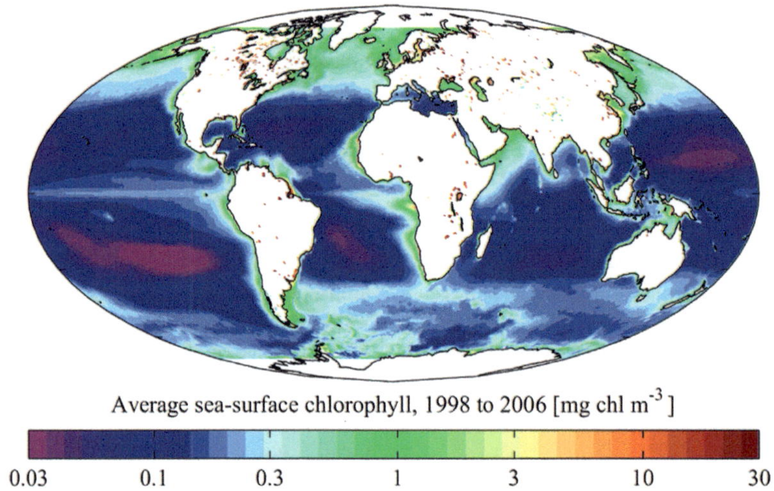

Fig. 4.6 Average sea-surface chlorophyll-a, 1998 to 2006 (mg chlorophyll m^{-3}). Chlorophyll-a is used as an indicator of phytoplankton biomass. Source: SeaWIFS Project [https://oceancolor.gsfc.nasa.gov/SeaWiFS/]. SeaWIFS (Sea-Viewing Wide Field-of-View Sensor) was a satellite-borne sensor, active from September 1997 to December 2010, designed primarily to quantify chlorophyll produced by marine phytoplankton

Take light down: Due to the aggressiveness of ultraviolet radiation from the sun in the upper sea layers, and also the constant adaptation of the controlled buoyancy of the phytoplankton, the optimal wavelength for marine algal photosynthetic activity is within the blue range of the spectrum, because this has the highest water penetration coefficient (Fig. 4.7) (Mascarenhas and Keck 2018).

Luckily, thanks to today's LED technology, we are able to produce exactly the blue light that is needed, with acceptable levels of energy demand (Fig. 4.8).

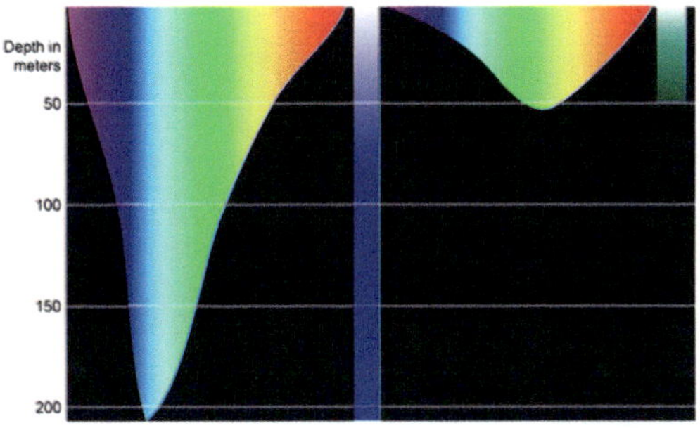

Fig. 4.7 Light penetration in the sea; open ocean at left, coastal waters at right. This diagram offers a basic illustration of the depth at which different colours of light penetrate ocean waters. Water absorbs warm colours like reds and oranges (long wavelength light) and scatters the cooler colours (short wavelength light).. Image courtesy of Kyle Carothers, NOAA-OE. *Source* NOOA Ocean Explorer at: [https://oceanexplorer.noaa.gov/explorations/04deepscope/background/deeplight/deeplight.html]

Furthermore, LEDs can be deployed easily in high-pressure environments, because, unlike light bulbs, they are not hollow and cannot implode. But, even if we manage to illuminate properly a portion of the dark aphotic zone (below 100 m depth) with arrays of blue LEDs, we still need to survey, handle and fix a production process in these depths. This is feasible but too constraining.

Bring nutrients up: Unfortunately, nutrients are diluted in great amounts of water, far too much to be pumped up with external (expensive!) energy to create an artificial upwelling. It would not be economically viable to spend so much energy. So can we bring up colossal amounts of nutrient-richdeep-sea water without energy? Yes we can!

Antarctic Intermediate Water (AAIW) is a nutrient-rich, and low salinity water body, which characteristics are due, among other factors, to the mixture of seawater with mineral-rich and fresh meltwater from the southern continental ice cap. These are the waters where the Antarctic life explosion takes place every summer. AAIW then flows slowly northwards in every ocean. In the South Atlantic, it meets warmer and saltier sub-Antarctic water at the convergence zone, 50–60° S. There, it sinks to a depth of approximately 1000 m (3280 feet), gliding northerly over cold, salty and very dense bottom water.

A large part of this water flows northeasterly to the South Atlantic Gyre, where it loses its characteristics by mixing. But a small part of it flows due north on the west side of the Atlantic, until it crosses over the submarine volcanic chain known as Vitória-Trindade (a volcanic hotspot chain), off the eastern coast of Brazil. AAIW lays at this depth, and does not mix with the water layers above and below, even in this tropical region, where surface water becomes denser, due to evaporation causing increased saltiness (Fig. 4.9).

The sea stays stratified because the diffusion between the different water masses is low, no

Fig. 4.8 Super Bright LEDs producing the blue light that is essential for marine algal photosynthesis. Image from the *GIZMODO* website [https://io9.gizmodo.com/the-2014-nobel-prize-in-physics-goes-to-the-inventors-o-1643316553] © 2020 G/O Media Inc.

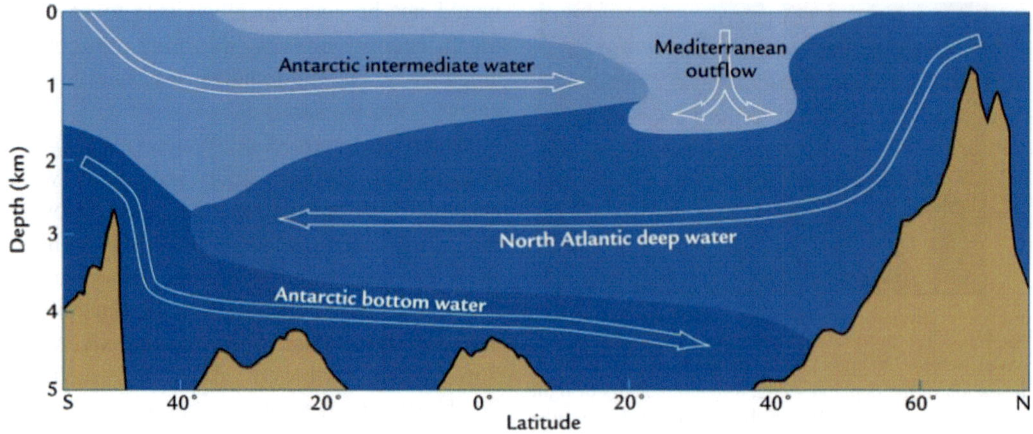

Fig. 4.9 Antarctic Intermediate Water currents (Courtesy: Ruddiman 2014)

storms are strong enough in this region to mix the layers, and no constant wind from landforms an upwelling.

These conditions are ideal to set up a **Perpetual Salt Fountain**, an 'ocean curiosity' which was first described by Hank Stommel in 1956 (Stommel et al. 1956). This can be made to work where you have a warm and salty water mass above a colder and fresher one. The technique is to insert a vertical duct between these two layers, and then pump it out until the pipe is filled with the deep water. You can then stop pumping. The upflow from the lower layer will last perpetually, without any other external energy expenditure. This is due to the fact that the heat energy difference between the water masses is conducted through the pipe walls, but the ***salinity difference*** remains unchanged, and it is the density difference caused by the salinity difference that drives the upward flow (Fig. 4.10).

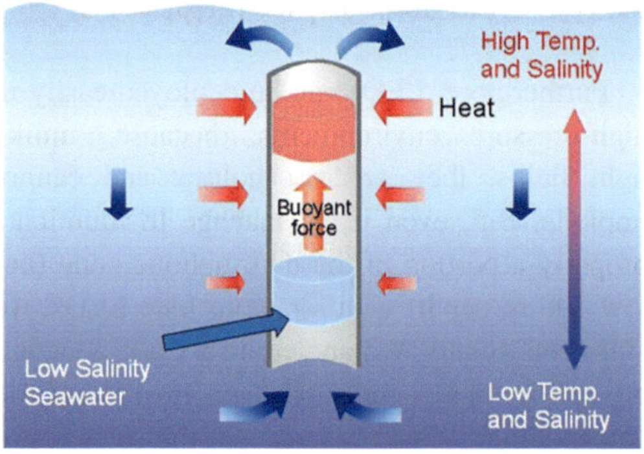

Fig. 4.10 Schematic diagram of a Stommel self-driven perpetual salt fountain. A duct links the very salty and warm water above to the low salinity colder water below. The salinity has the greatest influence on density so the cold water below has greater buoyancy and the salt fountain can be very strong. The system allows nutrients to be brought to the surface without the expenditure of energy. Image courtesy of Prof. Shigenao Maruyama, Institute of Fluid Science, Tohoku University, Sendai, Japan [http://www.ifs.tohoku.ac.jp/maru/english/research/research/laputa/index.html]

This property has been validated recently in open ocean experiments in the Mariana Trench area of the Philippines Sea. A 0.3 m diameter, 280 m long soft pipe made of PVC sheet was used in the experiment and gave an upwelling velocity of 212 m day^{-1}. Subsequently, it was demonstrated that the chlorophyll concentration around the pipe outlet was much greater than that in the surrounding seawater, providing evidence for increased primary production in the ocean by Stommel's perpetual salt fountain (Maruyama et al. 2004, 2011, 2013).

Similar experiments in Chinese ***coastal*** waters were also successful (Fan et al. 2020). These experiments were conducted in the oligotrophic Aoshan Bay, in the Shandong Province of China. In this case, a solar-powered, air-lift system was used to lift the nutrient-rich bottom water to the surface and the results show that the growth of seaweeds in a kelp farm was stimulated by this artificial upwelling. The authors calculate that an extra carbon removal potential of 14.8 thousand tons in Chinese coastal waters might be expected if their system were to be applied along the Chinese coast.

Furthermore, AAIW, found today at 1000 m depth in the tropical zone of South West Atlantic, was in contact with the atmosphere about 300 years ago, in other words, long before the Anthropocene, and the wide dissemination of anthropogenic pollutants. Even with the inevitable transfers between water layers, caused for example by the daily vertical migration of plankton, **the concentration of anthropogenic pollutants in AAIW should remain insignificant**. By contrasting the present state of seafood contamination with heavy metals and persistent organic pollutants (POPs), these waters constitute a very attractive aquafeed production environment.

For several years now, the World Health Organisation has recommended we eat fish for its healthy omega-3 content, ***but not more than twice a week, because of its pollutant content***. Just before Christmas 2016, a leading French consumer association revealed also, that organic farmed salmon has higher contaminant levels than conventionally farmed fish (both, luckily, well below WHO standards), due to a higher fishmeal inclusion in their feed (issue No. 521 of the magazine ***60 millions de consommateurs***, in a report entitled *Saumon: carton rouge sur le bio* [https://www.60millions-mag.com/2016/11/24/saumon-le-bio-n-est-pas-irreprochable-10800]).

From my point of view, the (still relatively low) contamination of fishmeal is not due directly to wild forage fish, but to the 30% sourcing of trimmings from higher trophic level fishes, in which pollutants accumulate most heavily when packaged in the filleting factories. It is time to take serious steps to reduce the currently observed contamination levels of farmed fish with heavy metals and POPs.

Another issue is to build a floating infrastructure with, at least, a 1000 m deep anchorage, or with dynamic positioning. This is quite an outrageous plan if you consider the scale of the engineering needed to reach the seafloor and the production levels we are aiming for. But the plan is made more realistic by seamounts, and more specifically guyots (also known as table mounts) with a flat top; extinct volcanoes rising up from the seafloor, sometimes almost to the surface. Such a guyot is perfectly suited **to support an infrastructure on its top, and pipes along its slopes**. In addition, due to Taylor columns effects (arising as a result of the Coriolis forces), seamounts have the particularity to let the isotherms rise and form a vortex that retains the surrounding waters. This phenomenon allows us to find AAIW at less deep levels and, after being raised through the Perpetual Salt Fountain pipes, to keep it above the seamount for a sufficient time for exploitation.

A last issue is the provision of the **required energy supply** so far away from the continental shores. Leaving aside nuclear power, which I consider a lethal activity for mankind, and fossil fuels which have no real future anymore, we must consider renewable power sources.

- Solar power: solar panels, deployed on a large area, are certainly too vulnerable in an open sea environment (at the moment).
- Wind power: wind turbines have already been anchored offshore, but may provide only an intermittent power supply.
- Wave power: Wave Energy Converters are fragile mechanical devices and are also unable to provide a regular supply.
- Ocean current power: marine turbines need a very strong current to be efficient and it is better to avoid that for our purposes.
- Power from osmosis: semipermeable membranes are expensive; cleaning chemicals are needed and the system is very complex to handle.
- Power from temperature gradients: Ocean Thermal Energy Conversion plants are also very sophisticated systems and need chemical refrigerants.

If none of these are acceptable candidates, what else is there? I suggest we think in terms of geothermal energy. If we locate our infrastructure just above an extinct volcano, we could bring the Earth's internal heat as a power source rising up from the seafloor almost to the surface.

4.6 Where Could This Take Place?

The Brazilian continental margin includes several volcanic islands and submerged volcanic seamounts, and there is a suitable location for our purposes in the Vitória-Trindade Seamount Chain, which is located off the central coast of Brazil. Starting 175 km off the coast of Espírito Santo State and extending for 950 km eastward, the seamounts are disposed almost linearly at 20° and 21°S (Fig. 4.11) (Alves 1998; Almeida 2006; Mohriak 2020).

Many of these seamounts rise higher than 2.5 km from the ocean floor, with more than half reaching the surface layers that receive enough light for photosynthesis to occur (the euphotic zone) (Motoki et al. 2012). The mechanism of formation of these seamounts and islands has been linked to volcanic episodes dated from the Late Cretaceous (100.5 to 66 million years ago) to the Pleistocene (2.58 to 0.012 million years ago), with the Trindade Archipelago, at the easternmost end of the Vitória-Trindade Seamount Chain, being the youngest volcanic eruptions (2.8 to 1.2 million years ago)(Geraldes et al. 2013).

I have located a suitable guyot in the Vitória-Trindade Chain, which is called the **Davis Bank**. In the tropical southwest Atlantic, it is the only suitable seamount that does not belong (yet) to a national Exclusive Economic Zone. It rises from 4000 m on the seafloor to less than 50 m depth (160 feet), and has a very large flat top of around 90,000 hectares (222,000 acres) (Figs. 4.11 and 4.12).

The Brazil Current, a warm water current that flows south along the Brazilian coast towards the Río de la Plata, is the western boundary current of the South Atlantic subtropical gyre. It transports warm water polewards and as it passes through this region it has relatively low nutrient availability. These environmental conditions favor mixotrophs (flexible organisms that can kill and eat other plankton as well as photosynthesising), heterotrophs (consumer organisms unable to photosynthesise), or diazotrophs (nitrogen-fixing microbes). A plankton survey near the Vitória-Trindade Seamount Chain

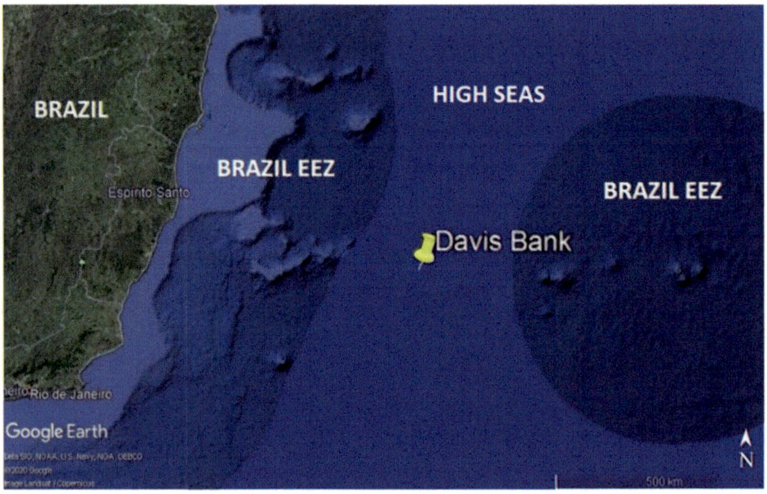

Fig. 4.11 Davis Bank, a guyot (or seamount) situated in the centre of the Vitória Trindade Chain, between two of Brazil's EEZs (Exclusive Economic Zones), the continental one and that enclosing the islands Trindade and Martim Vaz. (*Source* KML file of Flanders Marine Institute (2020), Maritime Boundaries Geodatabase: High Seas, version 1, available online at http://www.marineregions.org created on Google Earth with the author's annotations)

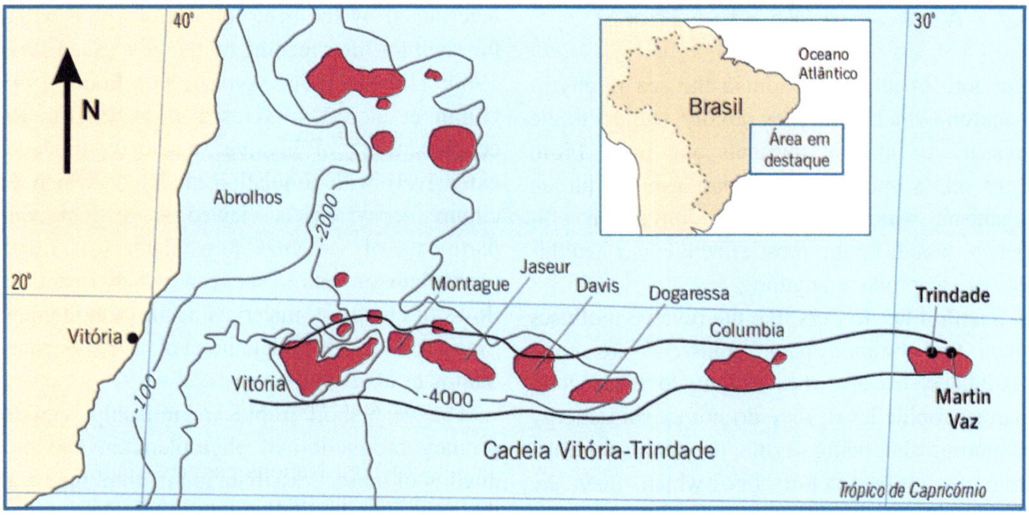

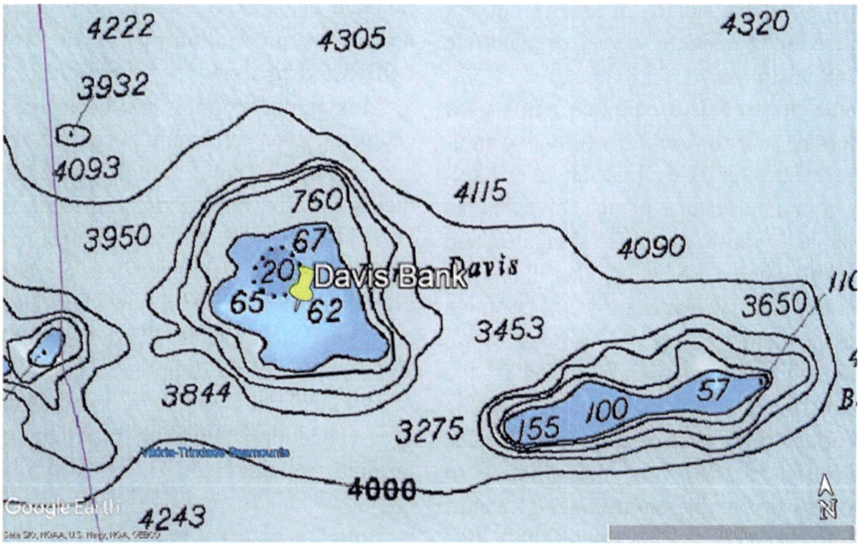

Fig. 4.12 **Top**: Geographical location of the Vitória-Trindade Seamount Chain, which is located off the central coast of Brazil. From Almeida, 2006. **Bottom**: Details of bathymetric contours around Davis Bank, the seamount (or guyot) situated just off Brazil's continental EEZ. This map was generated on Google Earth with a KML file specifically created for my purpose by Marine Geogarage (with the data of Centro de Hidrografia da Marinha do Brazil) to show the bathymetry of Davis Bank, from 4000 m on the seafloor to between 20 m and 60 m from sea level

identified 175 taxa, representing Cyanobacteria (photosynthetic bacteria, some nitrogen-fixing), Bacillariophyta (diatoms), Dinophyta (dinoflagellates), and Ochrophyta (brown and golden-brown algae). The greatest species diversity was seen among the dinoflagellates (Lubiana and Dias 2016). These seamounts therefore appear to be hotspots of bacterial and alga-like primary productivity. The waters around them also contain large stocks of commercially important fish (Meirelles et al. 2015; Pinheiro et al. 2015). Indeed, Meirelles et al. (2015) conclude that the Vitória-Trindade Seamount Chain "... represent important hotspots of biodiversity that deserve further conservation actions". I couldn't agree more!

4.7 What Shall We Raise There?

The core of all production in the sea is phytoplankton, which multiplies rapidly, subject to the presence of mineral nutrients and light. From there on, a short trophic relationship with an organism, which has a good conversion efficiency, would be the most effective and healthiest way to produce anything.

Sessile filter feeders like the bivalve molluscs given the common name 'mussels' are good candidates, because they belong to the second marine trophic level, they do not expend energy swimming and, being sessile, they cannot escape from any infrastructure into which they are introduced. Before going further I should point out that by using the name 'mussel' I am not limiting attention to a single species or even one genus of organism.

The name 'mussel' is a common name used for members of several ***families*** of bivalve molluscs, from freshwater as well as saltwater habitats. The common feature of the bivalves so named is a shell whose outline is elongated and **asymmetrical**, whereas shells of other clams are rounded or oval but ***symmetrical***. I will continue here to use the name 'mussel' in this broad sense of a ***sessile*** bivalve species best suited to the conditions of the location in which it is to be farmed. Variability of shell shape in bivalves is an adaptive feature allowing the animals to respond to less favourable environmental conditions with thicker and heavier shell valves than normal, that can be closed tightly to protect the body. The shell length, a factor often measured to represent the animal's size, does not always accurately reflect the soft tissue content because growth of shell and soft tissue do not occur simultaneously; generally shell growth precedes the growth of soft tissue. During spawning or food shortage, internal energy reserves are consumed while the shell may continue to grow. Salinity, temperature, and food supply have the greatest influence on this variability. In intertidal bivalves, shell weight is supported by a smaller biomass in animals located high on the shore than in those at the low water mark. A response determined by the need to provide space within the shell for filter feeding by the gills (Seed 1968, 1980; Hilbish 1986; Aypa 1990; Franz 1993; Gimin et al. 2004; Telesca et al. 2018). The ecophysiology of bivalves has been modelled extensively with simulation models in which the culture ecosystem is viewed as distinct compartments of variables (e.g. shell, soft tissue, phytoplankton, etc.), between which quantified flows of energy or materials occur (Scholten and Smaal 1998, 1999; Filgueira et al. 2015; Fuentes-Santos et al. 2019).

The very short trophic relationship between primary production of phytoplankton and production of mussels assimilating it, combined with the very pure AAIW quality, are the unquestionable guarantee of the **lowest level of anthropogenic pollutants at the end of the cultivation process**.

Mussel aquaculture has been practiced for centuries, even millennia, because it is rather an easy culture. They multiply profusely, attach themselves on any rough surface, natural or artificial, and feed on any organic matter suspended in the water they filter. Several cultivation techniques are used nowadays and practical advice is widely available (Lovatelli 1990; Utting and Spencer 1991; Helm et al. 2004).

Wikipedia [https://en.wikipedia.org/wiki/Mussel#Culture_methods] lists the following farming methods:

- **Bouchot culture**: Intertidal growth technique, or bouchot technique (also known as pole culture or stake culture): wooden (or other) pilings, known in French as bouchots, are planted in the shore; ropes, on which the mussels grow, are tied in a spiral on the pilings; some mesh netting prevents the mussels from falling away. This method needs an extensive tidal zone. Bouchot-grown mussels have a longer shelf life, as the animals are habituated to being out of the water at low tide.
- **On-bottom culture**: On-bottom culture is based on the principle of transferring mussel 'seed' (spat) from areas where they have settled naturally to areas where they can be

placed in lower densities to increase growth rates, facilitate harvest, and control predation, or which are simply owned by the farmer. This method requires minimum investment. It is effectively 'free-range' cultivation but has the disadvantages of the natural environment: heavy predation by oyster drills (whelks), starfish, crabs, as well as siltation (accumulation of fine sand and clay particles in the animal, which reduces market value), poor growth and relatively low yields per unit culture area.

- **Raft culture**: Raft culture is a commonly used method throughout the world. Lines of rope-mesh 'socks' are seeded with young mussels and suspended vertically from a raft. The specific length of the socks depends on water depth and food availability. Principal advantages of this type of culture are reduced predation, utilisation of planktonic food at all levels of water, and minimum siltation.
- **Longline culture (rope culture)**: Mussels are cultivated this way extensively in Galicia (Spain) and Chile, as well as in New Zealand. The conventional method is to attach mussels to ropes that are hung from a rope back-bone supported by large plastic floats. The most common species cultivated in New Zealand is the New Zealand green-lipped mussel. Longline culture is the most recent development for mussel culture and is often used as an alternative to raft culture in areas that are more exposed to high wave energy. A long line is suspended by a series of small anchored floats and ropes or socks of mussels are then suspended vertically from the line.

Mussel production costs are highly variable, depending on temperature (which affects growth rate), carrying capacity (availability of nutrients), cultivation method, technical automation, labour costs and productivity, rate of predation, availability of mussel 'seed' (spat), and offshore transport costs. Using long lines is the most productive technique for mussel aquaculture, because it takes advantage of the volume in the water column, instead of a surface. However, this production method, intended for human consumption, is too expensive and needs to be adapted to suit aquafeed production. Several sorting and cleaning stages essential for harvests intended as human food can also be abandoned, freeing a lot of time and money. The long lines themselves are too expensive in their present form, and also need to be re-designed (Fig. 4.13).

Despite the reservations just expressed, **long lines mussel farming is by far the world's most productive breeding method**, currently yielding 60 to 70 metric tonnes of mussel flesh, per hectare per year. To put these figures into perspective, ***beef*** production is only around 0.340 tonne per hectare per year, around two hundred times less! With mussels on long lines, we can reasonably forecast (if the carrying capacity is given) a production between 3 to 6 million tonnes of mussel flesh in a square of 90,000 ha, like the flat top of Davis Bank.

This figure for the potential yield can also be compared to the 5.5 million tons of Peruvian anchovies caught in 2018, which is the world's largest fishery dedicated to the production of fishmeal and fish oil. These were their highest landings since 2011 and are unlikely to be surpassed, due to government conservation policies that limit the total allowable catches. Moreover, 2018 did not experience the famous El Niño event, which periodically occurs and results in a temporary collapse in yield of this fishery.

However, as a consequence of the relative ease of culture, some aspects of the biology of mussels, such as their most efficient diet, have only been poorly documented. We know that mussels filter and ingest plankton and other organic particles from 5 to 15 microns. Diatoms, the main intake, provide DHA (the omega-3 fatty acid that is a primary structural component of membranes in nerves in the human brain—see above) and flagellates provide EPA (eicosapentaenoic acid, another omega-3 fatty acid that is good for our heart). Control of the diet of mussels, neglected until now, is all important here, especially if it can be managed.

Each mussel filters nearly 100 L of water per day. Retaining all the particles present in the water, its food conversion factor can fluctuate between 30 and 80%. That means mussels produce between 20

Fig. 4.13 Rope cultures of blue mussel (*Mytilus edulis*). *Source* The MARICULT Research Programme [https://kodu.ut.ee/~olli/eutr/html/htmlBook_111.html#id7]. See also Olsen, 2002

and 70% faeces (which pass through the digestive system), or pseudofaeces (materials that are immediately rejected). These sink and pollute the surroundings because the organic content is not degraded by aerobic bacteria during sedimentation. This is especially important when the seafloor is not very deep, which is the case in most coastal areas where mussels are farmed.

When the faeces accumulate on the seafloor, digestion depends on anaerobic bacteria, which have a negative impact on ecosystem health. That's why knowledge of the most efficient diet for mussels is so important. Especially if we are able to influence this food conversion factor and the nature of the metabolised substances, by using the appropriate phytoplankton mix to feed the mussels. In our plans, the most favourable phytoplankton species could be injected into the rising nutrient-rich AAIW passing through the salt fountain pipes, to let them multiply on their way up, thanks to strings of blue LEDs inside the pipes. In this way, it should be possible to control the diet of our cultivated mussels, in order to minimise faecal output and **maximise omega-3 fatty acid production**.

Now we have the framework. But it will not work as described so far because of the environmental footprint of such a huge mussel farm. To avoid contamination, it still needs to be distributed between several places, or at least trophically isolated in some way, by the cultivation of macroalgae ('kelp forests') for example, to establish greater biodiversity among the species cultivated so that the 'mussel farm' becomes a self-sufficient ecological community or biotope. This approach is called ***Integrated Multi-Trophic Aquaculture*** where the waste products of one species are recycled as feed for another (Fig. 4.14). The integration of fish aquaculture with the cultivation of rice, aquatic plants, vegetables, fruit, and even leaves for silkworm cultivation (Ruddle and Zhong 1988; Yang 1994) has been practised in China for over 2,000 years (discussed in Chapter 3, above). But

Fig. 4.14 This illustration depicts a conceptual model for an Integrated Multi-Trophic Aquaculture (IMTA) System. Small orange dots and orange arrows show the flow and uptake of inorganic dissolved nutrients from the salmon finfish net pen (centre) towards the Kelp rafts (at left). White arrows show the direction of the water currents within an IMTA system. Green dots and arrows show the flow and uptake of organic particulate nutrients by filter feeders such as scallops and mussels (at right) as well as deposit feeders such as sea cucumbers, sea urchins and sea worms. Organic nutrients are shown for both fine particles (represented by smaller and lighter green dots) and large particles (represented by larger and darker green dots). *Source* © Fisheries and Oceans Canada, drawn by Joyce Hui) [https://dfo-mpo.gc.ca/aquaculture/sci-res/imta-amti/index-eng.htm]

the term Integrated Multi-Trophic Aquaculture (IMTA) was coined by Thierry Chopin in 2004 and the concept is now widely accepted (Milhazes-Cunha and Otero 2017; Knowler et al. 2020; van Beijnen and Yan 2021).

Mussel faeces cause pollution problems in most of today's ***monoculture*** farm locations. To avoid this, the soluble faeces can be assimilated by kelp or other macroalgae, and the solids can be assimilated by scavengers on the sea bottom like sea cucumbers (see Fig. 4.14 legend and the SeafoodSource website [https://www.seafoodsource.com/, and view https://tinyurl.com/bdcprywx]). These will also strongly contribute to the project's viability, because of their importance in Asian cuisine and medicine. You can read more about it on the website of Fisheries and Oceans Canada's IMTA Research Laboratory at this URL: https://www.dfo-mpo.gc.ca/aquaculture/sci-res/imta-amti/imta-amti-eng.htm [and view http://www2.unb.ca/chopinlab/index.html].

However, despite IMTA techniques, it is commonly assumed that such a mussel farm could lead to a huge threat for the existing biodiversity of this rhodolith covered seamount, which counts several species of algae, corals, crustaceans, sponges, fishes, some of which are endemic (not found elsewhere) (Pinheiro et al. 2015; ***and view*** Hudson Pineiro's YouTube video at [https://www.youtube.com/watch?v=ZsV3AkDvvvE]). Eutrophication is a major threat to coastal environments due to the run-off of minerals and nutrients from the land. Mussel farming might be expected to aid in extracting nutrients from coastal marine environments as mussels feed on the algae and particles suspended in the water; thereby potentially reducing eutrophic ecological conditions. On the other hand, it is also argued that mussel faeces may so increase the deposition of organic matter as to offset the benefits of nutrient management by these filter feeders. Studies of a newly re-established mussel farm on the east coast of the Danish peninsula of Jutland in the Baltic Sea found that there were immediate changes to the biogeochemistry of the sediments beneath the farm, including 4 to fivefold increases (relative to reference conditions) in the release of dissolved inorganic nitrogen and dissolved inorganic phosphorus underneath the farm. Impacts of the mussel farm were measurable during the first year of its establishment (Hylén et al. 2021).

Other research, though, suggests that the prospects may not be so bleak. We have recently been informed that an offshore mussel farm in UK waters that encounters waves of up to 9 m on a regular basis has been documented to promote a huge increase in biodiversity and seabed regeneration (https://offshoreshellfish.com/sustainability/). It is clear that only specific trials at Davis Bank would be able to reveal what kind of environmental consequences a mussel farm would have in its specific local conditions.

In short, the Vitória-Trindade Chain that includes Davis Bank is a rich reef biotope, but not without its threats. To quote Pinheiro et al. (2015):

> … The structure of fish assemblages was similar between islands and seamounts, not differing in species geographic distribution, trophic composition, or spawning strategies. Main differences were related to endemism, higher at the islands, and to the number of endangered species, higher at the seamounts. Since unregulated fishing activities are common in the region, and mining activities are expected to drastically increase in the near future (carbonates on seamount summits and metals on slopes), this unique biodiversity needs urgent attention and management.

Since this was written, carbonates mining to provide fertilisers for Brazilian sugar cane plantations has occurred there, together with mining trials of cobalt-rich crusts. Moreover, for several years now, fleets of fishing vessels from China, South Korea, Portugal and Spain commonly work there (Fig. 4.15). If these fleets use bottom trawls, the reef might already be in pretty bad shape. I don't know how this paradise looks today, but if it is still pristine despite these damaging attacks, we may perhaps choose a less space-requiring marine organism than bivalve molluscs farms for our purposes.

Copepods are also a good candidate, especially species of *Calanus* (Fig. 4.16). From 13,000 species of known copepods, 10,000 are marine and 5,000 of them are nonparasitic and free living zooplankton (like *Calanus*). Populations of these 1–2 mm crustaceans are so large that they represent considerable marine biomass and are, in nature, a crucial link between energy-producing phytoplankton and fish. Like mussels, they belong to the second marine trophic level. In spring, they aggregate together in huge swarms near the surface, until seawater becomes like a syrup that the baleen whales scoop into huge mouthfuls.

A single copepod can catch and consume a few hundred thousand phytoplankton cells per day, clearing about a million times their own body volume of water. They feed at the surface at night to avoid visual predators and eventual toxic emanations from phytoplankton. During daytime, they sink several hundred metres down to stay protected. In Polar Regions, they sink to a thousand metres depth to hibernate during the cold season and live on their reserves, accumulated during the summer.

This is why their metabolism allows them to build up a very high protein and LC-PUFA content. To be able to sink, they change their oils into denser fats/wax esters. They also accumulate

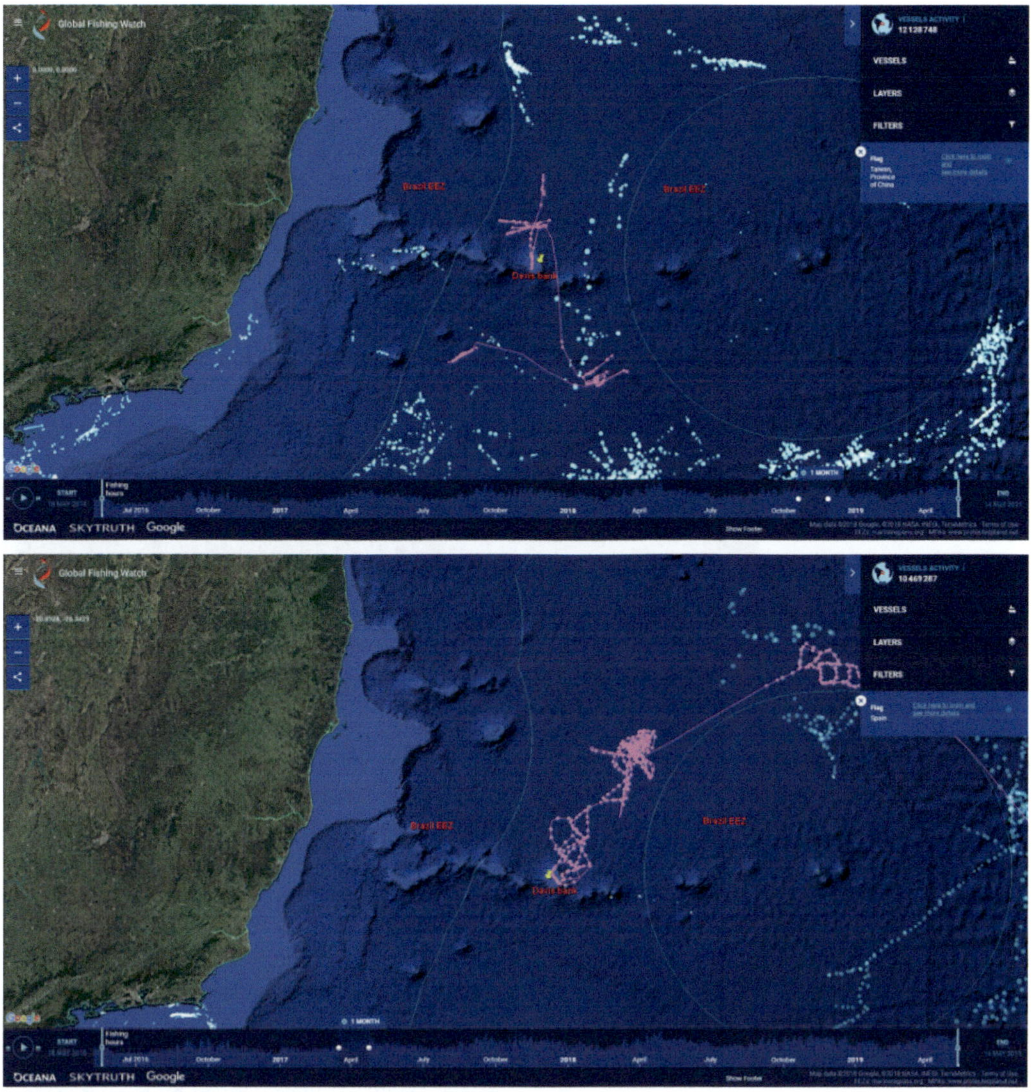

Fig. 4.15 Top: Chinese fishing vessel tracks on Davis Bank in November 2018; bottom: Spanish fishing vessel tracks on Davis Bank in April 2018. Screen captures made of observations on Global Fishing Watch [https://globalfishingwatch.org/]

astaxanthin, a lipid soluble carotenoid, well known for its antioxidant properties, which enables them to hibernate with preserved stocks of nutriments. Astaxanthin is also the substance that gives the orange-red colour to wild salmon flesh. Fish feed is currently supplemented with synthetic astaxanthin to avoid rancidness and give the right colour to farmed salmon flesh.

With a view to breeding *Calanus* spp. for fish feed, some very interesting properties can be highlighted:

- They are already natural fish feed, a wholesome feed for aquaculture.
- They aggregate together naturally and can thus be easily harvested.

Fig. 4.16 The copepod *Calanus* sp. Photo courtesy of Terje van der Meeren, Institute of Marine Research, Bergen, Norway (Escobar-Lux et al. 2019)

- They have a very high protein and LC-PUFA content, naturally well balanced with all needed micronutrients.
- Bound as wax esters, the LC-PUFA is much better assimilated than those of fish oil and lead to very pronounced positive effects on important metabolic parameters.
- Thanks to their astaxanthin content which acts as a 'self-preservative', they may not need to be processed into meal and oil to be stored at room temperature.

My expectation is that we should be able to breed swarms of *Calanus* spp. inside the Perpetual Salt Fountain pipes, harvesting them at the top outflow to avoid the threat to the environment. A proportion of the harvest would be returned to the bottom of the pipes as inoculum for the next growth cycle.

Obviously, a great deal more research needs to be done but the scientific work to which I refer here suggests my project has a solid base to counteract the challenges of fish farming sustainability and preservation of the biological productivity of the open oceans. Pragmatic ecology and oceanology suggest we must start these efforts by installing perpetual salt fountain pipes on Davis Bank.

4.8 My Vision about the High Seas

In my opinion, in the very near future, providing food for the whole of humanity will become the main challenge we face as a global species. Not gold, diamonds, oil or gas, not rare earth elements nor even personal data; the overriding problem that is rushing towards us is just the provision of food for all of us on earth. Jewels are not a necessity, neither are gluttonous cars or the latest smartphones, for us to live well. On the other hand, food shortages happen now and cause populations to migrate and create deadly conflicts. Driven by hunger, each of us is able to steal and to fight. The vital need for food supplants everything else. Because of the expected global resource depletion, food supplies will become the 21th century's first means of exercising political power and influence. The United Nations Organisation (UNO) should be able to count on a supranational food production, and I believe my project can provide this. **Food sufficiency for all is the first guarantee of peace**.

I would like to see the huge amounts of animal proteins that my project is likely to produce when fully developed directed first towards family-based and small scale fish farms, in a manner that allows redirection easily and quickly to communities in need. At the same time, mussels are essentially an excellent food for humans. As the programme expands, a part of the production could be put at the disposal of the *World Food Program* (WFP). To ensure that no national power can interfere, this production needs to be directly accountable to UNO in the framework of **"the common heritage of mankind"** (Bollier and Helfrich 2013; and view: http://wealthofthecommons.org/home).

This concept was first defined legally in 1982 at Montego Bay by the *Convention on the Law Of the Sea* (UNCLOS), giving birth to Exclusive Economic Zones (EEZ), inside 200 nautical miles off the coasts, in which resources belong to the contiguous country, and also to the *International Seabed Authority* (ISA), which was charged with managing the mining resources outside EEZ for humanity (***"the common good"***). The ISA gives mining licenses to countries or companies and surveys their activities in terms of pollution, together with their benefits in term of sharing. Some mining licenses are already active, but the resulting obligations remain unclear, though is true that the opportunity to exploit polymetallic nodules, hydrothermal vents or cobalt crusts is technically and economically not yet mature. Seabed mining will really start first inside EEZs, on the continental shelf, in which locations there is no certainty of 'sharing for the common good'.

But outside EEZs, new marine activities, such as deep-sea bottom trawling or fish gathering devices, are already widely exploited, and plunder the sea without any legal framework to protect marine biodiversity. This remaining free area (on a first-come, first-served basis) called *The Area*, or *The Commons*, covers half of our planet!

The *High Seas Alliance*, which counts more than 40 environmental NGOs, mobilised since 2011 to initiate a competent authority, which is able to regulate *The Commons* in the interest of the public good. An UN ad-hoc and open-ended informal working group was first charged to find issues for the governance of marine Biological diversity *Beyond areas of National Jurisdiction* (BBNJ). A major step was taken in January 2015 when the United Nations agreed to begin negotiating a legally binding treaty to conserve marine biodiversity in the high seas (Sala et al. 2021). The first round of negotiations started at the end of March 2016 for two weeks, but the talks are limited to the definition of Marine Protected Areas and the global sharing of biological patents derived from marine organisms. It is still good news, but what will be the state of the oceans when a binding resolution is adopted, which, at normal rates of progress in such matters is unlikely to happen in less than 10 years? For my part, I think that market forces are likely to act much faster.

Beyond EEZ, everybody is free today to build infrastructure, an artificial island, the status of which is ruled by the owner's home nation, or a flag of convenience. For myself, being already a dual French and German citizen, I lack only the other 191 UN member nationalities to be an international representative! To assure the international global status of the high seas as a Common Heritage of Humanity, we need a supranational ***High Seas Authority***.

Davis Bank, the perfect guyot in my view for my project, is situated between two separate Brazilian EEZs, the main continental shelf EEZ, and the one around the islands of Trindade and Martin Vaz. In the year 2004, as it had the right to, Brazil submitted to the Commission on the Limits of the Continental Shelf (CLCS) an extension of its EEZs beyond 200 nautical miles. The term 'Blue Amazon' was created to describe this extension to territorial limits specifically to call attention to the immense riches of the oceanic area under Brazilian jurisdiction. In aggregate, the proposal encompasses a maritime area equivalent to more than 50% of the total land area of Brazil (Wiesebron 2011, 2013).

My concern is that this *Blue Amazon* submission by Brazil attempts to join its two EEZs, by extending territorial 'continental shelf' limits from the Abrolhos Archipelago to Trindade and Martin Vaz, in order to include all the Vitória-Trindade Seamount Chain, including Davis Bank (Fig. 4.17). This has been requested despite the fact that the raising of these submarine volcanoes is obviously due to a magmatic hotspot, and has nothing to do with the continental shelf.

The submission was partially rejected by CCLS in April 2007, but Brazil submitted new

Fig. 4.17 Map in Portuguese of the proposed limits of the Continental Shelf in order to extend the Exclusive Economic Zone (EEZ) of Brazil. The ***Blue Amazon*** submission by Brazil attempts, among other things, to join its two current EEZs by extending territorial 'continental shelf' limits from the Abrolhos Archipelago to Trindade and Martin Vaz, in order to include all the Vitória-Trindade Seamount Chain, including Davis Bank. *Source* la Marine brésilienne at http://f.i.uol.com.br/folha/mercado/images/1024957.jpeg, via https://commons.wikimedia.org/w/index.php?curid=27752731

study papers in April 2015 to try to reverse the original decision (Wiesebron 2011, 2013). The consideration of this submission was programmed for the 38th session of CLSC from July to September 2015, but as far as I am aware, the resulting recommendations have not yet been published. However, as far as Brazil's hopes for ongoing oil prospecting beyond the existing EEZ is concerned, the section between the continental EEZ and Trindade-Martin Vaz is still preserved. I am convinced that ecology should be placed above politics and business, but I am pragmatic enough to realise that multinational commercial interests control events in this world.

Our global economic decision-makers could show a little more wisdom and a little less acquisitiveness, but I suppose this is wishful thinking. That's why I believe that only an internationally supported sustainable and profitable business (truly ***healthy*** fish ***farming***), which markets the same product as the wild fisheries around the world, will be able to slow and finally stop today's overfishing. **The better our aquaculture, the less we fish; and let the oceans live**. There is a lot of ocean to manage responsibly for the Common Good. The global High Seas covers half of our planet (Fig. 4.18).

I estimate my project is likely to cost ***about*** one billion US dollars to make it become a reality, but this is little more than a guestimate. What is clear is that the need for very substantial financial facilities is an inescapable feature. Each simple operation is tremendously expensive on the open sea. It's not the sort of operation that could be started off in my garage, and I am not at the head of a personal fortune. Furthermore, this project is not of the sort that could start small and grow; production would have to begin on a large scale to allow to guarantee this high-quality product at an affordable price. Further, it will have to be a philanthropic operation, unlikely to make profits at least until the world's population is nourished properly and the oceans have regained their biological balance.

Today's aquaculture major players have been able to expand thanks to strong capital investments from the likes of the oil industry (Norway), capture fishing industry (Chile) or ship owners (Greece). Perhaps the fish farming industry itself, or its trade associations or common interest groups, like the *Global Salmon Initiative* [https://globalsalmoninitiative.org/en/], could invest in the project. They have the most to gain by developing a sustainable omega-3 rich feed (which does not depend on terrestrial agriculture) to maintain and expand their own sustainable and healthy production of farmed fish.

I think also of all those billionaires, who have decided to give half of their fortune to

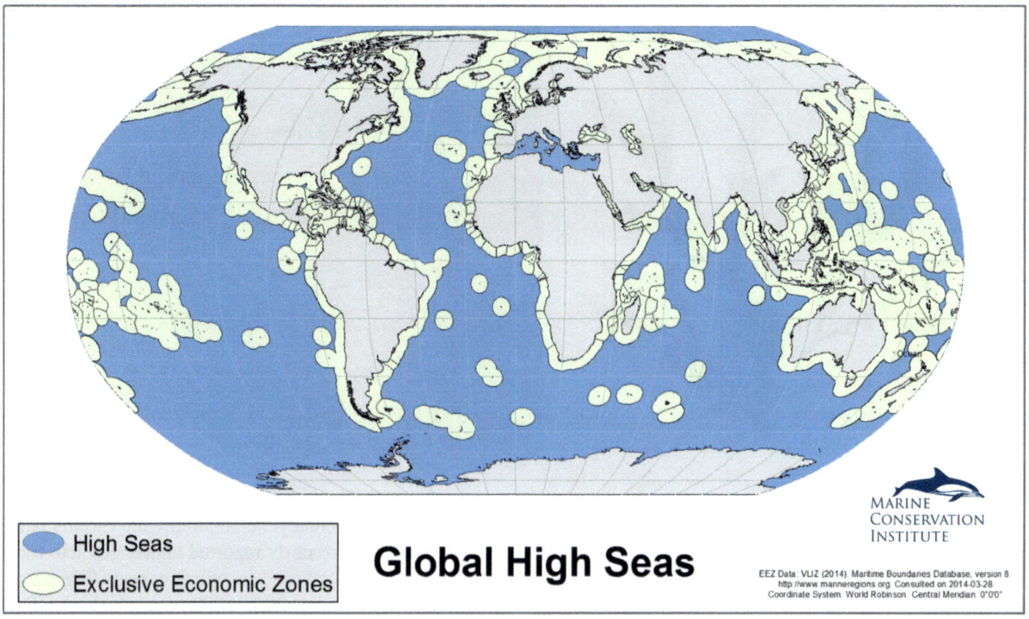

Fig. 4.18 The ***High Seas*** covers half of the total surface of our planet. Here the High Seas are compared with the world's Exclusive Economic Zones which are subject to the jurisdiction of individual nation states. *Source* Marine Conservation Institute (2020), MPAtlas [On-line]. Seattle, WA. Accessed 25 August 2020, URL: http://www.mpatlas.org/data/map-gallery/

philanthropy during their lifetime. To date, more than 200 of them have made their commitment within the framework of the Giving Pledge [https://givingpledge.org/], founded by Bill Gates and Warren Buffet. I am sure that most of them are sincere and not involved in philanthro-capitalism (the 'smart way' to give—for its own interests).

A third perspective is the involvement of UNO itself, to create a supranational food supply, and make a start establishing the Common Heritage of Mankind by initiating a global economy for all of us, and perhaps **the beginning of a universal income**.

4.9 My Dream of a Shared Half World

Alongside the production of healthy aquafeed, which is the initial aim of my project and the intended purpose of the high seas infrastructure, the latter is also a wonderful stepping stone to develop other opportunities for sharing industries.

Cosmetics and medicines: Thanks to the biological affinity between our blood and seawater, between our cells and marine ones, cosmetics containing active agents from marine sources provide some of the most sympathetic interactions with the skin. The market for natural cosmetics is constantly growing and already faces shortages in supply of uncontaminated ingredients.

The understandable tendency for consumers to seek natural ingredients in their cosmetics no longer allows the industry to artificially reproduce natural molecules they find interesting. In order to be both accepted and appreciated, ingredients need to be produced naturally in a healthy and sustainable way. Chemically synthesised molecules or even cultures in stainless steel or plastic tanks will not be able to compete with cultures in natural seawater originating from before the Anthropocene (Fig. 4.19).

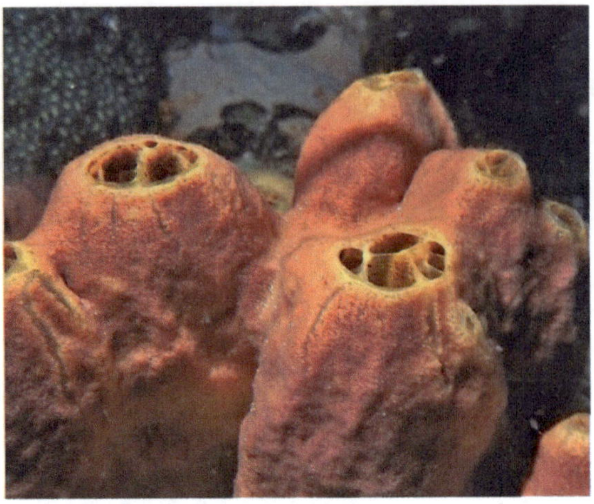

Fig. 4.19 Marine sources for cosmetics and, potentially, medicines. **LEFT**: *Dictyopteris membranacea*, herbarium specimen (collected by B. Navez, July 1982 near Villefranche-sur-mer (France), imaged 2010-02-01 (full size = 12 cm). URL: https://commons.wikimedia.org/wiki/File:Dictyopteris_membranacea_herbarium_item.jpg . This is a very widely distributed marine alga. Extracts have been shown to have anti-inflammatory, antioxidant and antimicrobial activities (Aoun et al. 2010) and have been used in anti-ageing cosmetic formulations. **RIGHT**: *Acanthostrongylophora ingens*, a common sea sponge growing in Indonesian waters produces a chemical called manzamine A, a pentacyclic alkaloid with various bioactivities, including recently reported anticancer activity on pancreatic, colorectal and cervical cancer (Lin *et al.*, 2018; Karan *et al.*, 2020). Photographed by B.W. Hoeksema, *in situ*, South of Bangka Island, Indonesia. Image taken from http://www.marinespecies.org/porifera/porifera.php?p=image&tid=166797&pic=40123 on the World Porifera Database [http://www.marinespecies.org/index.php] under a Creative Commons Share Alike 4.0 License

Moreover, many chemicals produced by marine organisms are useful as medicines. Manzamine A, for example, a pentacyclic alkaloid with various bioactivities, including recently reported anticancer activity on pancreatic, colorectal and cervical cancer (Lin et al. 2018; Karan et al. 2020) (Fig. 4.19). It is produced by a sponge common in Indonesian waters. *Acanthostrongylophora ingens* is able to grow well in poor-quality polluted water, but heavy metals and other contaminants could bind to manzamine A, making it more difficult and expensive to extract. Yet another argument is to prefer the pristine waters from the AAIW upwelled with the Perpetual Salt Fountain pipes for cultivation of marine species.

Fishing with dolphins: The oasis of life that could be created on Davis Bank will also result in numerous interactions with wild animals and inspire other challenges. Sea bream, for example, will come and crunch mussels. Dolphins, seals or other marine mammals will also be attracted and could provide the final delights. Indeed, these particular species can be educated, trained, or encouraged to assist capture fishermen by selecting wild fish shoals in their territories using their echolocation and driving the shoals of fish into the fishermen's nets.

There are several places on Earth where dolphins collaborate in hunting fish together with mankind, in Brazil, but also in Mauritania and Myanmar as well. These cooperative and mutually beneficial bonds have lasted for generations, even since the fifteenth century in Mauritania, and are based on trusting relationships between individuals of the two species who both do well out of it. The cooperation between the bottlenose dolphins and the fishermen of Laguna, Brazil is illustrated in the YouTube video at this URL: https://youtu.be/GjuW6xODzw4.

Combined with the dophins' ability to recognise geometric figures and colours (demonstrated on a daily basis in marine parks the world over), their goodwill could certainly allow us to herd schools of targeted fish species towards prepared fishing nets around the Davis Bank facility, which would avoid or greatly reduce the almost inevitable 'bycatch', which is the unwanted fish and other marine creatures trapped by commercial fishing nets which are so often thrown back into the sea dead (in Asiatic prawn fisheries the bycatch is currently 95% of the total catch!). We need to change our old hunter habits and imagine new possibilities to take advantage of our oceans and preserve them at the same time.

Carbon sequestration: Another opportunity is to develop a possible atmospheric carbon sequestration sink. This is based on the fact that bivalve shells are made of carbon dioxide permanently removed from the atmosphere. If the Davis Bank installation produces 100 tonnes of fresh mussels on long lines per hectare per year (expected to be a low production average), 50% of that harvest will be shell, representing 50 t of calcium carbonate containing 12% of carbon, **that is 6 t of carbon per hectare per year, being permanently removed each year**. Representing a total of **540,000 t of carbon** being drawn down from the atmosphere each year by the planned 90,000 ha installation.

It is difficult to compare this performance with the highly varied terrestrial data "... due to the inconsistent use of terms, geographic scope, assumptions, programme definitions, and methods. For example, there are at least three distinct definitions for a 'ton of carbon'..." (Richards and Stokes 2004). We calculate the range of estimates to be between 0.27 and 9.55 tonnes of carbon per hectare per year, with an estimate of about 4 t ha^{-1} y^{-1} being a fair average (Richards and Stokes 2004; Toochi 2018; Le Quéré et al. 2018; Pugh et al. 2019). Comparing with this 'fair average' the ***mussel farm sequesters three times as much carbon as terrestrial ecosystems retain***. Though, of course, mussel shell sequestration is an immediate permanent removal from the atmosphere, whereas terrestrial ecosystems retain their carbon sinks only transiently, while the plants are alive and growing.

This, and the scientific debate about the relative importance of the different carbon fluxes occurring during shell production are discussed in detail in **Chapter 2** (Moore et al. 2021a, b). At first sight, it seems that the amount of carbon sequestered in the shell is less important than that released by the animal's energy metabolism during shell calcification, although this last is merely a component of the metabolic carbon cycle (Figs. 4.20 and 4.21, below) and ***not a net contribution*** to environmental CO_2. This is why shellfish cultivation, unlike reforestation, has not been taken into account as a carbon offset in the international carbon trading system first implemented by the Kyoto Protocol and then taken over by the Paris Agreement on a voluntary basis. In my opinion, however, these isolated carbon fluxes have to be understood in correlation with other related natural cycles (Figs. 4.20 and 4.21).

The chemistry involved in the process of shell-making, called marine ***calcification*** or ***biomineralisation***, relies on a set of ionic (or electrolytic) dissociations and associations in water, which are in equilibria governed by local conditions (Fig. 4.20). $CaCO_3$ and CO_2 are produced from calcium and bicarbonate ***ions*** in solution as described by the following scheme:

$$Ca^{2+} + 2HCO_3^- \rightleftharpoons CaCO_3 + CO_2 + H_2O$$

(in words : one divalent calcium ion
+ two monovalent bicarbonate ions
⇌ calcium carbonate + carbon dioxide + water)

One molecule of CO_2 from the bicarbonate ions of seawater is released, together with a molecule of water, during the calcification (biomineralisation) reaction. Other aspects of marine carbonate system chemistry are illustrated in Fig. 4.20.

Seawater is over-saturated with calcium and its concentration of bicarbonates largely dominates those of carbonates and dissolved free CO_2. In these conditions, the molecule of CO_2 released during the biomineralisation of shells (if it is not used directly by the surrounding phytoplankton

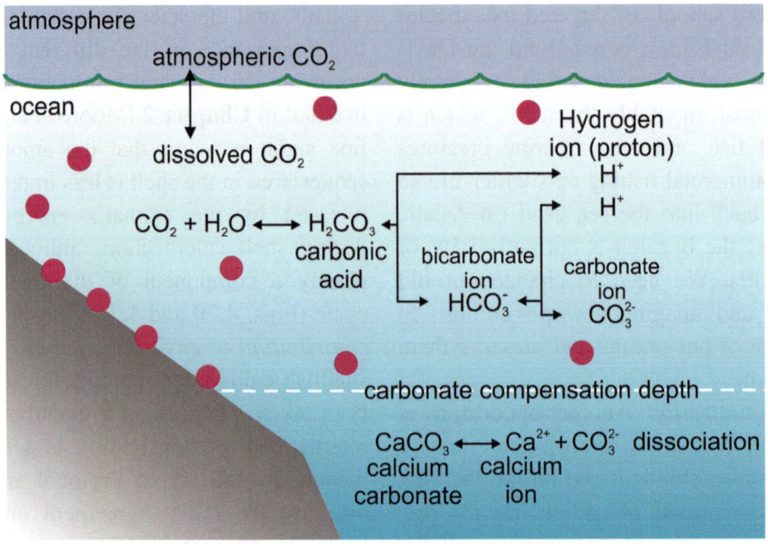

Fig. 4.20 Summary of the reactions between carbon dioxide (CO_2) with water (H_2O) in the oceans. Calcifying organisms, plant and animal, sessile and floating (represented here by magenta circles) form calcium carbonate ($CaCO_3$) in shallow water, using bicarbonate ions (HCO_3-) and calcium ions (Ca^{2+}) according to the reversible reaction $2HCO_3\text{-} + Ca^{2+} \leftrightarrow CaCO_3 + H_2O + CO_2$. Today, the calcium ***ion*** concentration, [Ca^{2+}], in the oceans is essentially a function of salinity, and is fairly constant in the ocean. At normal temperatures and pressures the ***salt***, $CaCO_3$, is essentially insoluble in water but solubility increases with pressure and decreases with temperature. In the ocean depths $CaCO_3$ dissolves (and the salt dissociates into $Ca^{2+} + CO_3^{2-}$) below the ***carbonate compensation depth*** (CCD; also known as the carbonate saturation horizon). The CCD varies between 3000 to 5000 m depth in different marine regions (shallower at high latitudes). If the solubilised calcium carbonate is swept to lesser depths it will recrystallise and the crystals will have the opportunity to grow until they reach a size that prompts their sedimentation to the sea floor at those lesser depths. Figure adapted and redrawn after an image in *Encyclopædia Universalis France* [https://www.universalis.fr/media/DE120411/]

for its photosynthesis), will bind with water, forming carbonic acid which will dissociate forming bicarbonate ions and protons that would be available for marine calcifiers to form more calcium carbonate. Alternatively, the carbonic acid can dissociate to form a carbonate ion and two protons. These electrolyte dissociations and associations are illustrated in Fig. 4.20 and described by these schemes:

$$CO_2 + H_2O \rightleftharpoons H_2CO_3 \rightleftharpoons H^+ + HCO_3^- \rightleftharpoons 2H^+ + CO_3^{2-}.$$

(in words : one molecule of carbon dioxide +
one molecule of water ⇌ one molecule of carbonic acid
⇌ one hydrogen ion (= proton)
+ one monovalent bicarbonate ion
⇌ two hydrogen ions + one divalent carbonate ion).

Release of hydrogen ions (protons) will clearly cause acidification of seawater. Acidity of a solution is measured in terms of the pH, a logarithmic scale of the inverse of the concentration of hydrogen ions (1/[H^+]). A neutral solution has a pH of 7, strongly acid, a pH of 5, and a strongly alkaline solution, a pH of about 9 [https://en.wikipedia.org/wiki/PH]. The *Encyclopædia Universalis France* states that.

> … Since the industrial era, the ocean's basic [alkaline] pH has fallen from 8.2 to 8.1. This drop of 0.1 unit corresponds to an ***increase in acidity of about 25%*** [because the scale is logarithmic]… [https://www.universalis.fr/encyclopedie/acidification-des-oceans/].

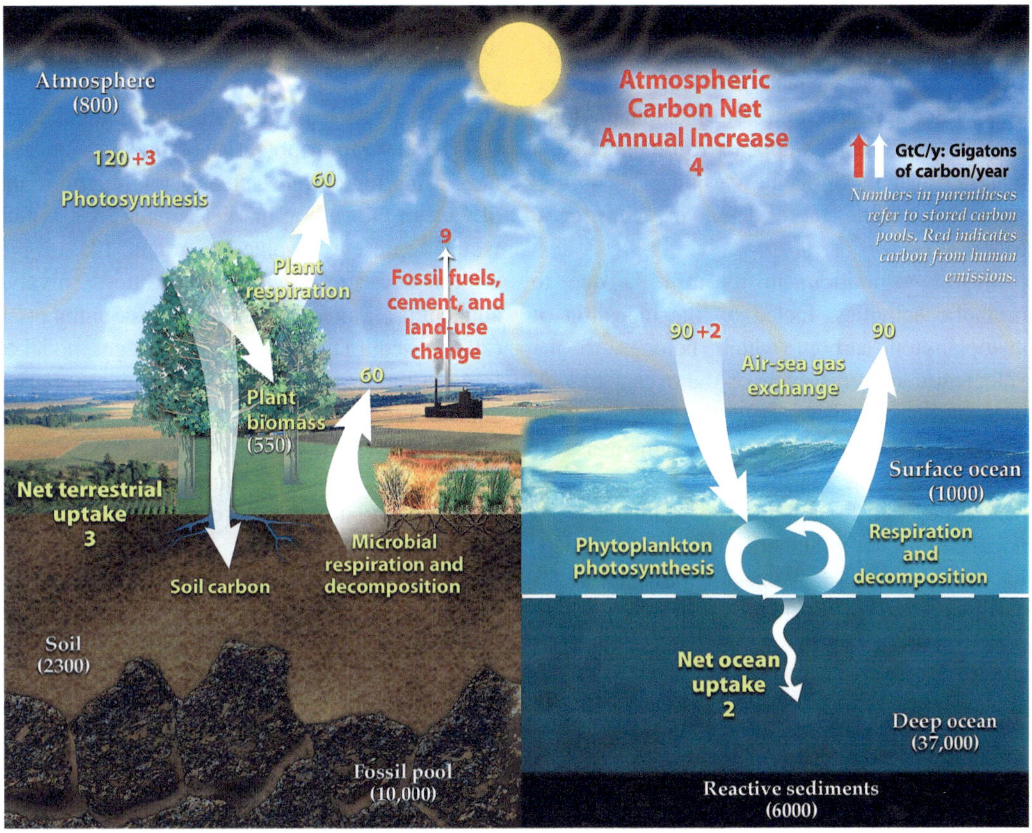

Fig. 4.21 Movement of carbon between land, atmosphere, and ocean; numerals show amount of carbon in billions of metric tonnes per year (GtC/y). Yellow numbers are natural fluxes, red are human contributions, white are stored carbon. The effects of volcanic and tectonic activity are not included. Image from the Report: US DOE, (2008). *Carbon Cycling and Biosequestration: Report from the March 2008 Workshop*, DOE/SC-108, U. S. Department of Energy Office of Science (https://genomicscience.energy.gov/) (Riebeek, 2011)

The concern with ocean acidification is the fear that formation of calcium carbonate by calcifier organisms will be disturbed. Fitzer et al. (2016) have demonstrated significant changes in the hydrated and dehydrated forms of amorphous calcium carbonate in the crystalline layers of mussel (*Mytilus edulis*) shells cultured under acidification conditions. However, there is evidence that, in *Mytilus*, acidification eases the negative effects of increased sea temperatures on biomineralization, suggesting a complex relationship between calcification and the various components of climate change (Knights et al., 2020).

Adverse effects of present-day ocean acidification are clearly seen to impact the viability of symbiotic algae of coral and giant clams, and in those cases, too, are interwoven with elevated temperatures and light levels in relatively shallow tropical waters.

The present levels of elevation of marine CO_2 concentration are more likely to encourage calcification than discourage it due to a consequential increase in proton concentration. The calcification process is thought to have originated when large amounts of excess calcium occurred in seawater at the Precambrian-Cambrian boundary, about 550 million years ago. The organisms of the time had already evolved sophisticated mechanisms for the maintenance of cellular calcium homeostasis. It is theorised that the environmental calcium excess produced conditions favouring natural selection for calcification in protists and invertebrates as a mechanism to detoxify extracellular Ca^{2+} and avoid intracellular precipitation of phosphate ions. Now the same process appears to be effective at sequestering anthropogenic carbon. This is discussed in a little more detail in **Chapter 6** (Moore 2021).

Mussel respiration, in its turn, is due to metabolic activities, fueled by the ingestion of phytoplankton carbon, which belongs to the biological carbon reservoir of the ocean, like all other marine organisms. After more or less residence time there, only a fraction of this reservoir carbon reaches the sea bottom sediments and is then sequestered for a long time—millennia and more. When phytoplankton is ingested by shellfish, the carbon track is exactly the same as for other plankton eaters, but involves only the animal's soft body with its metabolism, not its shell formation. That's why mussel respiration is neutral in this budget.

Shellfish shells, including those of crustacea, are not made of living cells and are produced outside the animal's body. Bivalve shell calcium carbonate, is elaborated by the mollusc's mantle using calcium bicarbonate from seawater, the carbon of which originates ultimately from the atmosphere. The important thing is what remains after the animal's death. The shell carbon is effectively and permanently sequestered in a crystalline mineral form, indigestible and ***chemically stable for geological periods of time***.

Among recent scientific publications, Zhang et al. (2017) identify recalcitrant dissolved organic carbon produced by microbial decomposition of bivalve faeces as an additional CO_2 sink that has been neglected in the past. Alonso et al. (2021) reviewed the capacity of bivalve aquaculture to mitigate global warming in a circular economy concept to introduce it into the international carbon market. They point out that bivalve aquaculture is already an activity with a low carbon footprint, quoting 'cradle-to-farm gate' carbon footprints ranging from 500 kg CO_2 equivalents t^{-1} of mussels to around 1500 kg CO_2 eq t^{-1} of cultured oysters. This is generally low compared with finfish aquaculture and compares with the carbon footprint of beef cattle, which ranged between 8,000 and 22,000 kg CO_2 t^{-1} carcass weight (Desjardins et al. 2012; Rotz et al. 2015). Alonso et al. (2021) estimate that the CO_2 sequestration potential of bivalve aquaculture, using the current value of **one** metric tonne of CO_2 in the carbon market is over 25 €, which would represent a value of around 125 to 175 million € y^{-1} to the European Union's bivalve aquaculture industry alone. Fuentes-Santos et al. (2021) used a Dynamic Energy Budget model to forecast the impact of climate change on marine aquaculture production in the NW Iberian coastal upwelling system. They focused on long lines mussel cultivation, this being the main aquaculture industry in Spain, producing 40% of cultured mussels in the EU. They found that the predicted impact of climate change on mussel growth in their model was low compared to the influence of the seeding time parameter (this being the time, usually when water temperature reaches around 15 °C in nature, when mussels spawn and the mussel farmer deploys the lines on which the swimming larvae settle and attach). Modelling revealed that the response of mussels varied between climate models, ranging from a minor growth decline to a moderate growth increase. The variability is linked to both farm management strategies and climate uncertainty.

The same pattern of carbon fluxes occurs with forests. In the same way as the marine phytoplankton, terrestrial green plants are photosynthetic primary producers that fix carbon dioxide out of the atmosphere into their carbohydrates. The carbon reservoir of a forest is represented, of course, by its biomass of wood, leaves, fruit and roots, but also by all of the animals, large and small, that depend on the plants for food, all of the microbes, bacteria, fungi and protozoa that digest the wastes of the forest, and all of the carbon accumulated over the years in the humus of the forest soil.

The most important point about sequestration of carbon by the forest biome is that it is ***temporary***, because the plants (and the animals they feed) will die all too soon and their subsequent decay releases their carbon back into the atmosphere again. This is discussed in detail in

Chapter 2 and will not be repeated here, but Fig. 4.21 illustrates the general features of the global carbon cycle. It is enough to say that forests are relevant carbon sinks only as long as they are in active growth. So, forest industries are ***not*** good long-term candidates for the international carbon trade despite the fact that reforestation is believed so widely to be a respected opportunity for industries who need to improve their carbon footprint. **There is a better alternative**.

In contrast to forests, if the expansion of shellfish cultivation were to be accepted as a carbon sink within the framework of the carbon trading system, it would be much more sustainable, easier to implement and, most importantly, offer permanent removal of carbon from the atmosphere (**Chapter**2). It is easier to implement because cultivating a marine species does not challenge the use of scarce agricultural land for conventional terrestrial farming for food. Further, shellfish cultivation can be combined with conservation projects (see Chapters 2, 3 and 55; Moore et al. 2021a, b) and restoration of extinct shellfish fisheries which would tackle the problem of undernourishment around the globe. Polluting states, cities and industries wishing to improve their carbon footprint (on a permanent basis), could thus make a real contribution, not only to climate mitigation but also to conservation of coastal areas and the fight against malnutrition by funding the research, infrastructure, equipment and teaching required to spread enhanced shellfish cultivation around the globe. Healthy marine proteins would be in the gift of these contributors against climate change, especially for developing countries, where most coastal sea areas are still free from commercial exploitation and where around a billion people suffer from hunger. But at the same time, their involvement would result in the permanent removal of carbon from the atmosphere. A fact that ***must*** be taken into account as a carbon offset in the international carbon trading system.

How my project can contribute here is that apart from the planned mussel production intended for aquafeed on Davis Bank (which is centred on harvested mussel ***meat***), a carbon sequestration programme (focussed on the mussel ***shell***) could also be deployed easily, and on a massive scale, from this location towards the high seas. Given the large biomass of mussel larvae (each female spawns millions of eggs), it would be feasible to envisage producing small biodegradable floatation devices already spawned with fixed juvenile mussels produced in the Davis Bank facility that could be released into the passing Brazil Current (BC). This surface current then flows south-easterly towards the South Atlantic Gyre.

This idea can also be expanded in an even more scalable way without my fixed installation on Davis Bank (intended for meat production), with several factory ships (intended for shell production), equipped with mussel hatcheries and producing those biodegradable floatation devices, already spawned with fixed juvenile mussels that could be released in all ocean currents and ocean gyres.

In both cases, the shellfish will grow (even in oligotrophic waters where the animal will make proportionately more shell than flesh) and, after a while, will sink under their own weight. When the animals die, the carbon locked in their shells as crystalline calcium carbonate will be sequestered in the ocean's depths, at least until reaching the Carbonate Compensation Depth, or CCD, which is defined as the depth in the oceans below which the solid carbonate crystals dissolve again. As long as the ocean floor lies above the CCD, carbonate particles will accumulate in bottom sediments, but below that depth, there is no net accumulation (Fig. 4.20).

This effect is due to the influences of pressure, temperature and seawater composition on the ***solubility*** of $CaCO_3$, not on its stability as a salt. Calcium carbonate is essentially insoluble in sea surface waters at the present time and the CCD varies between 3000 to 5000 m in different marine regions (shallower at high latitudes). If the solubilised calcium carbonate is swept to lesser depths it will recrystallise and the crystals will have the opportunity to grow until they reach a size that prompts their sedimentation to the seafloor at those lesser depths.

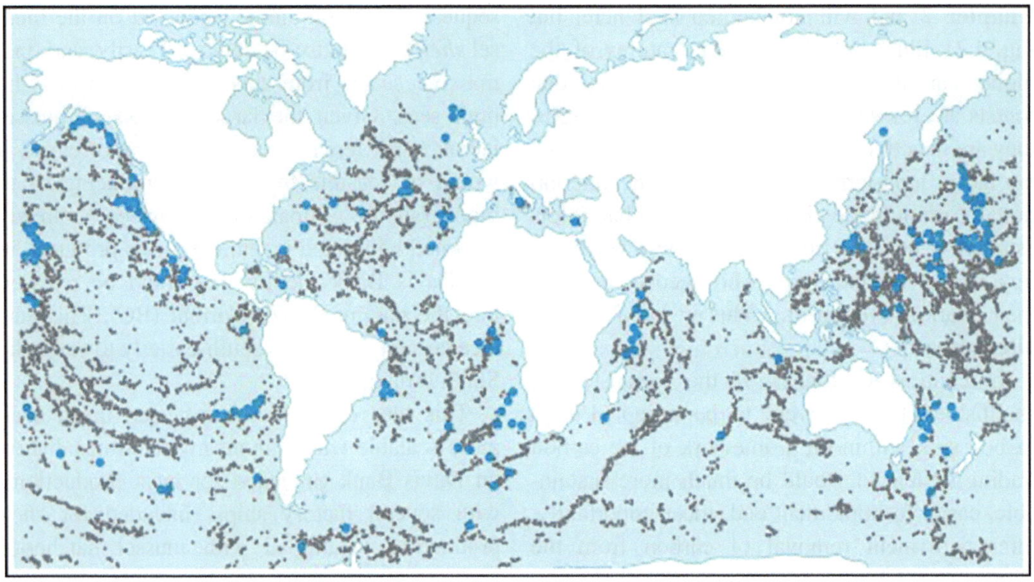

Fig. 4.22 Global Distribution of seamounts (known seamounts in blue, approximate location of other seamounts in grey). The spatial distribution of seamounts describes where and when seafloor volcanism has occurred in the past. The vast majority of Earth's volcanism occurs on the seafloors and the majority of the seafloor volcanism that forms seamounts occurs in the Pacific Ocean basin. Seamounts are not uniformly, or randomly, distributed in the ocean basins; they are spatially clustered. Most linear chains of seamounts are formed by plates moving over hotspots. Image based on data from http://www.seaaroundus.org/large-seamount-areas/

Even in the worst circumstances, the carbon sequestered in shellfish shell is permanently removed from the atmosphere. If it enters solution at or below the CCD it will be carried by the global thermohaline circulation, and is likely to take a good thousand years to surface again. We can make a start on this carbon sequestration process on Davis Bank because of its ideal environment, but there are many seamounts (and other ocean gyres) awaiting us when the technology has been developed and validated (Fig. 4.22).

Considering all these arguments, it seems that shellfish farming is the only industry able to scale up massively to provide protection to us ***against climate change***, while ***improving global food supply*** at the same time. The most harmful effect of climate change being the undermining of the ecological basis of food production, my proposal to include shellfish cultivation in the carbon trading system may hit two targets with one bullet and would support at least five of the United Nations Sustainable Development Goals Goals, namely (Fig. 4.23):

- **Goal 2** End hunger, achieve food security and improved nutrition and promote sustainable agriculture (*to which* ***we*** *would add* **'and aquaculture'**) [https://sdgs.un.org/goals/goal2].
- **Goal 3** Ensure healthy lives and promote well-being for all at all ages [https://sdgs.un.org/goals/goal3].
- **Goal 12** Ensure sustainable consumption and production patterns [https://sdgs.un.org/goals/goal12].
- **Goal 13** Take urgent action to combat climate change and its impacts [https://sdgs.un.org/goals/goal13].

Fig. 4.23 We support the sustainable development goals of the United Nations

- **Goal 14** Conserve and sustainably use the oceans, seas and marine resources for sustainable development [https://sdgs.un.org/goals/goal14].

References

Almeida F (2006) Ilhas oceânicas brasileiras e suas relações com a tectônica atlântica [The Brazilian oceanic islands and relationships with the Atlantic tectonics]. *Terrae Didatia* 2:3–18. https://doi.org/10.20396/td.v2i1.8637462

Alonso AA, Álvarez-Salgado XA, Antelo LT (2021) Assessing the impact of bivalve aquaculture on the carbon circular economy. J Cleaner Prod 279:123873. https://doi.org/10.1016/j.jclepro.2020.123873

Alves RJV (1998) *Ilha da Trindade & Arquipélago de Martin Vaz: Um Ensaio Geobotânico*. Rio de Janeiro: Serviço de Documentação da Marinha, Brasil. pp 144. ISBN: 8570470549. https://www.marinha.mil.br/secirm/sites/www.marinha.mil.br.secirm/files/documentos/protrindade/producao/ensaio-geobotanico.pdf

Aoun ZB, Said RB, Farhat F (2010) Anti-inflammatory, antioxidant and antimicrobial activities of aqueous and organic extracts from *Dictyopteris membranacea*. Bot Mar 53:259–264. https://doi.org/10.1515/BOT.2010.027

Aypa SM (1990) Mussel culture. Chapter 4 In: Selected papers on mollusc culture, Regional Seafarming Resources Atlas: Volume II, SF/WP/90/2, (ed A. Lovatelli). UNDP/FAO Regional Seafarming Development and Demonstration Project (RAS/90/002), National Inland Fisheries Institute, Kasetsart University Campus, Bangkok, Thailand. http://www.fao.org/3/AB737E/AB737E04.htm

Bernasconi AA, Wiest MM, Lavie CJ, Milani RV, Laukkanen JA (2020) Effect of omega-3 dosage on cardiovascular outcomes: an updated meta-analysis and meta-regression of interventional trials. Mayo Clinic Proceedings, online ahead of publication. pp 10. https://doi.org/10.1016/j.mayocp.2020.08.034

Berners-Lee M (2019) There Is No Planet B: A Handbook for the Make or Break Years. Paperback, Cambridge University Press, UK, pp 302. ISBN-13: 978–1108439589

Bollier D, Helfrich S (eds) (2013) The wealth of the commons: a world beyond market & state. Levellers Press, Amherst, MA, USA, pp 442. ISBN: 978–1937146146. http://wealthofthecommons.org/home

Desjardins RL, Worth DE, Vergé XPC, Maxime D, Dyer J, Cerkowniak D (2012) Carbon footprint of beef cattle. Sustainability 4:3279–3301. https://doi.org/10.3390/su4123279

Escobar-Lux RH, Fields DM, Browman HI, Shema SD, Bjelland RM, Agnalt A-L, Skiftesvik AB, Samuelsen OB, Durif CMF (2019) The effects of hydrogen peroxide on mortality, escape response, and oxygen consumption of *Calanus* spp. FACETS J 4:626–637. https://doi.org/10.1139/facets-2019-0011

Fan W, Zhang Z, Yao Z, Xiao C, Zhang Y, Zhang Y, Liu J, Di Y, Chen Y, Pan Y (2020). A sea trial of enhancing carbon removal from Chinese coastal waters by stimulating seaweed cultivation through artificial upwelling. Appl Ocean Res 101(102260). https://doi.org/10.1016/j.apor.2020.102260

FAO (2020) The state of world fisheries and aquaculture 2020: sustainability in action. Published by the Food and Agriculture Organization of the United Nations, Rome, open access under licence CC BY-NC-SA 3.0 IGO (https://creativecommons.org/licenses/by-nc-sa/3.0/igo). ISBN 9789251326923. www.fao.org/3/ca9229en/ca9229en.pdf

Filgueira R, Comeau L, Guyondet T, Mckindsey C, Byron C (2015) Modelling carrying capacity of bivalve aquaculture: a review of definitions and methods. In: Meyers RA (ed) Encyclopedia of sustainability science and technology. Springer Science+Business Media, New York, USA. ISBN: 9781493924936. https://doi.org/10.1007/978-1-4939-2493-6_945-1

Fitzer SC, Chung P, Maccherozzi F, Dhesi SS, Kamenos NA, Phoenix VR, Cusack M (2016) Biomineral shell formation under ocean acidification: a shift from order to chaos. Sci Reports 6(21076). https://doi.org/10.1038/srep21076

Franz DR (1993) Allometry of shell and body weight in relation to shore level in the intertidal bivalve *Geukensia demissa* (*Bivalvia: Mytilidae*). J Exp Mar Biol Ecol 174:193–207. https://doi.org/10.1016/0022-0981(93)90017-I

Fuentes-Santos I, Labarta U, Álvarez-Salgado XA (2019) Modelling mussel shell and flesh growth using a dynamic net production approach. Aquaculture 506:84–93. https://doi.org/10.1016/j.aquaculture.2019.03.030

Gammone MA, Riccioni G, Parrinello G, D'Orazio N (2019) Omega-3 polyunsaturated fatty acids: benefits and endpoints in sport. Nutrients 11(46). https://doi.org/10.3390/nu11010046

Geraldes MC, Motoki A, Costa A, Mota CE, Mohriak WU (2013) Geochronology (Ar/Ar and K-Ar) of the South Atlantic post-break-up magmatism. Geological Society, London, Special Publications 369:41–74. https://doi.org/10.1144/SP369.21

Gimin R, Mohan R, Thinh LV, Griffiths AD (2004) The relationship of shell dimensions and shell volume to live weight and soft tissue weight in the mangrove clam, *Polymesoda erosa* (Solander, 1786) from northern Australia. NAGA, WorldFish Center Quarterly, 27(3 & 4). https://digitalarchive.worldfishcenter.org/bitstream/handle/20.500.12348/2026/na_2318.pdf?sequence1=

Helm MM, Bourne N, Lovatelli A (2004) Hatchery culture of bivalves. A practical manual. FAO Fisheries Technical Paper. No. 471, pp 203. Rome: Food and Agriculture Organization of the United Nations. ISBN 9251052247. http://www.fao.org/3/y5720e/y5720e02.htm. PDF: https://www.fao.org/documents/pub_dett.asp?pub_id=166536

Hilbish TJ (1986) Growth trajectories of shell and soft tissue in bivalves: seasonal variation in *Mytilus edulis* L. J Exp Mar Biol Ecol 96:103–113. https://doi.org/10.1016/0022-0981(86)90236-4

Holmgren D (2011) Permaculture principles & pathways beyond sustainability. 2nd edn. Paperback, Published by: Permanent Publications: Hampshire, UK, pp 320. ISBN: 978–1856230520

Hylén A, Taylor D, Kononets M, Lindegarth M, Stedt A, Bonaglia S, Bergström P (2021) *In situ* characterization of benthic fluxes and denitrification efficiency in a newly re-established mussel farm. Sci Total Environ 782(146853). https://doi.org/10.1016/j.scitotenv.2021.146853

Karan D, Dubey S, Pirisi L, Nagel A, Pina I, Choo Y-M, Hamann MT (2020) The marine natural product manzamine A inhibits cervical cancer by targeting the SIX1 protein. J Nat Prod 83:286–295. https://doi.org/10.1021/acs.jnatprod.9b00577

Knights AM, Norton MJ, Lemasson AJ, Stephen N (2020) Ocean acidification mitigates the negative effects of increased sea temperatures on the biomineralization and crystalline ultrastructure of *Mytilus*. Front Mar Sci 7(567228). https://doi.org/10.3389/fmars.2020.567228

Knowler D, Chopin T, Martínez-Espiñeira R, Neori A, Nobre A, Noce A, Reid G (2020) The economics of integrated multi-trophic aquaculture: where are we now and where do we need to go? Rev Aquac 12:1579–1594. https://doi.org/10.1111/raq.12399

Le Quéré C, Andrew RM, Friedlingstein P, Sitch S, Pongratz J, 72 others. (2018) Global carbon budget 2017. Earth Syst Sci Data 10:405–448. https://doi.org/10.5194/essd-10-405-2018

Lin LC, Kuo TT, Chang HY, Liu WS, Hsia SM, Huang TC (2018) Manzamine A exerts anticancer activity against human colorectal cancer cells. Mar Drugs 16(252). https://doi.org/10.3390/md16080252.

Lovatelli, A. (1990). Artificial propagation of bivalves: techniques and methods. Technical paper prepared as part of the Mollusc-Shellfish Culture Course of the 7th NACA Training Course for Senior Aquaculturists in Asia and in the Pacific. FAO/UNDP Regional Seafarming Demonstration and Development Project (RAS/90/002) Bangkok, Thailand: National Inland Fisheries Institute, Kasetsart. http://www.fao.org/3/AB739E/AB739E00.htm

Lubiana KMF, Dias C (2016) The composition and new records of micro- and mesophytoplankton near the Vitória-Trindade Seamount Chain. Biota Neotrop 16. https://doi.org/10.1590/1676-0611-BN-2016-0164

Maruyama S, Behnia M, Chisaki M, Kogawa T, Okajima J, Komiya A (2013) Large eddy simulation of the diffusion process of nutrient-richup-welled seawater. Front Heat Mass Transf 4(023002). https://doi.org/10.5098/hmt.v4.2.3002

Maruyama S, Tsubaki K, Taira K, Sakai S (2004) Artificial upwelling of deep seawater using the perpetual salt fountain for cultivation of ocean desert. J Oceanogr 60:563–568. https://doi.org/10.1023/B:JOCE.0000038349.56399.09

Maruyama S, Yabuki T, Sato T, Tsubaki K, Komiya A, Watanabe M, Kawamura H, Tsukamoto K (2011) Evidences of increasing primary production in the ocean by Stommel's perpetual salt fountain. Deep Sea Res Part I 58:567–574. https://doi.org/10.1016/j.dsr.2011.02.012

Marza E, Long T, Saiardi A, Sumakovic M, Eimer S, Hall DH, Lesa GM (2008) Polyunsaturated fatty acids influence synaptojanin localization to regulate synaptic vesicle recycling. Mol Biol Cell 19:833–842. https://doi.org/10.1091/mbc.e07-07-0719

Mascarenhas V, Keck T (2018) Marine optics and ocean color remote sensing. In: Jungblut S, Liebich V, Bode-Dalby M (eds) YOUMARES 8—Oceans across boundaries: learning from each other, Proceedings of the 2017 conference for YOUng MARine RESearchers, Kiel, Germany. pp 41–54. Open access. Switzerland: Springer Nature. ISBN: 978–3030066307. https://doi.org/10.1007/978-3-319-93284-2_4.

Meirelles PM, Amado-Filho GM, Pereira-Filho GH, Pinheiro HT, de Moura RL, Joyeux J-C, Mazzei EF, Bastos AC, Edwards RA, Dinsdale E, Paranhos R, Santos EO, Iida T, Gotoh K, Nakamura S, Sawabe T, Rezende CE, Gadelha LMR, Francini-Filho RB,

Thompson C, Thompson FL (2015) Baseline assessment of mesophotic reefs of the Vitória-Trindade Seamount chain based on water quality, microbial diversity, benthic cover and fish biomass data. PLoS ONE 10(e0130084). https://doi.org/10.1371/journal.pone.0130084

Milhazes-Cunha H, Otero A (2017) Valorisation of aquaculture effluents with microalgae: the Integrated multi-trophic aquaculture concept. Algal Res 24:416–424. https://doi.org/10.1016/j.algal.2016.12.011

Mohriak W (2020) Genesis and evolution of the South Atlantic volcanic islands offshore Brazil. Geo-Mar Lett 40:1–33. https://doi.org/10.1007/s00367-019-00631-w

Moore D (2020) A biotechnological expansion of shellfish cultivation could permanently remove carbon dioxide from the atmosphere/Una ampliación biotecnológica del cultivo de moluscos bivalvos podría eliminar permanentemente el dióxido de carbono de la atmósfera. Mex J Biotechnol 5:1–10. https://doi.org/10.29267/mxjb.2020.5.1.1

Moore D (2021).Saving the planet with appropriate biotechnology: 4. Coccolithophore cultivation and deployment/Salvando el planeta con biotecnología apropiada: 4. Cultivo de cocolitóforos e implementación. Mex J Biotechnol 6:129–155. https://doi.org/10.29267/mxjb.2021.6.1.129

Moore D, Heilweck M, Petros P (2021a) Saving the planet with appropriate biotechnology: 1. Diagnosing the Problems/Salvando el planeta con biotecnología apropiada: 1. Diagnóstico de los problemas. Mex J Biotechnol 6:1–30. https://doi.org/10.29267/mxjb.2021.6.1.1

Moore D, Heilweck M, Petros P (2021b) Saving the planet with appropriate biotechnology: 2. Cultivate Shellfish to Remediate the Atmosphere/Salvando el planeta con biotecnología apropiada: 2. Cultivar mariscos para remediar la atmósfera. Mex J Biotechnol 6:31–91. https://doi.org/10.29267/mxjb.2021.6.1.31

Motoki A, Motoki KF, Melo DP (2012) Caracterização da morfológia submarina da cadeia Vitória-Trindade e áreas adjascentes-ES, com base na batimetria predi-tada do topo versão 14.1 [Submarine morphology characterization of the Vitória-Trindade chain and the adjacent areas, State of Espírito Santo, Brazil, based on the predicted bathymetry of the topo version 14.1]. *Revista Brasileira de Geomorfologia* 13:151–170. https://tinyurl.com/4393d426

Olsen Y (2002) MARICULT research programme: background, status and main conclusions. Hydrobiologia 484:1–10. https://doi.org/10.1023/A:1021376517916

Opsahl-FerstadH-G, Rudi H, Ruyter B, Refstie S (2003) Biotechnological approaches to modify rapeseed oil composition for applications in aquaculture. Plant Sci 165:349–357. https://doi.org/10.1016/S0168-9452(03)00194-8

Pauly D, Christensen V, Dalsgaard J, Froese R, Torres F (1998) Fishing down marine food webs. Science 279:860–863. https://doi.org/10.1126/science.279.5352.860

Pinheiro HT, Mazzei E, Moura RL, Amado-Filho GM, Carvalho-Filho A, Braga AC, Costa PAS, Ferreira BP, Ferreira CEL, Floeter SR, Francini-Filho RB, Gasparini JL, Macieira RM, Martins AS, Olavo G, Pimentel CR, Rocha LA, Sazima I, Simon T, Teixeira JB, Xavier LB, Joyeux JC (2015) Fish biodiversity of the Vitória-Trindade seamount chain, Southwestern Atlantic: an updated database. PLoS ONE 10(e0118180). https://doi.org/10.1371/journal.pone.0118180

Pugh TAM, Lindeskog M, Smith B, Poulter B, Arneth A, Haverd V, Calle L (2019) Role of forest regrowth in global carbon sink dynamics. Proc Natl Acad Sci USA 116:4382–4387. https://doi.org/10.1073/pnas.1810512116

Richards KR, Stokes C (2004) A review of forest carbon sequestration cost studies: a dozen years of research. Clim Change 63:1–48. https://doi.org/10.1023/B:CLIM.0000018503.10080.89

Riebeek, H. (2011). The carbon cycle. NASA: earth observatory. http://earthobservatory.nasa.gov/Features/CarbonCycle/?src=eoa-features

Rotz CA, Asem-Hiablie S, Dillon J, Bonifacio H (2015) Cradle-to-farm gate environmental footprints of beef cattle production in Kansas, Oklahoma, and Texas. J Anim Sci 93:2509–2519. https://doi.org/10.2527/jas.2014-8809

Ruddiman WF (2014) Earth's climate: past and future, 3rd edn. W.H. Freeman & Co.: NY, USA, pp 464. ISBN: 978–1319154004

Ruddle K, Zhong G (1988) Integrated agriculture-aquaculture in South China. The Dike-Pond system of the Zhujiang delta. Cambridge, UK: Cambridge University Press. pp 192. ISBN: 9780521341936. https://www.researchgate.net/publication/258212339_Integrated_Agriculture-Aquaculture_in_South_China_The_Dike-Pond_System_of_the_Zhujiang_Delta

Ruyter B, Sissener N, Østbye T, Simon C, Krasnov A, Bou M, Sanden M, Nichols PD, Lutfi E, Berge GM (2019) N-3 Canola oil effectively replaces fish oil as a new safe dietary source of DHA in feed for juvenile Atlantic salmon. Br J Nutr 122:1329–1345. https://doi.org/10.1017/S0007114519002356

Sala E, Mayorga J, Bradley D, Cabral RB, Atwood TB et al (2021). Protecting the global ocean for biodiversity, food and climate. Nature (2021). https://doi.org/10.1038/s41586-021-03371-z

Scholten H, Smaal AC (1998) Responses of *Mytilus edulis* L. to varying food concentrations: testing EMMY, an ecophysiological model. J Exp Mar Biol Ecol 219:217–239. https://doi.org/10.1016/S0022-0981(97)00182-2

Scholten H, Smaal AC (1999) The ecophysiological response of mussels (*Mytilus edulis*) in mesocosms to a range of inorganic nutrient loads: simulations with the model EMMY. Aquat Ecol 33:83–100. https://doi.org/10.1023/A:1009995823741

Seed R (1968) Factors influencing shell shape in the mussel *Mytilus edulis*. J Mar Biol Assoc UK 48:561–584. https://doi.org/10.1017/S0025315400019159

Seed R (1980) Shell growth and form in the Bivalvia. In: DC Rhoads (ed) Skeletal growth of aquatic organisms: biological records of environmental change, pp 23–68. New York: Springer Science+Business Media. pp 750. ISBN 9780306402593

Shepard M (2013) Restoration agriculture: real world permaculture for farmers. Paperback, pp 339. Published by Acres U.S.A. Inc.: Austin TX, USA. ISBN: 978–1601730350

Stommel H, Arons AB, Blanchard D (1956) An ocean curiosity: the perpetual salt fountain. Deep-Sea Res 3:152–153. https://doi.org/10.1016/0146-6313(56)90095-8

Telesca L, Michalek K, Sanders T, Peck LS, Thyrring J, Harper EM (2018) Blue mussel shell shape plasticity and natural environments: a quantitative approach. Sci Rep 8(2865). https://doi.org/10.1038/s41598-018-20122-9

Toochi EC (2018) Carbon sequestration: how much can forestry sequester CO_2? Forestry Res Eng Int J 2:148–150. https://doi.org/10.15406/freij.2018.02.00040

Utting SD, Spencer BE (1991) The hatchery culture of bivalve mollusc larvae and juveniles. Laboratory Leaflet Number 68. pp 32. Lowestoft, UK: Ministry of Agriculture, Fisheries and Food, Directorate of Fisheries Research. Download PDF: https://www.cefas.co.uk/publications/lableaflets/lableaflet68.pdf

van Beijnen J, Yan G (2021) The multi-trophic revolution: a deep dive with IMTA guru Thierry Chopin. Article on *The Fish Site website*. https://thefishsite.com/articles/the-multi-trophic-revolution-a-deep-dive-with-imta-guru-thierry-chopin

Wiesebron ML (2011) L'Amazonie bleue et les enjeux et contraintes maritimes du Brésil. *La Revue Défense Nationale*, Revue n° 738:45–51. https://www.seer.ufrgs.br/austral/article/viewFile/35039/23931&lang=en

Wiesebron ML (2013) Blue Amazon: thinking the defense of Brazilian maritime territory. AUSTRAL Braz J Strategy Int Relat 2:101–124. https://seer.ufrgs.br/index.php/austral/article/viewFile/35039/23931

Yang H (1994) Integrated fish farming. In: S Li, J Mathias (eds) Freshwater fish culture in China: principles and practice. Developments in Aquaculture and Fisheries Science, vol 28. Amsterdam, Elsevier, pp 219–270. ISBN: 9780444888822

Zhang Y, Zhang J, Liang Y, Li H, Li G, Chen X, Zhao P, Jiang Z, Zou D, Liu X, Liu J (2017) Carbon sequestration processes and mechanisms in coastal mariculture environments in China. Sci China Earth Sci 60:2097–2107. https://doi.org/10.1007/s11430-017-9148-7

Farming Giant Clams in 2021: A Great Future for the 'Blue Economy' of Tropical Islands

5

David Moore

5.1 In this Chapter...

A specific and dramatic example for the tropics is detailed to avoid too much attention being diverted to Northern Hemisphere shellfish cultivation. There is nothing more dramatic than Giant Clams, which have been fished out to extinction in many Indian Ocean and Pacific waters, but elsewhere contribute to a still thriving industry, though clam dredging is now doing immense damage to coral reefs in many areas. The topic of giant clam cultivation covers conservation and restocking of clams, but with the potential bonus of rehabilitating coral reefs degraded by bleaching induced by climate change, as well as food production, and development of remunerative local industry for local Pacific Island communities. It's not just the food value of the animal; the shells are used for carving (large!) ornaments, and several species are traded around the world for marine aquariums. Work towards 'seeding' and recolonising has been going on in the Pacific region for more than 30 years. Much of this work has been published and many of the faults in approach and problems of governance identified. In addition, though, several local enterprises have developed methods to produce economically large numbers of young giant clams for restocking tropical seas. The conservation and educational programmes that have resulted deserve wider attention and greater investment as they tie-in well with our call to 'cultivate shellfish to remediate the atmosphere'.

5.2 Introducing Giant Clams

So far, our **Chapters** have instanced Northern Hemisphere shellfish cultivation, but here I want to detail a dramatic example for the tropics. There is nothing more dramatic than Giant Clams, which have been fished out to extinction in many Pacific and Indian Ocean waters, a practice which has done immense damage to coral reefs in many areas. Several replenishment efforts have made successful contributions both to a still thriving clam fishing industry and to coral reef conservation and restoration in general.

There are a number of large clam species which are native to the shallow coral reefs of the South Pacific and Indian oceans, the South China Sea and the shores of the Philippines and Borneo. They have a long history of traditional and cultural use in the region and their European documentation goes back to 1521 when the chronicler of Ferdinand Magellan's circumnavigation, the Venetian scholar Antonio Pigafetta, documented Giant Clams in his journal.

Giant Clams have always been, and remain, an important resource throughout the tropical Indo-Pacific region, from Mauritius to the South China Sea. Traditionally, the meat of the animals has been an important subsistence food and the shell has been used to make dishes, tools, jewellery and ornaments (Ellis 1997; Morris et al. 2019; Fig. 5.1).

D. Moore et al., *Aquaculture: Ocean Blue Carbon Meets UN-SDGS*,
Sustainable Development Goals Series, https://doi.org/10.1007/978-3-030-94846-7_5

Fig. 5.1 Exquisitely carved giant clam shells (*Tridacna gigas and T. derasa*) produced by numerous small workshops in the city of Tanmen (Hainan, China). This industry is based on old discarded giant-clam shells recovered by fishermen from the seabed of the South China Sea and left there several decades ago by poachers harvesting clams for their adductor muscle destined to the Taiwan seafood market. The shells, of course, are solidified atmospheric carbon dioxide. Image from The Tridacna Mariculture Development Center website [http://lagoonclams.com/]

Markets for these animals have changed a great deal more recently but they are still heavily harvested. The meat is widely sold in Asian and Pacific markets as a delicacy, rather than a staple food, so it maintains a premium price level. Chinese Traditional Medicine believes that the clam's adductor muscle has aphrodisiac powers. The most recent use for the more brightly coloured species of giant clam is as a living decoration in home and public marine aquariums (Fig. 5.2).

Fig. 5.2 Brightly coloured giant clams wanted by the aquarium industry as living decoration in home and public marine aquaria. Image from The Tridacna Mariculture Development Center website [http://lagoonclams.com/]

5.3 Biology and History of Giant Clams

The first bivalve molluscs occur in the Early Lower Cambrian fossil record, about 500 million years ago. Their diversification, both taxonomic and ecological, surges in the fossil record from about 450 million years ago to the present day. They are today the second most diverse group of molluscs on the planet (second only to gastropods, the snails and slugs) with well over 10,000 described species of bivalve, living as important members of most marine and freshwater ecosystems. Bivalves are characterised by their two-halved shell. They can live on the ocean floor or burrow into seafloor sediment (or, indeed, into wet timber, like *Teredo navalis*, the naval shipworm, which is a marine bivalve mollusc, not a worm). Some (like the scallops, Family *Pectinidae*) can even swim through the water by snapping their shell open and shut. A few bivalves have evolved a reduced shell or have completely lost the shell [https://ucmp.berkeley.edu/taxa/inverts/mollusca/bivalvia.php].

Giant clams (Phylum *Mollusca*: Class *Bivalvia*: Family *Tridacnidae*) are the largest marine bivalve molluscs of the present day. They are found in coastal areas of the Indo-Pacific region; ranging eastwards from Cape Agulhas, the southern tip of the African continent and the dividing line between the Atlantic and Indian Oceans (Syukri bin Othman et al. 2010).

Giant bivalves occur in the fossil record. The largest found to date is a specimen of *Inoceramus* measuring 187 cm (74 inches) across its longest diameter [https://en.wikipedia.org/wiki/Inoceramus]. This extinct bivalve resembles the pearly oysters of the present-day genus *Pteria*. *Inoceramus* lived from the Early Jurassic (200 million years ago) to close to the end of the Cretaceous (about 66 million years ago). Another interesting group of extinct bivalves, the rudists (Order *Rudista* or *Hippuritida*) also became extinct at about the same time (Johnson 2002). Some of these bivalves were also very large. They arose during the Late Jurassic (about 150 million years ago), diversifying during the Cretaceous to become the major reef-building organisms in the Tethys Ocean, which, during the Mesozoic Era (251 to 65.5 million years ago), separated the supercontinent of Laurasia in the north (consisting of today's North America and Eurasia north of the Alpine-Himalayan mountain ranges), from Gondwana in the south (consisting of today's South America, Africa, India, Australia, Antarctica and Eurasia south of the Alpine-Himalayan mountains). Today, their fossils are found throughout in the Mediterranean region, the Middle East, the Caribbean and Southeast Asia.

It is thought that a decline of species of these giant bivalves is seen just ***before*** the asteroid collision that caused the Chicxulub crater in Mexico and is blamed for the extinction of the dinosaurs (among other extinctions). The final mass extinction of all rudist species was caused by the drastic environmental impact of this asteroid collision, however, the earlier decline of giant bivalve species has been ascribed to temperature changes in the then existing tropical ecosystems (Arthur et al. 1996) with this quotation:

> … With simultaneous poleward movement of surface and subsurface waters on sea-level highstands, the superheated middle Cretaceous tropics cooled, the reef line contracted, diversity decreased, and reef ecosystems collapsed, leading to mass extinction … (Arthur et al. 1996).

With the proviso that this quotation describes a cooling rather than warming of the ocean water, notice anything familiar in that melancholy description?

In the present day, eight species of giant clam of varying size and habitat preference have been described (*Tridacna gigas*, *T. derasa*, *T. squamosa*, *T. maxima*, *T. crocea*, *T. tevora*, *Hippopus hippopus* and *H. porcellanus*). A ninth species, *Tridacna rosewateri* has been described more recently and is endemic to Mauritius.

T. maxima and *T. crocea* are smaller but more colourful clams, and are found within limestone substrata; free-living species (*T. squamosa*, *T. derasa* and *T. gigas*) are larger and usually occur near reefs or on sandy sea bottoms. Similarly, *Hippopus* spp. are usually found in seagrass beds. All of these bivalves are unusual in that their mantle tissues act as a habitat for symbiotic single-celled dinoflagellate algae (zooxanthellae; *Symbiodinium* spp.) from which the adult clams get most of their nutrition (although the clams are also filter feeders). By day, the clam opens its shell and extends its mantle tissue so that the algae receive the sunlight they need to photosynthesise; the animal benefits from the products of photosynthesis. Giant clams are a highly prized food source, and both subsistence fishing and commercial fishing, which exports clam meat to many Asian markets, have been responsible for stock depletion across their range. The clams are also harvested for their shells and for live export for the marine aquarium trade (Teitelbaum and Friedman 2008).

Tridacna gigas, as its specific name indicates, is the species which truly merits the description 'Giant'; these are very large animals (Fig. 5.3). According to the website of the Mongabay Conservation and Environmental Science News Service, the world's biggest specimen of *Tridacna gigas* was found off the coast of Sumatra. The two shells of this specimen weighed 230 kg (rather more than an average female grizzly bear) and measured 1.4 m across [https://news.mongabay.com/2012/06/forgotten-species-the-wonder-inducing-giant-clam/].

The other species are smaller than *T. gigas*, but still large enough to be called giants. With their large size and impressive mantle colours, ranging from electric blue, through green, pink, and purple to gold, giant clams have been described as 'charismatic megafauna' that act as ***flagship taxa***, which, unfortunately, serve to direct attention to the continuing destruction of coral reefs (Soo and Todd 2014).

It is likely that clam harvesting by subsistence fishing was sustainable for centuries until

Fig. 5.3 *Tridacna gigas*, as its specific name indicates, is the species which truly merits the description '***Giant***'. Here, a young islander, is determined not to allow his find to escape from Fizroy Island, Australia. Image from The Tridacna Mariculture Development Center website [http://lagoonclams.com/]

commercial fishers took the animals from the waters to satisfy an increasing demand for them as food and crafts, but also for the aquarium trade and traditional medicine. So, by the end of the twentieth century, many giant clam species had become overharvested and were even locally extinct in some regions.

Frias-Torres (2017) describes how giant clam populations have been depleted due to overfishing for meat, shells and the aquarium trade, and how eutrophication and reef degradation contribute to their decline in spite of local management efforts, including mariculture and restocking. She goes on to state:

> "… Recently, bans on the elephant ivory trade have increased giant clam fishing, so they fulfill the demand in Asian markets for jewelry and house ornaments (clam shells) and aphrodisiacs (clam meat) … Increased giant clam shell trade and poaching also result in widespread coral reef destruction in the South China Sea, as poachers use boat propellers to loosen the giant clams, dragging them through the reef and carving up long stretches of lifeless rubble …" For references to all these assertions, see Frias-Torres (2017).

Today, four of the giant clam species are listed as Vulnerable by the **IUCN Red List** (view: https://www.iucnredlist.org/), including *Tridacna gigas*, while others are listed as Lower Risk or Conservation Dependent with commercial trade regulated by the Convention on International Trade in Endangered Species of Wild Fauna and Flora, see Appendix II (CITES 2013). *Hippopus hippopus* is also in severe danger because it is often harvested for its decorative shell as well as its meat. The problem is the extremely slow growth and reproduction rate of this species, which together mean that stocks can take several decades to recover from a single harvest. Not surprisingly, the species has been overfished in many countries (Wells 1996).

Conservationists have responded to this decline in the giant clam populations with research and conservation efforts. Over the past 40 years or so, scientists from all over the Indo-Pacific region have closely studied giant clam biology and cultivation methods to support programmes for giant clam restocking, conservation, farming, public education, and, because giant clams live in close association with coral reefs throughout the Indo-Pacific, the management and conservation of coral reefs. However, even if the conservation succeeds in protecting giant clams, climate change is the new threat, as it is elsewhere, and could have long-term impacts on the giant clam family. I will describe some of this work in the next sections, but first there are three aspects of recently studied giant clam general biology that contribute to their husbandry. These are: giant clam ***behaviours***, their ***ecological roles*** in coral reef ecosystems, and their ***distributions and population sizes*** in their habitats. Better understanding of these aspects of giant clam biology will greatly aid recovery of depleted clam populations, conservation and sustainable exploitation.

First, despite the size and apparent immobility of mature clams, the animals exhibit a varied repertoire of **behaviours** during their life cycle. Soo and Todd (2014) reviewed close to 100 papers that included behavioural observations published between 1865 and 2014, which they sorted into the following four general themes.

- ***Spawning***, giant clams display diel (day + night 24-h periodicity) and lunar periodicity in reproduction, and for some species, peak breeding seasons have been recognised.
- ***Locomotion***, giant clam larvae have considerable mobility, ranging from swimming and gliding; juveniles and adults are able to crawl across their substratum. The animals are chemotactic (moving in response to chemical signals) and gravitactic (moving in response to the gravity vector). Giant clams are not phototactic (no movement towards or away from a source of light). At least one species exhibits clumping behaviour, which may enhance physical stability, assist reproduction or provide protection from predators (the 'selfish herd' [https://en.wikipedia.org/wiki/Selfish_herd_theory]).
- ***Feeding***, giant clams go through several changes in mode of nutrition; starting with gaining nourishment from the yolk originally contained within its egg (= lecithotrophic) and feeding on plankton (planktotrophic) as larvae, switching to pedal feeding after metamorphosis (in which pulsations of the animal's foot in the juvenile bivalve drives the flow of water through the shell) followed by the transition to a dual mode of filter feeding (water is drawn over the gills through the beating of cilia) and phototrophy (using the products of their symbiont's photosynthesis) once symbiosis with zooxanthellae (*Symbiodinium* spp.) is established.
- ***Anti-predation and Stress Responses***. Adult giant clams cannot escape rapidly from threats using locomotion. The young giant clam secretes byssus (proteinaceous filaments; also called 'sea silk' [https://en.wikipedia.org/wiki/Sea_silk]) to attach itself to a solid surface on the seabed, with the shell hinge pointing downwards. Subsequently, the byssus regresses and the animal rests on his own weight; over time, corals, sponges and algae surround it and contribute to its fixation. Their anti-predation behaviours include sudden contraction of the mantle, closing the valves of the shell together, and squirting water through the siphon that is normally used in water intake for filter feeding.

The **ecological contributions** of giant clams to coral reefs have been studied by Neo et al. (2015) using data from the literature and their own studies. They described the following.

- Giant clam tissues are food for a wide array of predators and scavengers, while their discharges of live symbiotic zooxanthellae, faeces and their own gametes are eaten by opportunistic feeders.
- Giant clams increase heterogeneity of the structure of the reef; act as reservoirs of zooxanthellae (*Symbiodinium* spp.), and their filter feeding activity potentially offsets any excessive growth of algae induced by eutrophication (excessive enrichment of the water with minerals and nutrients).
- The shells of giant clams provide substrate for colonisation by epibionts, while commensal and ectoparasitic organisms live within their mantle cavities.
- Finally, dense populations of giant clams produce large quantities of calcium carbonate shell material that is eventually incorporated into the reef framework.

Another important feature that has been reviewed are the population densities and distribution of giant clams (Syukri bin Othman et al. 2010). Combining density records from 15 countries, from both literature reports and reef-monitoring data. These authors showed that while some populations have giant clam densities in excess of 100 individuals per square ***metre*** (= 100 million individuals per km^2) the density more typically ranges from 10^{-3} to 10^{-5} individuals per square metre (= respectively, 1000 individuals per km^2 to 10 individuals per km^2). *Tridacna maxima* had the most cosmopolitan distribution; almost encompassing the entire geographical range of all the other giant clams species. In contrast, discovered *T. costata*, *T. rosewateri*, *T. teveroa* and *H. porcellanus* have the most restricted geographical ranges. Reef-monitoring data includes records of giant clams beyond previously defined geographical boundaries; extending their known occurrence slightly west to near Cape Agulhas, South Africa.

Syukri bin Othman et al. (2010) cite overfishing, habitat loss, pollution, and increases in sea surface temperatures as contributing to decreases in giant clam populations, and state that restocking efforts provide an opportunity to redress this decline. Though, without enhanced protection and enforcement, these efforts may be unsuccessful.

A more recent study of giant clam populations at Dongsha Atoll, the largest northernmost atoll of the South China Sea, examined the mitochondrial cytochrome oxidase subunit I (COI) gene sequences of *Tridacna maxima* and *T. noae*, to assess the genetic structure of their populations (Neo et al. 2018). They found that the four species assessed had an overall density of 3.14 per 100 m^2. This is approximately equal to 30,000 individuals per km^2, which puts this density firmly into the ***upper end*** of the range observed by Syukri bin Othman et al. (2010) in their study area. Neo et al. (2018) considered it likely that overharvesting had depleted populations of *T. squamosa* and *Hippopus hippopus*, and concluded that these species may not be reproductively viable on the Dongsha Atholl. On the other hand, they described populations of *T. maxima* and *T. noae* as:

> … thriving and replenished by recruits …", but these two species "… showed low levels of mitochondrial genetic diversity that could reduce adaptability and may become further impacted by exploitation and global warming.

Low genetic diversity is not often a recipe for evolutionary success, so this is a cause for concern even though these two species are **apparently** thriving, particularly as the other two species seem to be sliding down the tragic road to extinction. These findings must be used to inform the design of conservation strategies. Significantly, the haplotype networks for *T. maxima* and *T. noae*, which show the relationships between different haploid genotypes found in a dataset, showed population structuring correlated with geographic boundaries between Dongsha Atoll, Taiwan and the Philippines. So, there is genetic diversity to be found across the wider geographic region and this should be a crucial consideration in any planning for restocking these species.

Purcell et al. (2020) assessed abundances of four species of giant clams across 20 barrier reef and 30 lagoon reef sites across 600 km of coastline in New Caledonia (a French territory in the South Pacific) with emphasis on sustainability and protection strategies. *Tridacna maxima* was the most common species, though *T. derasa* was significantly more abundant in marine reserves than at sites open to fishing. There was no such effect of marine reserves on total abundance when data for all four species were pooled. *T. squamosa* was significantly more abundant on the barrier reef (which surrounds the main island, Grand Terre, and is a major scuba-diving destination), while *Hippopus hippopus* was found only on lagoon reefs.

- Implementation of daily commercial catch limits to five giant clams per vessel per trip resulted in a marked reduction in catches of giant clam.
- However, despite daily limits of two giant clams per recreational fishing trip, the recreational harvest was much greater than the commercial catch.
- The effectiveness of marine reserves on giant clam conservation is species specific, and further fishery restrictions applying to the less common species might be needed to ensure their persistence in the Indo-Pacific.

5.4 Giant Clams and Coral Reefs

There has been only limited success for programmes aimed at restocking giant clams by replacing clams in fished-out coastal environments. Projects have been carried out in Australia, Philippines and the Pacific Islands (Palau, Solomon Islands, Vanuatu, Tonga, Marshall

Islands and Cook Islands). Teitelbaum and Friedman (2008) conclude that most restocking projects have been only partially successful. The reasons for these mixed results include the following.

- The high costs and lengths of time required to produce 'seed' clams have been problems for many operations. High mortality of juvenile clams has also reduced success rates.
- In the initial stages, lack of knowledge about rearing and growing clams was a problem for many of the participating countries.
- Lack of consistently committed involvement of local communities in the projects. In some cases, projects were not matched to what the local community needed or wanted.
- Poor survey and reporting protocols, together with poor funding for monitoring, have limited assessment of some reintroduction and restocking programmes even to the point of failing to report successful results.

Coral reef bleaching, resulting from increasing water temperatures and water acidity (both generated by the anthropogenic increase in atmospheric carbon dioxide) is the reef ailment that is most commonly reported in the public media. But these seawater changes have also been demonstrated to have adverse effects on at least two species of giant clams, *Tridacna squamosa* and *T. crocea* in Thailand (Junchompoo et al. 2013). This study showed that exposure to temperatures over 30ºC for longer than two weeks could result in the expulsion of the symbiotic living zooxanthellae from the mantles. As these dinoflagellates make such a significant contribution to the nutrition of their host, their loss must have a direct impact on survival of the giant clam. However, giant clam survivors located in shallow waters did re-colour, indicating that they could regain the symbionts. Indeed, Morishima et al. (2019) have shown that zooxanthellae expelled in the fecal pellets of *Tridacna crocea* are viable and able to infect *T. squamosa* juveniles. Although thermally stressed coral expel partially digested zooxanthellae (Fujise et al. 2014), giant clams expel live and active zooxanthellae with no signs of digestion in their faeces, leading to the suggestion that giant clam fecal pellets are a ***source*** of zooxanthellae in coral reefs (Morishima et al. 2019).

Ramah et al. (2018) have investigated population declines between 1999 and 2016 in the giant clams *Tridacna maxima* and *T. squamosa* at seven reef sites around the island of Mauritius (three marine protected areas (MPAs) and four non-MPAs) in the western Indian Ocean. Their data revealed significant decline in the population densities of both species from 1999 to 2016 at all seven sites. That significant decline in densities occurred irrespective of protection levels at the sites surveyed indicates that overfishing is not the sole cause of giant clam losses and implies that seawater changes resulting from global warming may have contributed. But most coastal nations contain priority areas that can contribute substantially to achieving the three objectives of biodiversity protection, food provision and carbon storage (Sala et al. 2021).

There are several examples of the use of mariculture linked to restoration (restocking) of reefs being attempted as a solution to reverse giant clam extinctions locally. Some captive bred giant clam restoration efforts with *Tridacna maxima* (Waters et al. 2013), *T. derasa* (Heslinga et al. 1984), *T. squamosa* (Guest et al. 2008) and *T. gigas* (Gomez and Mingoa-Licuanan2006) have focused on juvenile individuals, demonstrating a relationship between size at transplant and mortality. Escape from predation in natural habitats requires that the juvenile exceeds a minimum size, resulting in variability of survival depending on size at transplant. In the most extreme cases, the mortality rate could be close to 100%. However, despite these early failures, it was concluded nearly 40 years ago (in an article that describes methods developed over a 4-year period at the *Micronesian Mariculture Demonstration Center* in Palau for mass culture of tridacnid clams from egg to maturity) that:

> … no serious biotechnical constraints would prevent commercial or subsistence farming of these autotrophic animals in the Indo-West Pacific … (Heslinga et al. 1984).

For over 30 years, the Marine Science Institute (MSI) at the University of the Philippines' Bolinao Marine Laboratory in the Province of Pangasinan, northern Philippines, has also cultured giant clams with the intention of restoring depleted populations, basing their approach on the protocols developed by Braley (1992a). Restocking activities were done in collaboration with local groups by providing training in the culture and ocean rearing of giant clams, promoting giant clam farming as a sustainable livelihood (Gomez and Mingoa-Licuanan2006; Gomez et al. 2006; Mingoa-Licuanan and Gomez 2007). These authors emphasise the care that needs to be given to:

- selection of release sites;
- arrangements with participating community groups to safeguard the released clams;
- transfer of technology to collaborators;
- transport procedures for large clams from nursery areas to release sites (and see Braley 1992b).

Gomez and Mingoa-Licuanan (2006) also report their experience of testing the viability of supplying giant clams for the aquarium trade to create new sources of income for local fishers. Global demand for the aquarium trade was about 200,000 pieces in the year 2007 and 69,000 pieces of the demand was exported from the Pacific region (Ponia 2010). Unfortunately, this initiative encountered legal obstacles when the government regulatory agency prohibited the export of ***cultured*** clams, regarding this as a threat to the conservation of wild individuals. Clearly, as well as everything else, there needs to be better understanding between regulatory authorities and the cultivation and conservation communities.

Widening the conservation goals towards improving biodiversity and productivity of stressed coral reef habitats by coral transplantation and giant clam restocking would be a contribution towards United Nations Sustainable Development Goal 14.2 ("By 2020, sustainably manage and protect marine and coastal ecosystems to avoid significant adverse impacts, including by strengthening their resilience, and take action for their restoration in order to achieve healthy and productive oceans" [https://sdgs.un.org/goals/goal14]).

Gomez et al. (2006) completed a 5-year study of 10 selected demonstration sites in the Philippines. These being intended to serve as models for other communities. Transplantation of corals was done by taking fragments from nearby large coral colonies to be transplanted to the target sites along with any solitary forms that were available, taking care to minimise damage to the source reefs. Cultured *Tridacna gigas* giant clams of 20–30 cm shell length were used. Monitoring focussed on macro-invertebrates and fish, as well as the assessment of the survival and growth of the transplanted animals. These experiments contributed data and experience which were published in 2007 as a set of guidelines with the title *Reef Restoration Concepts & Guidelines: making sensible management choices in the face of uncertainty* (Edwards and Gomez 2007), which offered communities, managers and decision-makers with clear guidance about coral reef restoration (now also available in French, Bahasa Indonesia and Spanish). These Guidelines were followed in 2010 by a *Reef Rehabilitation Manual* (Edwards 2010). More recently, Schmidt-Roach et al. (2020) have reviewed proposals to restore coral reef populations, and point out that the scale of the work required to implement these concepts in habitats on an ecosystem-wide management remains a major limitation for logistical and, more importantly, financial reasons. Their solution is to suggest implementation by including land-based coral gardening into architectural elements to enhance and beautify coastal development sites.

They see this as encouraging the necessary investment by providing additional value and rationale for ecotourism stakeholders. The overall conclusion of Schmidt-Roach et al. (2020) is that a:

> … global reduction of greenhouse gas (GHG) emissions would be the most effective and economically viable long-term strategy to mitigate climate change effects and protect vulnerable ecosystems such as coral reefs ….

Further, they express the fear that restoration efforts may have lessened effects in mitigating climate change impacts globally and that relying solely on decarbonisation to counteract the degradation of tropical habitats is unrealistic and will not suffice to conserve coral reefs or restore their previous abundance, even if the global community fully complies with the internationally agreed control strategies for the GHG emission pathway. They also bring attention to these other recent publications:

This chapter (entitled *Changing ocean, marine ecosystems, and dependent communities*) of the IPCC *Special Report on the Ocean and Cryosphere in a Changing Climate* [download from this URL: https://www.ipcc.ch/srocc/chapter/chapter-5/] (Bindoff et al. 2019), from which I take the following quotations:

> … There is clear evidence for observed climate change impacts throughout the ocean with consequences for human communities and require options to reduce risks and impacts. Coastal blue carbon can contribute to mitigation for many nations but its global scope is modest ... The survival of some keystone ecosystems (e.g., coral reefs) are at risk, while governance structures are not well-matched to the spatial and temporal scale of climate change impacts on ocean systems. Ecosystem restoration may be able to locally reduce climate risks … but at relatively high cost and effectiveness limited to low emissions scenarios and to less sensitive (p. 454…).
> … There are a broad range of identified barriers and limits for adaptation to climate change in ecosystems and human systems … Limitations include the space that ecosystems require, non-climatic drivers and human impacts that need to be addressed as part of the adaptation response, the lowering of adaptive capacity of ecosystems because of climate change, and the slower ecosystem recovery rates relative to the recurrence of climate impacts, availability of technology, knowledge and financial support and existing governance structures … (p. 455).

The only part of this report (Bindoff et al. 2019) with which I find fault is that unfortunate phrase "…*blue carbon can contribute to mitigation for many nations but its global scope is modest*…". Blue carbon science is relatively young, but has revealed the importance of aquatic ecosystems in the carbon balance and ecosystem services (specific examples are the monetary value of mangroves and seagrasses in ecosystem services and the monetary value of the seafood industry) but it deserves significantly increased attention (Macreadie et al. 2019). Perhaps it is appropriate here to remind the reader that the central thrust of ***our*** argument in this book is that the physiological chemistry of a few ocean creatures (coccolithophores, corals, crustacea, molluscs) enables them to extract carbon dioxide from the atmosphere. And that biotechnologies to make use of this to ***permanently sequester*** atmospheric carbon are already available. Unfortunately, this is another restorative activity that also requires huge effort, central governance, huge scale and huge finance. But more than anything else it requires the ***recognition*** that cultivation of coccolithophores, corals, crustacea and molluscs on a massive scale would make a massive and continued ameliorative contribution to the planetary ecosystem.

The review paper entitled *Rebuilding marine life* (Duarte et al. 2020), which concludes that achieving the United Nations Sustainable Development Goal 14 (to "… conserve and sustainably use the oceans, seas and marine resources for sustainable development …" [https://www.un.org/sustainabledevelopment/oceans/]) "… will require rebuilding the marine life-support systems that deliver the many benefits that society receives from a healthy ocean …". But they finally conclude that "… Rebuilding marine life represents a doable Grand Challenge for humanity, an ethical obligation and a smart economic objective to achieve a sustainable future …". In the opinion of Duarte et al. (2020), recovery rates seen in past studies of conservation interventions suggest that "… substantial recovery of the abundance, structure and function of marine life could be achieved by 2050, if major pressures—including climate change—are mitigated …".

The political limitations of conventional ecosystem governance are discussed by Morrison et al. (2020), focussing on coral reefs and the need to reassess the long-standing assumptions about coping with climate change caused by human activity, particularly the assumption that strong ***local*** institutions can maintain ecological and social resilience through management of the ecosystem adaptation and restoration. They conclude that a new governance paradigm applicable to all ecosystems is required. Governance that serves local needs for conservation and traditional livelihoods is not sufficient; it must be changed to encompass the interests of a broader range of stakeholders, investors and sponsors. As examples they cite governance of the Great Barrier Reef and governance of the Pacific Islands.

Governance of the Great Barrier Reef "… has evolved over the last decade from a local assemblage of social actors (dominated historically by fishing and tourism stakeholders, local conservation groups, and traditional owners) to a more complex polycentric regime, including mining lobbyists, UNESCO, and large international environmental NGOs…". Similarly, "… governance of the Pacific islands now involves international banks, coastal engineers, and property lawyers, among others …" (Green and Hale 2017; Javeline 2014; Keohane 2015). So this is the paradigm shift that is necessary, from local to regional or global governance:

> "… to establish and test the political legitimacy and effectiveness of proposed interventions, to measure political feasibility and modify interventions accordingly, and to guide the development of completely new interventions that are often overtly political. Indeed, securing a future for coral reefs under climate change is a political challenge as much as an ecological or social one …" (Morrison et al. 2020).

There is a published case study that makes the point about the valuable roles that adaptive management strategies involving international, national and local regulatory frameworks can play in mitigating economic impacts from climate-related events (Andréfouët et al. 2017). This study reviews the adaptive management response of nine main groups of stakeholders in the small-scale mariculture of the endangered giant clam species *Tridacna maxima* in French Polynesia which suffered an El Niño-induced giant clam bleaching event in the Tuamotu atolls in 2015–2016. The authors point out that while this case study deals specifically with giant clam farming in the islands of French Polynesia, its general lessons could be applied on other islands.

In their letter to the journal *Science*, Gordon et al. (2020) claim that "… Marine restoration projects are undervalued …". In their final paragraph they conclude:

> … Political agreements for global reductions in atmospheric carbon have been slow to emerge. Relying on their implementation as the only solution to the degradation of tropical habitats is a major gamble. In the meantime, restoration projects could help maintain species survival and ecosystem services, ultimately providing humanity with the breathing space to stabilize the climate … (Gordon et al. 2020).

Overall, then, there is no shortage of scientific and practical knowhow about what could be done to rescue our suffering coral reef ecosystems. But there is insufficient effort due to poor financial, legislative and political support to make sure that what ***could*** be done ***is*** put into effect over the necessary time and geographic scales. This divide between what is possible and what we are doing is a theme to which I will return at the end of this **Chapter**.

5.5 Giant Clam Cultivation and Restocking

Moorhead (2018) discusses how aquaculture of giant clams was developed in the Pacific region, during the 1970s and 1980s, as a community farming effort to develop opportunities to create high-value exports. She points out that:

> … giant clams hold other values, notably cultural value, socio-ecological value, value to tourism, and value through building social capital, and these may explain the continuation of the aquaculture programmes …

Evidently, the value of the giant clam 'ecosystem service' extends way beyond the local coral reef or conservation project. Conservation is just the start of the valuation of a giant clam cultivation programme.

Crucial to any cultivation activity is an understanding of the reproductive biology of the chosen species. Reproduction in giant clams has been studied for over 40 years. Braley (1984) observed *Tridacna gigas* and *T. derasa* spawning sperm in nature during the Australian summer on the Great Barrier Reef. Diel (twenty-four hour) spawning periodicity was observed in *T. gigas*; spawning generally coinciding with incoming tides near the second, third and fourth phases of the moon. Heslinga et al. (1984) confirmed that *T. gigas* spawns with lunar and diel periodicity in Palau. In the central tropics there is no evidence in these early studies of seasonality in reproduction (Gwyther and Munro 1981; Beckvar 1981). Giant clams are 'protandric hermaphrodites' that is they mature as males during the first 2–3 years and later also develop female gonads (Ellis 1997). Braley (1984) did not observe eggs being spawned in nature.

Beckvar (1981) reared larvae of the giant clams *Tridacna gigas*, *T. derasa* and *T. squamosa* in the laboratory, the juveniles being subsequently cultivated outdoors in flowing seawater in open sunlight. Gametes were obtained from spontaneous laboratory spawnings and by inducing spawning with hydrogen peroxide. No supplemental food was added to the system.

> "... Laboratory reared *T. gigas* reached a mean length of 2.6 cm at 10 months post-fertilisation;*T. derasa* were 1.1 cm mean length at 5 months; and *T. squamosa* were 6.7 cm mean length at 2 years ..."

Beckvar's (1981) growth studies projected rates of: *T. gigas,* 8–12 cm y^{-1}; *T. derasa*, 3–6 cm y^{-1}; *Hippopus hippopus*, 3–5 cm y^{-1} and *T. squamosa*, 2–4 cm y^{-1}. Heslinga et al. (1984) measured rates of meat production for 1 to 3-year-old*Tridacna derasa* and found them to compare favourably with values reported for intensively managed mussel farms in Europe.

According to Singh and Azam (2013), the reported methods of inducing spawning in giant clams include:

- Beckvar (1981) induced spawning with hydrogen peroxide, by syringe injection of 10 to 20 cm^3 of a 3% solution into the incurrent siphon. Spawning takes place only occasionally, but the majority of individuals do not respond, yet still show many signs of stress (Fitt and Trench 1981; Gwyther and Munro 1981).
- Wada (1954), Fitt and Trench (1981), Jameson (1976), LaBarbera (1975) and (Heslinga et al. 1990) have induced spawning in tridacnids using fresh and macerated gonads as the stimulus. LaBarbera (1975) reports giant clams being stimulated to spawn by the addition of macerated gonads to the water, and that individuals of *Tridacna maxima* collected at Anae Island, Guam, spawned from November to March; on Palau, *Hippopus hippopus* spawned in June and *Tridacna crocea*, in July.
- Braley (1985) induced spawning by intra gonadal injection of serotonin, which, in most studies is the most successful method of inducing spawning in giant clams (Navneel et al. 2013; Singh and Azam 2013).
- Exposure to a rapid temperature change also induces spawning (Ellis 1997).
- As does mechanical irritation of the posterior abductor muscle (Gwyther and Munro 1981) and
- administering a mild electric shock (LaBarbera 1975) also induces spawning. These last three physical techniques are based on causing a stressful event. Thermal stress (heat shock) is also widely used in the oyster and scallop industries for induced spawning (Heslinga et al. 1990).

These methods have been reviewed by Mies and Sumida (2012) who also succinctly described the aquaculture of giant clams in a land-based hatchery, which is made up of several tanks with different purposes for larval rearing. Prior to reproductive events, broodstock are conditioned and gametogenesis stimulated by frequent food additions and increased photoperiod, while optimum physical–chemical conditions are strictly maintained. After application of the chosen method to induce spawning (e.g. intragonadal injection of serotonin):

> ... gametes are fertilized and a succession of planktonic larval stages is cultured in hatching tanks and raceways. Embryonic development lasts

> for approximately 12 hours until the hatch of free-swimming and non-feeding **trochophore** larvae. After 24 hours post-fertilization larvae morph into **veliger** stage, made evident by the presence of calcium carbonate shells and velum. Veliger larva are fed with live or preserved phytoplankton and must also acquire symbiotic zooxanthellae. The last stage is the **pediveliger** stage at approximately one week post-fertilization, when settlement takes place and metamorphosis is soon attained … (Mies and Sumida 2012).

A critical step in this progress is the acquisition of zooxanthellae, this being a colloquial term for single-celled dinoflagellates that are able to live in symbiosis with several marine invertebrates including demosponges, corals, jellyfish and nudibranchs, as well as clams. There are several genera with *Symbiodinium* being most commonly encountered. The symbiotic algae are not found in either fertilised eggs or trochophore larvae. Fitt and Trench (1981) found that all strains of *S. microadriaticum* introduced to veliger larvae of *Tridacna squamosa* clams were taken into the stomach through the mouth, with motile zooxanthellae being more readily ingested than non-motile ones.

> "… Within 2-9 days after metamorphosis, zooxanthellae moved … into the developing siphonal tissues. Survival and growth of veliger larvae and juveniles with zooxanthellae was greater than those without zooxanthellae. Juveniles with zooxanthellae can survive and grow in Millipore-filtered seawater with light as the sole energy source for over 10 months, illustrating the phototrophic aspect of the association …" (Fitt and Trench 1981).

Growth rates increase sharply after the acquisition of zooxanthellae (Jameson 1976). Juveniles of *Hippopus hippopus* established a symbiosis only with strains of *S. microadriaticum* when offered a variety of free-living and symbiotic species of algae. Other species of algae were digested if small enough to be ingested. The symbiotic dinoflagellate *Amphidinium klebsii* and the diatom *Phaeodactylum tricornutum* were not ingested by veliger or juvenile clams (Fitt et al. 1989).

Giant clams become giants by consuming the sugars and proteins produced photosynthetically by the billions of dinoflagellate algae that live in their tissues. In exchange, they offer the algae a safe home and behavioural patterns that provide regular access to sunlight for photosynthesis, basking by day with their shells open and their alga-containing mantles exposed. They also use a siphon to draw in water to collect and consume plankton by filter feeding. Klumpp and Griffiths (1994) compared the relative contributions of phototrophy (that is, translocation of photosynthates from zooxanthellae) and heterotrophy (filtered planktonic particles) towards the carbon requirements for tissue and shell growth, and metabolism in four species of giant clam (*Tridacna gigas*, *T. crocea*, *T. squamosa* and *Hippopus hippopus* from the Great Barrier Reef. The three species of *Tridacna* were able to satisfy all their growth and metabolic requirements for energy from the combination of photosynthate and particulate food, although the smallest *H. hippopus* did not obtain sufficient carbon from these sources (so their growth was nutrition-limited. However, in general, phototrophy is the most significant source of energy for clams, and the importance of the contribution of filter feeding decreases with increasing size of clam. Another limitation is probably the burrowing habit of *T. crocea*, the physical constraints of which caused this species to have the lowest growth rate overall.

LaBarbera (1975) made measurements of the developing larvae and juveniles, finding that:

- Fertilised eggs of *Tridacna crocea*, *T. maxima* and *Hippopus hippopus* had mean diameters of 93.1, 104.5 and 130.0 μm.
- Shell calcification starts at the transitional stage between the trochophore and the veliger larvae of *Tridacna squamosa* (LaBarbera 1974).
- *Tridacna crocea* larval mortality was exponential during the first 48 h of life declining significantly afterwards (Mies et al. 2012).
- The day-2 veligers of *T. crocea*, *T. maxima* and *H. hippopus* had mean shell lengths of 155.0, 168.0 and 174.4 μm, respectively.
- Mean growth ***rate*** was 11.3 μm day^{-1}, increasing after addition of zooxanthellae to 18.0 μm day^{-1}. Survival increased to about

75% after the addition of zooxanthellae (Mies et al. 2012).

- Settlement occurred 12, 11 and 9 days after fertilisation at a mean shell length of 168.0, 195.0 and 202.0 μm for *T. crocea*, *T. maxima* and *H. hippopus*, respectively.
- Metamorphosis was basically complete about 1 day after settlement.
- Juveniles of *T. crocea*, *T. maxima* and *H. hippopus* first acquire zooxanthellae after 19, 21 and 25 days, respectively.
- Juvenile shells show first signs of becoming opaque after 47 days for *T. maxima* and after 50 days for *H. hippopus* (except where indicated, bulleted data taken from LaBarbera 1975).

Coral reef restoration efforts usually focus on growth and reattachment of reef-building corals, but Frias-Torres (2017) has demonstrated that captive bred, adult giant clams survive restoration in the wild in the Seychelles (western Indian Ocean). Frias-Torres (2017) describes the giant clams as sharing:

> … the role of ecosystem engineers alongside hermatypic corals [= reef-building corals] by providing topographic relief and calcium carbonate to the reef framework … (see above; and see Neo et al. 2015).

In a similar study some years earlier, Guest et al. (2008) had shown that the giant clam, *Tridacna squamosa*, can be restored on Singapore's coral reefs. Seven reefs off Singapore's southern islands were surveyed and an experiment was conducted to determine if *T. squamosa* reared in aquaria could survive if transplanted to reefs around Singapore in an environment with high levels of sedimentation and turbidity resulting from massive coastal development projects and regular dredging of shipping lanes. Giant clams can, indeed, survive and grow well, encouraging the view that restocking efforts using maricultured clams might effectively enhance the dwindling local populations.

Teitelbaum and Friedman (2008) concluded from their assessment of attempts to reintroduce giant clams to fishing-depleted coral reefs in the Indo-Pacific region that most restocking projects can claim only partial success (for reasons which are discussed above at the start of the previous section, *Giant clams and coral reefs*).

Despite such disappointing outcomes, the attempts to restock giant clams continue. Several atolls and islands in French Polynesia have the world's highest stocks of giant clams in shallow, accessible waters, which are consequently highly vulnerable to fishing pressure. Van Wynsberge et al. (2013) used population spatial modelling to simulate the 30-year track of a *Tridacna maxima* stock under different management approaches for the local fishery authority (*Direction des Resources Marines*). For Tubuai, the largest island of the Austral archipelago, the model suggested that reducing fishing effort (through fixed quotas) and banning fishing below the 12 cm clam size limit (as currently implemented) were the most effective management actions to sustain *T. maxima* populations into the future. Interestingly, they found that implementing No-take-Areas was a poor strategy because although giant clam stocks increased inside the protected area, overfishing ***increased*** in the neighbouring areas, so, overall there was no net improvement in stocks.

Just a few of the numerous and wide variety of research studies that have been conducted over an extensive geographical area are reviewed in this **Chapter**. They have contributed to an equally wide variety of printed instruction manuals and guides in the past 27 years (Table 5.1) and, in the current fashion, YouTube videos and websites (Table 5.2).

My general conclusion as an outsider to this area of science is that for 40 years or more a wide range of academics and agencies have studied the decline of stocks of giant clams and their coral reef habitats due to commercial overfishing, climate change and growth in demand for aquarium supplies and recreational (tourist) SCUBA fishing. It is evident from Tables 5.1 and 5.2 (together with other citations in the text above) that (a) numerous well tested techniques and protocols exist that are able, within a reasonable timescale, to restore the

Table 5.1 Giant Clam Cultivation: conventionally published reviews and 'how to do it' manuals

References	Title	Click here
Andréfouët (2017)	Adaptive management for the sustainable exploitation of lagoon resources in remote islands: lessons from a massive El Niño-induced giant clam bleaching event in the Tuamotu atolls (French Polynesia)	https://doi.org/10.1017/S0376892917000212
Braley (1992a)	The giant clam: a hatchery and nursery culture manual	https://aciar.gov.au/publication/books-and-manuals/giant-clam-hatchery-and-nursery-culture-manual
Cabaitan and Conaco (2017)	Bringing back the giants: juvenile *Tridacna gigas* from natural spawning of restocked giant clams	https://doi.org/10.1007/s00338-017-1558-9
Castaños (1992)	Mariculture of giant clams	https://repository.seafdec.org.ph/handle/10862/2602
Ellis (1997)	Spawning and early larval rearing of giant clams (Bivalvia: Tridacnidae)	http://www.ctsa.org/files/publications/CTSA_1306316728608730954041.pdf
Edwards (2010)	Reef Rehabilitation Manual	https://ccres.net/images/uploads/publications/3/reef_rehabilitation_manual_web.pdf
Edwards and Gomez (2007)	Reef Restoration Concepts and Guidelines: making sensible management choices in the face of uncertainty	https://ccres.net/images/uploads/publications/430/ccres_2018_annual_report_final_.pdf
Fatherree (2006)	Giant Clams in the Sea and the Aquarium	https://www.amazon.co.uk/Giant-Clams-Sea-Aquarium-Identification/dp/0978619404
Fatherree (2019)	Giant Clams in the Reef Aquarium: Biology, Identification, and Care	https://www.amazon.co.uk/Giant-Clams-Reef-Aquarium-Identification/dp/0978619455/ref=tmm_pap_swatch_0?_encoding=UTFqid=sr=
Heslinga (2013)	Saving Giants: Cultivation and Conservation of Tridacnid Clams (E-book, free download from blurb.com)	http://store.blurb.com/ebooks/374835
Heslinga et al. (1990)	Giant Clam Farming	https://www.worldcat.org/title/giant-clam farming/oclc/26539012.
Knop (1996)	Giant Clams: A Comprehensive Guide to the Identification and Care of Tridacnid Clams	https://www.amazon.co.uk/Giant-Clams-Comprehensive-Identification-Tridacnid/dp/3921684234
Mies and Sumida (2012)	Giant Clam aquaculture: a review on induced spawning and larval rearing	https://www.researchgate.net/publication/260020319_Giant_Clam_Aquaculture_a_Review_on_Induced_Spawning_and_Larval_Rearing
Mingoa-Licuanan et al. (2007)	Giant clam hatchery, ocean nursery and stock enhancement	https://repository.seafdec.org.ph/handle/10862/2147
Moorhead (2018)	Giant clam aquaculture in the Pacific region: perceptions of value and impact	https://doi.org/10.1080/09614524.2018.1467378

(continued)

Table 5.1 (continued)

References	Title	Click here
Munro (1992)	Giant Clams	https://spccfpstore1.blob.core.windows.net/digitallibrary-docs/files/47/476dd96e30d654ffea944a5c2fe47bd9.pdf?sv=2015-12-11sr=bsig=WH8ywxSdgXKYxjNRTRIcGTabOoCdeNCYXRxDGwStKoU%3Dse=2021-03-11T10%3A26%3A39Zsp=rrscc=public%2C%20max-age%3D864000%2C%20max-stale%3D86400&rsct=application%2Fpdfrscd=inline%3B%20filename%3D%22FFA_1992_075.pdf%22
Neo (2017)	Mei Lin Neo's Conservation of Giant Clams - Part 1 (website)	https://meilin5giantclam.wordpress.com/2017/02/27/conservation-of-giant-clams-part-1-sustainable-mariculture/
Neo et al. (2017)	Giant Clams (*Bivalvia*: *Cardiidae*: *Tridacninae*): A comprehensive update of species and their distribution, current threats and conservation status: an annual review (301 pp., open access)	https://s3-us-west-2.amazonaws.com/tandfbis/rt-files/docs/Open+Access+Chapters/9781138197862_oachapter4.pdf
Schmidt-Roach et al. (2020)	Beyond reef restoration: next-generation techniques for coral gardening, landscaping and outreach	https://doi.org/10.3389/fmars.2020.00672
Teitelbaum and Friedman (2008)	Successes and failures in reintroducing giant clams in the Indo-Pacific region	https://pdfs.semanticscholar.org/afea/43fc57acbbe2d98b7ef1c15f39a15d72b7f0.pdf?_ga=2.245289798.235832041.1597489527-286070954.1597489527
Waters et al. (2013)	A methodology for recruiting a giant clam, *Tridacna maxima*, directly to natural substrata: a first step in reversing functional extinctions?	https://doi.org/10.1016/j.biocon.2012.12.036
Wynsberge et al. (2013)	Best management strategies for sustainable Giant Clam fishery in French Polynesia Islands: answers from a spatial modelling approach	https://doi.org/10.1371/journal.pone.0064641

biodiversity of coral reef systems to something close to normality; and (b) local efforts to implement these conservation schemes are in general only partially successful for a mixture of reasons, among which are limited time and limited funding both contributing to limited scale of the operations. But the greatest limitations emerge from conflicting demands between conservationists and local communities and conflicting politics between local, regional and even national and international administrations. Evidently the concern which has been expressed over the governance (Morrison et al. 2020) and undervaluation (Gordon et al. 2020) of marine restoration projects is fully justified. But whatever is achieved with ecologically protected reserves and even the most effective conservation measures can be severely threatened by determined illegal harvesting of giant clam shells, as is demonstrated by the news report illustrated in Fig. 5.4.

Work towards 'seeding' and recolonising has been going on around the Pacific for about 40 years, and much of this work has been published. Among the numerous detailed studies that have been made of the practical aspects of the

Table 5.2 Giant clam cultivation: YouTube videos and websites

Title and printed hyperlink	Click here
Farming Giant Clams in Palau	https://youtu.be/c8CrhtVrIkc
Palau Once Again Has the Largest Giant Clam Farm in the World	https://youtu.be/NelrsGWsD_A
Visiting Coral Heaven at Simon's Nature Preserve, Solomon Islands	https://youtu.be/EMYxo-hjx-w
Jonathan Bird's Blue World Giant Clams	https://www.youtube.com/watch?v=Z-32RfYNbOY
Weird Sea Creatures' Giant Clams in the South Pacific	https://youtu.be/tbK-TAJ5C44
Mei Lin Neo's The fascinating secret lives of giant clams	https://youtu.be/vGX3FA_rQq4
How to Grow Giant Clams	https://youtu.be/FSbrwwJCK6s
Gerald Heslinga's Sunlight, symbiosis, and sustainable seafood	https://youtu.be/nLie82rfCxU
Websites	
Mei Lin Neo's Conservation of Giant Clams - Part 1: Sustainable Mariculture	https://meilin5giantclam.wordpress.com/2017/02/27/conservation-of-giant-clams-part-1-sustainable-mariculture/
Philip Dor's Tridacna Mariculture Development Center	http://lagoonclams.com/
Giant clams. The giants of the seabed	https://blogs.ntu.edu.sg/hp3203-2017-23/
Global Aquaculture Alliance	https://www.aquaculturealliance.org/advocate/giant-clam-mariculture/
A Giant Clam stock survey and preliminary investigation of Pearl Oyster resources in the Tokelau Islands	http://www.fao.org/3/AC293E/AC293E00.htm
ReefBase Pacific	http://www.reefbase.org/pacific/default.aspx

biology of giant clams, their nutrition, reproduction, husbandry and conservation, there are several dedicated (even impassioned) individuals who have made impressive contributions over the years to the conservation of giant clams and their coral reefs (Tables 5.1 and 5.2).

In addition, though, several local entrepreneurs have developed methods to produce economically large numbers of young giant clams for restocking tropical seas. The conservation and educational programmes that have resulted deserve wider attention and fit well with our call to 'cultivate shellfish to remediate the atmosphere'. One such example of a private individual's efforts is Philip Dor's *Tridacna Mariculture Development Center* [http://lagoonclams.com/] which I will briefly describe next as a case study.

5.6 Giant Clam Aquaculture is More Than Just Science!

The Tridacna Mariculture Development Center (TMDC) has developed a special 'proprietary' farming technique for giant clams which offers a very economic production method for 2–3 mm seed-clams (Fig. 5.5) at an overall cost of a few cents each.

The method depends on a modular submerged floating long line nursery system that is economical, fast and easy to assemble being constructed from PVC sections. It is versatile and long-lasting and suitable for manual as well as mechanical maintenance and harvest. This allows production of large numbers (millions) of juvenile giant clam 'seeds' at cost levels similar to

BBC | Home | News | Sport | Weather | iPlayer | More | Search

NEWS

Home | Prince Philip | Coronavirus | Brexit | UK | World | Business | Politics | Tech | Science | Heal | More

Asia | China | India

Philippines: Giant clam shells worth $25m seized in raid

6 hours ago

EPA

Fig. 5.4 A report posted on the BBC News website on the 17th April 2021 describing the seizure in the Philippines of around **200 tonnes** of illegally harvested giant clam shells worth an estimated US$ 25 million. *Source* https://www.bbc.co.uk/news/world-asia-56784215

Fig. 5.5 At **left**, two to three millimetre clam 'seeds' in quantity. At **right** are 17-mm juveniles (about 4 months old) of *Tridacna derasa*. Image from The Tridacna Mariculture Development Center website [http://lagoonclams.com/]

other shellfish industries like edible oysters & mussels, so improving general profitability. The technique can be applied to six species of giant clams, and the nursery system also keeps the stock safe from natural predators and severe adverse weather conditions (Figs. 5.6, 5.7).

Fig. 5.6 Natural giant clam beds in Phoenix Island (Sanya, Hainan Province, China). Image from The Tridacna Mariculture Development Center website [http://lagoonclams.com/]

Fig. 5.7. 30-year-old *Tridacna gigas* clams at Orpheus Island in the Great Barrier Reef, Australia. Image from The Tridacna Mariculture Development Center website [http://lagoonclams.com/]

Overall, TMDC claim their process enables economically viable, sustainable and environmentally friendly (and crucially, profitable) mariculture of giant clams on a large scale in any protected tropical coastal location with good water conditions. They point out that farmed clams are also suited for restocking of fished-out areas and reef rehabilitation as well as for 'clam gardens' suitable for recreational SCUBA diving for low impact ecotourism.

The principle attraction of this natural farming method is that it eliminates maintenance because after the seed-clams are sown directly on suitable reef-flats they are left alone to grow in the normal way until the selected harvest time. TMDC calls this their 'Bypass Method' because mass production of seed-clams in their hatchery bypasses the first three months of the usual conservation strategy by raising densities to levels that increase the probability of natural recruitment to the resident population.

TMDC argue that giant clam farming is much more an 'art' than a 'science'.

> "… because conditions change with every location and season. So far giant clam hatcheries are still using intensive techniques developed decades ago by the early research pioneers and not suited to large scale commercial production …"

TMDC offer effective technology transfer and training with included pilot hatchery hardware set-up to any tropical island country to produce Giant Clam seeds that can be used to conserve their fringing reefs.

5.7 More a Matter of Vision and Governance Than Either Science or Art

TMDC is one of many practical scientific studies that make their cultivation under protected conditions a viable procedure. But still the decline in stocks of all species continues; and climate change just gets worse.

However, it's not all doom and gloom. In the course of this pioneering activity, many countries have gained knowledge of giant clam biology and a greater awareness of the resource value of these animals. In some places, this has resulted in increased protection of giant clams at both national and community levels as the methods used to rear and grow clams have been adopted, developed and transferred between countries. Introduction of simple hatchery and larval rearing methods have increased capacity in many countries, and progress towards more 'difficult' species like sea cucumbers (Friedman and Tekanene 2005).

The fundamental problem is that though many people bemoan the sad fact that giant clams (and the coral reefs to which so many of them contribute) are on an accelerating march towards extinction, there are too few people doing anything in practical terms on a sufficiently large scale to reverse their march to extinction.

Like many other conservation issues, some well-funded central authority needs to take it by the scruff of the neck, shake it free of self-serving contradictions and drive it into effective action on a worldwide basis.

Acknowledgements I acknowledge, with sincere thanks, the contributions of Philip Dor and The Tridacna Mariculture Development Center [http://www.tmdc.info/] for the inspiration to develop this **Chapter**.

References

Andréfouët S, Wynsberge SV, Kabbadj L, Wabnitz CC, Menkès CE, Tamata T, Pahuatini M, Tetairekie I, Teaka I, Scha TA, Teaka T, Remoissenet G (2017) Adaptive management for the sustainable exploitation of lagoon resources in remote islands: lessons from a massive El Niño-induced giant clam bleaching event in the Tuamotu atolls (French Polynesia). Environ Conserv 45:30–40. https://doi.org/10.1017/S0376892917000212

Arthur MA, Barron EJ, Fawcett PJ, Johnson CC, Kauffman EG, Yasuda MK (1996) Middle Cretaceous reef collapse linked to ocean heat transport. Geology 24:376–380. https://doi.org/10.1130/0091-7613(1996)024%3c0376:MCRCLT%3e2.3.CO;2

Beckvar N (1981) Cultivation, spawning and growth of the giant clams Tridacna gigas, Tridacna derasa and Tridacna squamosa in Palau. Aquaculture 21:21–30. https://doi.org/10.1016/0044-8486(81)90040-5

Bindoff NL, Cheung WWL, Kairo JG, Arístegui J, Guinder VA, Hallberg R, et al. (2019) Changing ocean, marine ecosystems, and dependent communities. Chapter 5 (pp 447–588) In: IPCC, 2019: IPCC special report on the ocean and cryosphere in a changing climate (eds Pörtner H-O, Roberts DC, Masson-Delmotte V, Zhai P, Tignor M, Poloczanska E, Mintenbeck K, Alegría A, Nicolai M, Okem A, Petzold J, Rama B, Weyer NM). Geneva: intergovernmental panel on climate change. Open access. URL Chapter 5 (142 pp) https://www.ipcc.ch/srocc/chapter/chapter-5/. URL Full Report (765 pp) https://www.ipcc.ch/site/assets/uploads/sites/3/2019/12/SROCC_FullReport_FINAL.pdf

Braley RD (1984) Reproduction in the giant clams Tridacna gigas and T. derasa in situ on the north-central Great Barrier Reef, Australia, and Papua New Guinea. Coral Reefs 3:221–227. https://doi.org/10.1007/BF00288258

Braley RD (1985) Serotonin-induced spawning in giant clams (Bivalvia: Tridacnidae). Aquaculture 47:321–325. https://doi.org/10.1016/0044-8486(85)90217-0

Braley RD (ed) (1992a) The giant clam: a hatchery and nursery culture manual. ACIAR Monograph No. 15, 144 pp. Open access. URL: https://www.aciar.gov.au/sites/default/files/legacy/node/2184/mn15_pdf_19505.pdf

Braley RD (1992) Improved method to ship Tridacna gigas seed. Aquaculture 102:193–199. https://doi.org/10.1016/0044-8486(92)90302-2

Cabaitan PC, Conaco C (2017) Bringing back the giants: juvenile *Tridacna gigas* from natural spawning of restocked giant clams. Coral Reefs 36: article 519. https://doi.org/10.1007/s00338-017-1558-9

Castaños MT (1992) Mariculture of giant clams. Southeast Asian Fisheries Development Center, Aquaculture Department. In: Aqua Farm News **10**(4):10–11, 13. https://repository.seafdec.org.ph/handle/10862/2602

CITES (2013) *Convention on international trade in endangered species of wild fauna and flora* (2013) Appendices I, II and III. https://cites.org/eng/app/appendices.php. [Free PDF download: https://cites.org/sites/default/files/eng/app/2020/E-Appendices-2020-08-28.pdf]

Duarte CM, Agusti S, Barbier E, Britten GL, Castilla JC, Gattuso JP, Fulweiler RW, Hughes TP, Knowlton N, Lovelock CE, Lotze HK, Predragovic M, Poloczanska E, Roberts C, Worm B (2020) Rebuilding marine life. Nature 580:39–51. https://doi.org/10.1038/s41586-020-2146-7

Ellis S (1997) Spawning and early larval rearing of giant clams (Bivalvia: Tridacnidae). Open Source Publication Number 130. Center for Tropical and Subtropical Aquaculture: Waimanalo, Hawai'i 96795, USA. http://www.ctsa.org/files/publications/CTSA_1306316728608730954041.pdf

Edwards AJ (ed) (2010) Reef rehabilitation manual. Coral reef targeted research & capacity building for management program: St Lucia, Australia. 166 pp. Open access. URL: https://ccres.net/images/uploads/publications/3/reef_rehabilitation_manual_web.pdf

Edwards AJ, Gomez ED (2007) Reef restoration concepts and guidelines: making sensible management choices in the face of uncertainty. Coral Reef Targeted Research & Capacity Building for Management Programme: St Lucia, Australia. 38 pp. ISBN: 9781921317002. Open access. URL: https://ccres.net/images/uploads/publications/

Fatherree JW (2006) Giant clams in the sea and the aquarium. Liquid Media Publications, Tampa, Fl, pp 277. ISBN: 978-0978619404

Fatherree JW (2019) Giant clams in the reef aquarium: biology, identification, and care. Liquid Media Publications, Tampa, Fl, pp 196. ISBN: 978-0978619459

Fitt WK, Fisher CR, Trench RK (1989) Contribution of the symbiotic dinoflagellate Symbiodinium microadriaticum to the nutrition, growth and survival of larval and juvenile tridacnid clams. Aquaculture 55:5–22. https://doi.org/10.1016/0044-8486(86)90051-7

Fitt WK, Trench RK (1981) Spawning, development and acquisition of zooxanthellae by Tridacna squamosa

(Mollusca, Bivalvia). Biol Bull 116:213–235. https://doi.org/10.2307/1540800

Frias-Torres S (2017) Captive bred, adult Giant Clams survive restoration in the wild in Seychelles, Indian Ocean. Front Mar Sci 4. article 97. https://doi.org/10.3389/fmars.2017.00097

Friedman K, Tekanene M (2005) White teatfish at Kiribati sea cucumber hatchery: "Local technicians getting them out again". *SPC Beche-de-mer Information Bulletin*, 21:32–33. Open access URL: https://tinyurl.com/yckv56zh

Fujise L, Yamashita H, Suzuki G, Sasaki K, Liao LM, Koike K (2014) Moderate thermal stress causes active and immediate expulsion of photosynthetically damaged zooxanthellae (Symbiodinium) from corals. PloS One **9**: article e114321. https://doi.org/10.1371/journal.pone.0114321

Gomez ED, Cabaitan PC, Vicentuan KC (2006) Coral culture and transplantation and restocking of giant clams in the Philippines. In: Proceedings of the regional technical consultation on stock enhancement for threatened species of international concern 13–15 July 2005, (Primavera JH, Quinitio ET, Eguia MRR, eds), Aquaculture Department, Southeast Asian Fisheries Development Center, Iloilo City, Philippines, pp 39–48. URL: https://repository.seafdec.org.ph/handle/10862/2932

Gomez ED, Mingoa-Licuanan SS (2006) Achievements and lessons learned in restocking giant clams in the Philippines. Fish Res 80:46–52. https://doi.org/10.1016/j.fishres.2006.03.017

Gordon TAC, Radford AN, Simpson SD, Meekan MG (2020) Marine restoration projects are undervalued. Science 367:635–636. https://doi.org/10.1126/science.aba9141

Green JF, Hale TN (2017) Reversing the marginalization of global environmental politics in international relations: an opportunity for the discipline. PS Polit Sci Polit 50:473–479. https://doi.org/10.1017/S1049096516003024

Guest J, Todd PA, Goh E, Sivalonganathan BS, Reddy KP (2008) Can giant clam (Tridacna squamosa) populations be restored on Singapore's heavily impacted coral reefs? Aquat Conserv-Mar Freshwater Ecosyst 18:570–579. https://doi.org/10.1002/AQC.888

Gwyther J, Munro JL (1981) Spawning induction and rearing of larvae of Tridacnid clams (Bivlavia: Tridacnidae). Aquaculture 24:197–217. https://doi.org/10.1016/0044-8486(81)90057-0

Heslinga G (2013) Saving giants: cultivation and conservation of Tridacnid Clams. E-book, download from blurb.com. URL: http://store.blurb.com/ebooks/374835

Heslinga G, Perron FE, Orak O (1984) Mass culture of giant clams (F. Tridacnidae) in Palau. Aquaculture 39:197–215. https://doi.org/10.1016/0044-8486(84)90266-7

Heslinga GA, Watson TC, Isamu T (eds) (1990) Giant Clam Farming. Pacific Fisheries Development Foundation (NMFS/NOAA): Honolulu, Hawaii, & Koror, Republic of Palau: MMDC Giant Clam Hatchery. pp 141–179. URL: https://www.worldcat.org/

Jameson SC (1976) Early life history of the giant clams *Tridacna crocea* Lamarck, *Tridacna maxima* (Röding) and *Hippopus hippopus* (Linnaeus). Pac Sci 30:219–233. URL: http://hdl.handle.net/10125/10783. Open access URL: https://scholarspace.manoa.hawaii.edu/bitstream/10125/10783/v30n3-219-233.pdf

Javeline D (2014) The most important topic political scientists are not studying: adapting to climate change. Perspect Polit 12:420–434. https://doi.org/10.1017/S1537592714000784

Johnson CC (2002) The rise and fall of rudist reefs. Am Sci 90:148–153. https://www.jstor.com/stable/27857629

Junchompoo C, Sinrapasan N, Penpian C, Patsorn P (2013a) Changing seawater temperature effects on giant clams bleaching, Mannai Island, Rayong province, Thailand. In: Proceedings of the design symposium on conservation of ecosystems 2013. The 12th SEASTAR2000 workshop, Kyoto University, Japan, pp 71–76. https://doi.org/10.14989/176186

Junchompoo C, Sinrapasan N, Penpian C, Patsorn P (2013b) Changing seawater temperature effects on giant clams bleaching, Mannai Island, Rayong province, Thailand. In: Proceedings of the design symposium on conservation of ecosystems (the 12th SEASTAR2000 workshop), Kyoto University, Japan (2013):71–76. https://doi.org/10.14989/176186

Keohane RO (2015) The global politics of climate change: challenge for political science. PS Polit Sci Polit 48:19–26. https://doi.org/10.1017/S1049096514001541

Klumpp DW, Griffiths CL (1994) Contributions of phototrophic and heterotrophic nutrition to the metabolic and growth requirements of four species of giant clam (*Tridacnidae*). Mar Ecol Prog Ser 115:103–115. https://doi.org/10.3354/meps115103

Knop D (1996) Giant clams: a comprehensive guide to the identification and care of Tridacnid Clams. Dähne Verlag, Ettlingen, pp 255. ISBN: 9783921684238

LaBarbera M (1974) Calcification of the first larval shell of Tridacna squamosa (Tridacnidae: Bivalvia). Mar Biol 25:233–238. https://doi.org/10.1007/BF00394969

LaBarbera M (1975) Larval and post-larval development of the giant clams Tridacna maxima (Röding) and *Tridacna squamosa* Lamarck (*Tridacnidae*: *Bivalvia*). Malacologia 15:69–79. Not found online; journal URL: https://bioone.org/journals/malacologia

Macreadie PI, Anton A, Raven JA, Beaumont N, Connolly RM, Friess DA, Kelleway JJ, Kennedy H, Kuwae T, Lavery PS, Lovelock CE, Smale DA, Apostolaki ET, Atwood TB, Baldock J, Bianchi TS, Chmura GL, Eyre BD, Fourqurean JW, Hall-Spencer JM, Huxham M, Hendriks IE, Krause-Jensen D, Laffoley D, Luisetti T, Marbà N, Masque P, McGlathery KJ, Megonigal JP, Murdiyarso D, Russell BD,

Santos R, Serrano O, Silliman BR, Watanabe K, Duarte CM (2019) The future of Blue Carbon science. Nat Commun 10: article 3998. https://doi.org/10.1038/s41467-019-11693-w
Mies M, Braga F, Scozzafave MS, Lemosand DELD, Sumita PYG (2012) Early development, survival and growth rates of the giant clam Tridacna Crocea (Bivalvia: Tridacnidae). Braz J Oceanogr 60:127–133. https://doi.org/10.1590/S1679-87592012000200003
Mies M, Sumida P (2012) Giant Clam aquaculture: a review on induced spawning and larval rearing. Int J Mar Sci 2:62–69. https://doi.org/10.5376/ijms.2012.02.0009
Mingoa-Licuanan SS, Gomez ED (2007) Giant clam hatchery, ocean nursery and stock enhancement. Tigbauan, Iloilo, Philippines: SEAFDEC Aquaculture Department. 109 pp. ISBN: 9789718511800. URL: https://repository.seafdec.org.ph/handle/10862/2147
Moorhead A (2018) Giant clam aquaculture in the Pacific region: perceptions of value and impact. Dev Pract 28:624–635. https://doi.org/10.1080/09614524.2018.1467378
Morishima SY, Yamashita H, O'Hara S, Nakamura Y, Quek VZ, Yamauchi M, Koike K (2019) Study on expelled but viable zooxanthellae from giant clams, with an emphasis on their potential as subsequent symbiont sources. PLoS One 14: article number e0220141 (open source). https://doi.org/10.1371/journal.pone.0220141
Morris JP, Backeljau T, Chapelle G (2019) Shells from aquaculture: a valuable biomaterial, not a nuisance waste product. Rev Aquac 11:42–57. https://doi.org/10.1111/raq.12225
Morrison TH, Adger N, Barnett J, Brown K, Possingham H, Hughes T (2020) Advancing coral reef governance into the Anthropocene. One Earth 2:64–74. https://doi.org/10.1016/j.oneear.2019.12.014
Munro JL (1992) Giant Clams. In: FFA Report 92/75. Suva, Fiji: Institute of Pacific Studies, Forum Fisheries Agency, International Centre for Ocean Development. 19 pp
Neo ML, Eckman W, Vicentuan K, Teo SLM, Todd PA (2015) The ecological significance of giant clams in coral reef ecosystems. Biol Conserv 181:111–123, open access. https://doi.org/10.1016/j.biocon.2014.11.004
Neo ML, Liu L, Huang D, Soong K (2018) Thriving populations with low genetic diversity in giant clam species, Tridacna maxima and Tridacna noae, at Dongsha Atoll, South China Sea. Reg Stud Mar Sci 24:278–287. https://doi.org/10.1016/J.RSMA.2018.09.001
Neo ML, Wabnitz CCC, Braley RD, Heslinga GA, Fauvelot C, Van Wynsberge S, Andréfouët S, Waters C, Tan ASH, Gomez ED, Costello MJ, Todd PA (2017) Giant Clams (Bivalvia: Cardiidae: Tridacninae): a comprehensive update of species and their distribution, current threats and conservation status: an annual review. In: Hawkins SJ, Evans AJ, Dale AC, Firth LB, Hughes DJ, Smith IP (eds) Oceanography and marine biology: an annual review, vol 55. CRC Press, Boca Raton. https://doi.org/10.1201/b21944
Ponia B (2010) A review of aquaculture in the Pacific Islands 1998–2007: tracking a decade of progress through official and provisional statistics. Secretariat of the Pacific Community, Noumea, New Caledonia. 38 pp. URL: https://tinyurl.com/ycwxx844
Purcell SW, Gossuin H, Ceccarelli DM (2020) Fishery management and spatial variation of giant clam stocks: lessons from New Caledonia. Ocean Coastal Manage 193: article number 105243. https://doi.org/10.1016/j.ocecoaman.2020.105243
Ramah S, Taleb-Hossenkhan N, Todd PA, Neo ML, Bhagooli R (2018) Drastic decline in giant clams (*Bivalvia*: *Tridacninae*) around Mauritius Island, Western Indian Ocean: implications for conservation and management. Mar Biodivers 49:815–823. https://doi.org/10.1007/s12526-018-0858-9
Sala E, Mayorga J, Bradley D, Cabral RB, Atwood TB and 21 others (2021) Protecting the global ocean for biodiversity, food and climate. Nature 2021. https://doi.org/10.1038/s41586-021-03371-z
Schmidt-Roach S, Duarte CM, Hauser CAE, Aranda M (2020) Beyond reef restoration: next-generation techniques for coral gardening, landscaping, and outreach. Front Mar Sci 7: article 672. Open accesss. https://doi.org/10.3389/fmars.2020.00672
Singh NK, Azam K (2013) Comparative study of available spawning methods of the Giant Clam *Tridacna squamosa* [*Bivalvia*: *Tridacnidae*] in Makogai, Fiji. World J Fish Mar Sci 5:353–357. URL: https://core.ac.uk/download/pdf/18621343.pdf
Soo P, Todd PA (2014) The behaviour of giant clams (*Bivalvia*: *Cardiidae*: *Tridacninae*). Mar Biol 161:2699–2717, open access. https://doi.org/10.1007/s00227-014-2545-0
Syukri bin Othman A, Goh GH, Todd PA (2010) The distribution and status of giant clams (family *Tridacnidae*) - a short review. Raffles Bull Zool 58:103–111. Published by the National University of Singapore, open accesss. URL: https://lkcnhm.nus.edu.sg/app/uploads/2017/06/58rbz103-111.pdf
Teitelbaum A, Friedman K (2008) Successes and failures in reintroducing giant clams in the Indo-Pacific region. SPC Trochus Information Bulletin #14, July 2008. Published by the Pacific Community (SPC). URL: https://www.researchgate.net/publication/242729422_Successes_and_failures_in_reintroducing_giant_clams_in_the_Indo-Pacific_region
Van Wynsberge S, Andréfouët S, Gilbert A, Stein A, Remoissenet G (2013) Best management strategies for sustainable Giant Clam fishery in French Polynesia Islands: answers from a spatial modeling approach. PLoS One 8: article number e64641, open access. https://doi.org/10.1371/journal.pone.006464

Wada SK (1954) Spawning in the Tridacnid Clams. Jpn J Zool 11:273–285 (not found online)

Waters CG, Story R, Costello MJ (2013) A methodology for recruiting a giant clam, Tridacna maxima, directly to natural substrata: a first step in reversing functional extinctions? Biol Cons 160:19–24. https://doi.org/10.1016/j.biocon.2012.12.036

Wells S (1996) *Hippopus hippopus*. the IUCN red list of threatened species. IUCN, 1996: article number e.T10105A3164917. https://www.iucnredlist.org/species/10105/3164917

Coccolithophore Cultivation and Deployment

6

David Moore

6.1 In this Chapter...

The potential for the cultivation of coccolithophore golden-brown algae for carbon sequestration is addressed in this chapter. Coccolithophores have been major calcium carbonate producers in the world's oceans for about 250 million years. Today they account for about a third of the total marine $CaCO_3$ production by coating their single cells externally with delicately sculptured plates of microcrystalline $CaCO_3$. The possibility that these algae could be used to trap atmospheric CO_2 with existing technology has not been widely recognised. There is scope, however, for both high technology cultivation in bioreactors and low technology cultivation in terraced raceway ponds or lagoons on tropical coastal sites. The latter could produce a sludge of pure $CaCO_3$ that could be harvested as a feedstock for cement production in place of the fossiliferous limestone that is currently used (cement production accounts for around 8% of industrial fossil CO_2 emissions). Bioreactor cultivation of genetically engineered coccolithophores could produce customised calcite crystals for nanotechnology industries. On the high seas coccolithophores naturally produce extensive blooms, and the blooms emit a volatile gas (dimethyl sulfide) to the atmosphere, where it promotes the formation of clouds that block solar radiation. Imagine aquaculture nurseries onboard factory ships, cultivating both coccolithophores and bivalve molluscs. During their open ocean cruises the ships could produce biodegradable floats already spawned with fixed juvenile bivalve molluscs and streams of coccolithophore algae that could be released into the ocean currents and ocean gyres nourished by artificial upwelling of nutrient-rich waters when the ship deploys its perpetual salt fountains. The dual aim to be creating and maintaining blooms of coccolithophores in the oceanic high seas to sequester carbon from the atmosphere, and generation of cloud cover to cool the immediate environment.

6.2 Introducing Calcifying Algae

As I have stated in Chap. 2, the '... only publication I have found that recognises the true potential for marine calcification to remove CO_2 from the atmosphere...' is Steve Connor's Science News article in *The Independent* newspaper entitled '*Can seashells save the world?*' (Connor 2008). In this article, Steve Connor explains that:

> "... coccolithophores are microscopic marine plants that convert carbon dioxide into chalk. It was thought that rising CO_2 and more acid oceans would curb their activity. Instead they are booming - and fighting global warming ..." and that "... these tiny photosynthetic organisms play a critical role in banking huge amounts of carbon by growing in huge numbers. Indeed, coccolithophore 'blooms' are so big they can even be seen from space ..." (Connor 2008).

Connor (2008) quotes Paul Halloran of Oxford University, a co-author of Iglesias-Rodriguez et al. (2008) as stating that

D. Moore et al., *Aquaculture: Ocean Blue Carbon Meets UN-SDGS*,
Sustainable Development Goals Series, https://doi.org/10.1007/978-3-030-94846-7_6

coccolithophores have thrived during the recent increases of atmospheric CO_2 since the start of the Industrial Revolution; adding:

> Our research has also revealed that, over the past 220 years, coccolithophores increased their mass of calcium carbonate by 40%. These results are in agreement with previous observations of coccolithophores being abundant in a period of ocean acidification 55 million years ago.

Coccolithophore abundance in a period of ocean acidification goes against what seems to be a commonly held view of today that ocean acidification will disturb the formation of calcium carbonate by calcifier organisms. However, this assumption is mainly based on laboratory experiments that are unrepresentative of natural ecosystems. For example, Fitzer et al. (2016) demonstrated significant changes in the hydrated and dehydrated forms of amorphous calcium carbonate in the crystalline layers of mussel (*Mytilus edulis*) shells cultured under acidification conditions.

However, these experiments used CO_2 concentrations that were 2½ times higher than today's observed natural levels. Present levels, which are attributed to anthropogenic CO_2 emissions, are just 25% increased in the ocean over pre-industrial levels. It is unreasonable to predict detrimental consequences for calcifiers on the basis of such extreme experimental procedures. Indeed, there seems to be more evidence for the contrary expectation, that the present levels of elevation of marine CO_2 concentration are more likely to ***encourage calcification*** than discourage it. A year-round monitoring of Subantarctic populations of the common coccolithophore *Emiliania huxleyi* revealed highly calcified morphotypes in high-CO_2 (more acidified) conditions (Rigual-Hernández et al. 2020). These results challenge the idea that ocean acidification will necessarily be detrimental to calcifiers even though it is clear that ocean acidification impacts the viability of symbiotic algae of coral and giant clams, and are interwoven with elevated temperatures and light levels in relatively shallow tropical waters.

Connor's final paragraph warns that:

> "… The coming century could see carbon dioxide levels in the atmosphere rising to 600 parts per million and beyond—which is unprecedented in terms of the human timescale on this planet. So the question of how marine calcifers will cope with this change will be critical in terms of whether the earth's oceans will continue to help us to deal with our carbon dioxide emissions …"

In this Chap. 1 will give a brief description of the nature and biology of coccolithophores and make some suggestions about how **we** could harness **them** to ***save the world***. Table 6.1 notes a few YouTube videos you might like to view.

6.3 The Nature, Biology and Ecology of Coccolithophores

Coccolithophores are eukaryotic phytoplanktonic algae that are predominantly found as single, free-floating haploid or diploid cells (Geisen et al. 2004). Originally assigned to the kingdom

Table 6.1 YouTube videos about coccolithophores

	CLICK LINK
Coccolithophores and Calcium. From coccolithophores to the White Cliffs of Dover, physicist Helen Czerski explains the amazing cycle that makes Calcium her favourite element.	https://www.youtube.com/watch?v=EMNuYOEBOWI
Aliza Fassler's Diatoms, Coccolithophores & Climate Change. This video is about how climate change will affect diatoms and coccolithophores. Changes in the abundance of diatoms and coccolithophores will affect carbon cycling and sequestration	https://www.youtube.com/watch?v=KfQz16LyPP4
American Geophysical Union's Giant algal bloom sheds light on formation of White Cliffs of Dover	https://www.youtube.com/watch?v=Ep5tcBXyFoE

Protista, they are now usually included in the subkingdom *Hacrobia*, phylum *Haptophyta*. *Hacrobia* is assigned to the Chromalveolate supergroup (though the status of this assemblage is uncertain as it may not be monophyletic). Haptophyte cells have two large golden-brown chloroplasts located on either side of the cell and surrounding the nucleus, mitochondria, golgi apparatus, endoplasmic reticulum and other organelles. The cells have two slightly unequal, smooth flagella and a unique organelle called a **haptonema**, for which the phylum is named.

A haptonema is a threadlike organelle, that extends from a position between the bases of the two flagella. Superficially similar to a flagellum, it differs in the arrangement of microtubules. It is more than 100 μm long in some species, and a variety of functions have been demonstrated: attachment and gliding on a substrate, formation of food aggregates, food capture and transport and reception of mechanical stimuli. The haptonema is capable of rapid coiling movements that occur within a few milliseconds following mechanical stimulation which is suggested to depend on Ca^{2+}-binding microtubule-associated proteins (Nomura et al. 2019).

The majority of known haptophytes occur as marine coastal, or open oceanic, planktonic organisms, although a few species thrive in freshwater (Saez et al. 2004). Many can form massive blooms, which in some cases are a hazard for commercial fisheries and other natural biota (Fig. 6.1). The best-known haptophytes are those that have an exoskeleton of calcareous plates called ***coccoliths*;** these are the coccolithophores and they account for 673 of the 762 described species of haptophytes (Foissner 2005).

The distinguishing feature of coccolithophores is that the algal cell is enclosed by a cage of intricate calcium carbonate plates (or scales), which make up the enclosing structure, which is called a coccosphere. Coccoliths are produced inside the cell under genetic control in special organelles, the coccolith vesicles. When a coccolith is completed, it is extruded and arranged outside the cell. Each *Emiliania huxleyi* cell, for example, is surrounded by 10–15 coccoliths, but the coccoliths are easily detached and new coccoliths are constantly built.

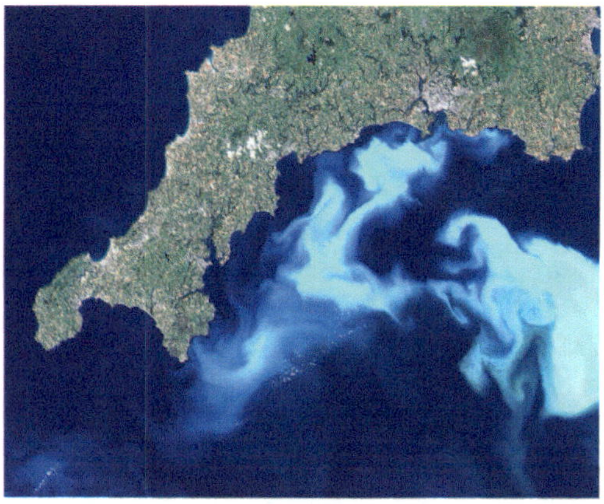

Fig. 6.1 LANDSAT Satellite image of *Emiliania huxleyi* bloom in the English Channel off the coast of Cornwall, 24th July 1999. What look like pale blue clouds in the water are, in fact, the reflected light from billions of coccoliths floating in the water-column. (Photo: NASA, text by Steve Groom, Plymouth Marine Laboratory; image source https://commons.wikimedia.org/wiki/File:Cwall99-lg.jpg under Creative Commons license CC-BY-SA 3.0)

The evolution of calcification in coccolithophore algae had a profound impact on ocean carbon cycling. It is thought that the most ancestral form of calcification produced simple coccoliths as plates of $CaCO_3$. Subsequently, the development of a silicon-dependent mechanism for crystal morphogenesis in the diploid life cycle stage drove the evolution of complex crystal morphology that promoted the ecological success of coccolithophores (Langer et al. 2021; Mock 2021) (and see Figs. 6.2 and 6.4).

The coccoliths are contructed from nanocrystals of $CaCO_3$, and are transparent so they do not shade the chloroplasts which need light for photosynthesis. In fact, the calcite in calcium carbonate allows coccoliths to scatter more light than they absorb, and this scattering enables satellite images to track coccolithospore blooms (Fig. 6.1).

A high concentration of coccoliths increases the temperature of surface water and decreases the temperature of deeper waters; resulting in greater stratification of the water column and decreased vertical mixing.

Fig. 6.2 Diversity of modern coccolithophores. All images are scanning electron micrographs of cells collected by seawater filtration from the open ocean. Species illustrated: **a** *Coccolithus pelagicus*, **b** *Calcidiscus leptoporus*, **c** *Braarudosphaera bigelowii*, **d** *Gephyrocapsa oceanica*, **e** *Emiliania huxleyi*, **f** *Discosphaera tubifera*, **g** *Rhabdosphaera clavigera*, **h** *Calciosolenia murrayi*, (**i**) *Umbellosphaera irregularis*, **j** *Gladiolithus flabellatus*, **k** and **l** *Florisphaera profunda*, **m** *Syracosphaera pulchra* and **n** *Helicosphaera carteri*. Scale bar, 5 μm. Image from Monteiro et al. (2016) under a Creative Commons Attribution 4.0 International License (https://creativecommons.org/licenses/by/4.0/)

However, recent estimates of the overall effect of coccolithophores on ocean temperatures is that it is less than that from anthropogenic sources (Morrissey et al. 2016). Consequently, rather than contributing to global warming, large blooms of coccolithophores cause a decrease in water column productivity in the deeper layers because less light penetrates to them. There seem to be no reports of ***coccolithophore*** toxicity, although closely related algae do produce haemolytic compounds that have been responsible for large fish kills and accumulate through the food chain. But toxicity tests in the laboratory with members of the oceanic coccolithophores *Emiliania*, *Gephyrocapsa*, *Calcidiscus* and *Coccolithus* and the coastal genus *Hymenomonas*, showed them to be non-toxic. Though the coastal genera *Pleurochrysis* spp. and *Jomonlithus* spp., were both toxic to the brine shrimp *Artemia* (Houdan et al. 2004).

One suggested function of the coccoliths is to act as lenses to focus illumination on the photosynthetic apparatus, enabling the cell to thrive in deeper zones where light levels are lower but nutrient levels higher than in surface waters, but several other potential functions have been suggested. These include isolating the intracellular environment from the marine; protection from osmotic, chemical and/or mechanical stress; protection from UV in sunlight; protection from predators among the zooplankton (coccolith appendages may hinder grazing by zooplankton); and it has also been proposed that coccoliths may allow the cell to control its buoyancy, perhaps enabling it to sink to deeper nutrient-rich levels in the water while avoiding descent to dangerous depths (Irie et al. 2010; Young et al. 2009).

Coccolithophores are almost exclusively marine and are found in large numbers throughout the sunlight zone of the world oceans and because of this production of calcite coccoliths, they are both (i) the largest atmospheric carbon sinks, and (ii) one of the largest primary producers on the planet, making them major contributors to global ocean calcification and long-term carbon fluxes. They can also form large amounts of lipids, especially long chain omega-3 polyunsaturated fatty acids (LC-PUFA or 'fish oils'), which have a high potential value as supplementary dietary nutrients. Consequently, in addition to their primary producer's algal photosynthesis and role as the ocean's major resource for calcification, they could also serve as a renewable fuel and alternative food source (Moheimani et al. 2012).

Morphologically, all coccolithophores share the same basic structure of a cell surrounded by the exoskeletal coccosphere, but coccosphere shapes range from spherical to cylindrical, with sizes ranging from about 3 to 30 μm. The number of coccoliths making up a coccosphere varies from as few as six to several hundred, in either one or many layers. The coccoliths themselves range from simple discs to those with elaborate ornamentations, including spines and other projections and delicate grilles. All of this results in a remarkable morphological diversity within the group (Fig. 6.2). However, environmental DNA sequencing shows even greater diversity in the coccolithophores of the marine plankton, many of which sequences are likely to represent novel species and lineages.

Coccolithophores are abundant in the marine phytoplankton, especially in the open ocean, and in the present day, sedimented coccoliths are a major component of the calcareous sediments that cover up to 35% of the ocean floor, being kilometres thick in some places (de Vargas et al. 2007). This abundance and wide geographic distribution is preserved into the fossil record resulting in coccoliths being the main component of the Late Cretaceous Chalk, a rock formation which outcrops widely in southern England, forming the ***White Cliffs of Dover***, and other similar rocks in many other parts of the world (Chimileski and Kolter 2017). Species diversity is believed to have peaked in the past and their presence is documented in the fossil record back to the Triassic, approximately 225 million years ago (Fig. 6.3). Some of their biomolecules are extraordinarily resistant to decay and are thus used by geologists as sedimentary representations of past climatic conditions (Eikrem et al. 2017).

The most abundant species of coccolithophore, *Emiliania huxleyi* (image **e** in Fig. 6.2) occurs in the plankton of almost all

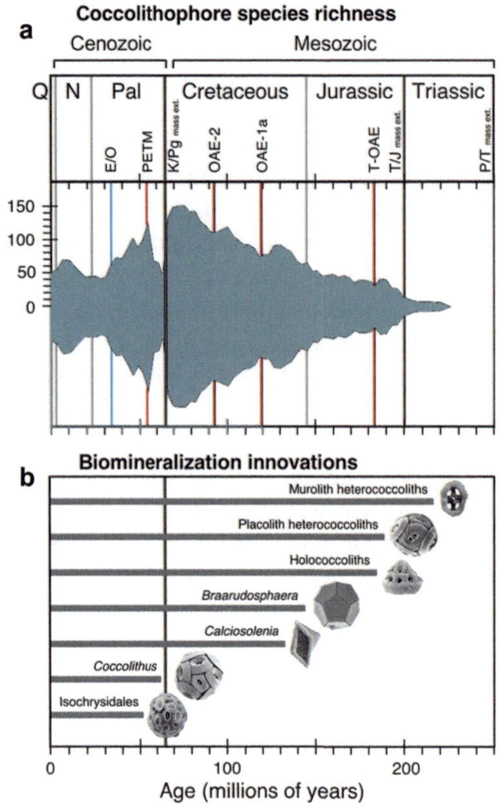

Fig. 6.3 Evolutionary history of coccolithophores. The top panel (**a**) shows species richness over time. **Q**, Quaternary; **N**, Neogene; **Pal**, Paleogene; **E/O**, Eocene/Oligocene glacial onset event; **PETM**, Paleocene/Eocene thermal maximum warming event; **K/Pg**, Cretaceous/Paleogene; **OAE**, oceanic anoxic event; **T-OAE**, Toarcian oceanic anoxic event; **T/J**, Triassic/Jurassic; **P/T**, Permian/Triassic; mass ext., mass extinction. Panel **b** summarises the fossil record of major coccolithophore biomineralisation innovations and morphological groups, including the first appearances of **muroliths** (simple coccoliths with narrow, wall-like rims), **placoliths** (coccoliths with broad shields that interlock to form strong coccospheres), **holococcoliths** (coccoliths formed from microcrystals in the haploid life cycle phase), *Braarudosphaera* (pentagonal, laminated nannoliths forming dodecahedral coccospheres); *Calciosolenia* (distinct, rhombic murolith coccoliths), *Coccolithus* (long-ranging and abundant Cenozoic genus), *Isochrysidales* (dominant order that includes *Emiliania*, *Gephyrocapsa*, and *Reticulofenestra*). Significant mass extinctions and paleoceanographic/paleoclimatic events are marked as horizontal lines. Data from Bown et al. (2004); graphic from Monteiro et al. (2016) under a Creative Commons Attribution 4.0 International License (https://creativecommons.org/licenses/by/4.0/)

ocean ecosystems from the equator to sub-polar regions, and from nutrient-rich upwelling zones to nutrient-poor oligotrophic waters, which makes *E. huxleyi* an important primary producer at the root of a great many marine food webs around the world (Foissner 2005).

Emiliania huxleyi has been widely studied as a model organism to understand physiological, biogeochemical and ecological processes in the oceans, because:

- It is easily cultured in vitro and, in fact, was the fastest growing coccolithophore among the six laboratory cultures studied by Buitenhuis et al. (2008).
- The extensive blooms it forms in nutrient depleted waters after the reformation of the summer thermocline (Chimileski and Kolter 2017) have been studied using floating laboratories with sea enclosures (Egge and Aksnes 1992).
- Long-term trends in surface winter nutrients and summer oxygen concentration of the euphotic zone, as well as seasonal and interannual variability in surface chlorophyll a (chl a) have been investigated for different shelf regions (depths less than 50 m) of the western Black Sea (Yunev et al. 2007). They showed that decrease in the silica to nitrogen ratio, caused by the numerous dams constructed on the River Danube, provoked a shift towards greater non-siliceous phytoplankton blooms (that is, from diatom blooms to haptophyte blooms). Phytoplankton needs sunlight and nutrients from the ocean to survive, so they thrive in areas with large inputs of nutrient-rich water upwelling from the lower levels of the ocean. The ratios between nitrogen, phosphorus and silicate concentrations determine competitive dominance between different phytoplankton communities by favouring either diatoms or other phytoplankton, such as coccolithophores. A low silicate to nitrogen and phosphorus ratio allows coccolithophores to outcompete diatoms, when silicate to phosphorus and nitrogen ratios are high coccolithophores are outcompeted by diatoms.

- Other sources of nutrients, such as inputs from shelf sediments and/or upwelling, and those related to the Danube River maximum discharge levels during spring, contributed to seasonal variations in chlorophyll-*a* measurements (used as a measure of phytoplankton concentration).
- It has been demonstrated that haptophyte pigments and C_{37}–C_{38} alkenones (long-chain biolipids) are synthesised at the seawater layers of highest primary production and therefore the C_{37} alkenone record reveals the temperature for the highest primary productivity of shallow (5 m) or deep (1100 m) waters. Due to their high resistance to chemical and microbial degradation these alkenone molecules are commonly used by earth scientists studying global climate change as a means to estimate past sea surface temperatures (Bentaleb et al. 1999).
- Today, coccolithophores contribute to ocean temperature regulation. They grow well in warm seas and algal blooms produce large amounts of dimethyl sulfide (DMS), a volatile gas that is emitted to the atmosphere, where it promotes the formation of clouds that block solar radiation. As the oceans then cool, the coccolithophore populations decrease and cloud cover also decreases because of the reduced levels of DMS. A classic feedback loop maintaining the temperature equilibrium of the seas (Keller 1989; Alcolombri et al. 2015).
- Coccolithophores, as calcifiers, have been such an important component of the Earth's carbon balancing system for hundreds of millions of years that they are naturally of particular interest for studies of contemporary global climate change. This is particularly true because of the widely expressed fear that as ocean acidity increases, coccolithophores may become less calcified. However, a detailed study of the most abundant coccolithophore species, *Emiliania huxleyi*, in the Bay of Biscay revealed a pronounced seasonality in the morphology of the individuals in the population. In summer, the heavily calcified morphology accounted for only 10% of the population, whereas in winter, when the waters were most acidic and $CaCO_3$ saturations were at their lowest, the population shifted to be 90% of the heavily calcified form. In other words, the most heavily calcified morphotype dominates when conditions are most acidic, which is contrary to the earlier fears and predictions for a high-CO_2 world, suggesting that even in more acidic conditions, coccoliths may be important as a carbon sink (Smith et al. 2012).

More recent work suggests, though, that *Emiliania huxleyi* is not ***one*** species, but a family group of closely related species (a *species complex*). Genomic analysis has shown that different *E. huxleyi* strains harbour extensive genome variability, implying that the strains are isolates from a species complex rather than a single species. The genome variability is reflected in the phenotypic variability, demonstrated as inter-strain variability in physiological and biogeochemical traits in strains maintained in different culture collections (Blanco-Ameijeiras et al. 2016), resulting from different metabolic capabilities that allow *E. huxleyi* to thrive in habitats ranging from the equator to the subarctic and enabling the 'species' to form large-scale episodic blooms under a wide variety of environmental conditions (Read et al. 2013) (and see ***PhycoCosm; the Algal Genomics Resource*** at this URL: https://mycocosm.jgi.doe.gov/Emihu1/Emihu1.home.html.

Emiliania huxleyi is the most abundant coccolithophore species in the Atlantic Ocean, but it is not the only one. *Umbellosphaera irregularis* was the next most abundant species, particularly in surface waters of the tropics. *Gephyrocapsa oceanica* was observed at lower frequencies in both surface and mid-depth samples of the tropics and subtropics but was relatively cosmopolitan, though patchily distributed. *Discosphaera tubifera*, was more commonly seen in surface samples of the subtropics and in only a few cases in the waters of the deep chlorophyll maximum (DCM).

In fact, the diversity and species richness of coccolithophore cells were usually greater in surface populations than in the deep DCM. (Balch et al. 2019). The DCM is a subsurface layer, often located tens of metres below the surface, that is enriched in chlorophyll-*a*. It forms near the nutricline and the bottom of the photic zone. Growth of phytoplankton in the DCM is limited by both nutrient and light availability and the location and formation of the DCM also depends on the season. The DCM cannot be observed using satellite-based remote sensing methods (Fig. 6.1) (Balch 2018; Moore et al. 2012). Estimates of primary productivity are often made using such remote sensing, which they obviously underestimate as they cannot detect the deepwater biomass.

At the surface, the upper photic zone is low in nutrient concentrations but high in light intensity and light penetration, and usually higher in temperature. The lower photic zone is high in nutrient concentration but low in light intensity because of low penetration through the overlying water, and relatively cool (Jordan and Chamberlain 1997). The abundance of deep-dwelling coccolithophore species is greatly influenced by nutricline and thermocline depths; increasing in abundance when the nutricline and thermocline are deep and decreasing when they are shallow (Kinkel et al. 2000; Boeckel et al. 2006).

Balch et al. (2019) reported the lowest concentrations of coccolithophores in equatorial waters; the highest concentrations of cells and coccoliths were associated with temperate, subpolar conditions. Coccolithophore species commonly found in deeper waters were: *Calciosolenia murrayi* (usually found at 40–100 m but also observed in several surface samples), *Florisphaera profunda* (at 50–200 m), *Michaelsarsia adriaticus* (in surface samples, otherwise seen between 50 and 130 m), *Rhabdosphaera clavigera* (also found in surface samples and otherwise between 40 and 200 m) and *Umbellosphaera foliosa* (at 50–200 m). Calcifier species in deeper waters are likely to be especially important for carbonate production at these depths. Although detached coccoliths were distributed to depths of about 300 m, coccolithophore ***cell*** concentrations in the subtropics were highest at less than 200 m depth in the South Atlantic and less than 100 m in the North Atlantic.

Living coccolithophores are also distributed widely in the North and South Pacific (Okada and Susumu 1973); indeed they occur in all large bodies of water, such as the Mediterranean Sea, and in all oceans, including the Southern Ocean, from tropical to polar regions (Thierstein and Young 2004; Winter and Siesser 2006; Winter et al. 2006; Saavedra-Pellitero et al. 2014; Foissner 2005; Chang and Northcote 2016; Menschel et al. 2016; Balch 2018).

It has been demonstrated in laboratory experiments that the process of calcification is an important physiological trait for coccolithophores (Walker et al. 2018). Coccolith production is an intracellular process and has been enabled by modifications to cell ultrastructure and metabolism; surveyed by Taylor et al. (2017).

> "... In addition to calcification, which appears to have evolved with a diverse range of functions, several other remarkable features that likely underpin the ecological and evolutionary success of coccolithophores ... include complex and varied life cycle strategies related to abiotic and biotic interactions as well as a range of novel metabolic pathways and nutritional strategies ..." (Taylor et al. 2017).

Though the benefits of calcification can be species specific there are some common features. In particular, the production of coccoliths requires uptake of dissolved bicarbonate and calcium.

Calcium carbonate and carbon dioxide are produced from calcium and bicarbonate ions by the following chemical reaction:

$$Ca^{2+} + 2HCO_3^- \rightleftharpoons CaCO_3 + CO_2 + H_2O$$

(Mackinder et al. 2010; Mejia 2011; Monteiro et al. 2016).

It is important here to emphasise the Kantian philosophy that the end of the process has more value in itself than the means to achieve it. The calcification reaction releases CO_2, as is shown in the reaction scheme above. This released CO_2 will be fixed in photosynthesis. All of these processes require energy and that energy is supplied by respiration, which returns CO_2 to the atmosphere. So, the ***means*** by which the calcification occurs involves the release of CO_2 to the atmosphere. However, the ***end*** of the process is that the bicarbonate ion that is converted to $CaCO_3$ is now ***permanently removed from the atmosphere***. When the cell dies, or the coccolith is shed, the crystalline $CaCO_3$ will eventually sediment to the seafloor. In the fullness of time, the sediment will become a layer of limestone and will remain as such until subducted into the Earth's mantle, through which the CO_2 will be vented eventually as volcanic gas. Coccolithophores have been major calcium carbonate producers in the world's oceans since the mid-Mesozoic era (Fig. 6.3).

Increase in Ca^{2+} concentrations at the Precambrian/Cambrian boundary has been related to the evolution of calcification in protists and invertebrates, linking formation of $CaCO_3$ to the need to detoxify excess ***calcium*** contained in the 'calcite seas' of the time (see below). It is argued that Ca^{2+} concentrations during the Precambrian era were a crucial promoter of the major steps in the evolution of early life such as photosynthesis, eukaryogenesis, multicellularity, origin of metazoans, etc.; all of which require close homeostatic control over intracellular Ca^{2+} levels. Elevated seawater Ca^{2+} concentrations in the Cretaceous and Jurassic created a need to detoxify extracellular Ca^{2+} to avoid intracellular precipitation of phosphate ions and maintain cellular calcium homeostasis (Simkiss 1977; Raven and Crawfurd 2012; Kazmierczak et al. 2013; Müller et al. 2015; Müller 2019). Consequently, evolutionary selection towards a mechanism to achieve the biochemical benefits of intracellular calcium fixation into the biological inert form of $CaCO_3$ also removed from the water and atmosphere the metabolic waste product and volcanic gas, CO_2. Coccolith fossils dating back to the Palaeocene–Eocene Thermal Maximum of 55 million years ago (labelled **PETM** in Fig. 6.3) are particularly interesting because this period is thought to correspond most directly to the current levels of CO_2 in the oceans (Lloyd et al. 2011; Self-Trail et al. 2012).

As the Precambrian eased into the past, the Palaeozoic Era began 541 million years ago with the Cambrian explosion, an extraordinary diversification of marine animal fossils in its rocks, and ended about 252 million years ago with the end-Permian extinction, which was the greatest extinction event in Earth history. Carbon dioxide levels reconstructed for the Palaeozoic show values around 2000–4000 ppm, which is more than ten times higher than our modern atmosphere. We have to admit that sea water in the Palaeozoic was probably totally different from that of the modern ocean. In particular, the Ordovician (485.4 million years ago (Mya) to 443.8 Mya) and Silurian (443.8 Mya to 419.2 Mya) periods were times of 'calcite seas' in which low-magnesium calcite is the main marine calcium carbonate precipitate. Calcite seas were coincident with times of rapid seafloor spreading which, by cycling seawater through hydrothermal vents transforms calcium-rich minerals in basalt to magnesium-rich clays. This selectively favours precipitation of the more stable calcite $CaCO_3$ crystals (rather than the metastable aragonite $CaCO_3$ crystal form that has a shorter lifetime) into the ocean sediments. Burial of calcite this way affects the acid/base buffering by calcium and bicarbonate ions and lowers seawater pH. Furthermore, the increased volcanism results in elevated levels of carbon dioxide in the atmosphere and oceans leading to global greenhouse climate conditions (Munnecke et al. 2010; Zahnle et al. 2010).

If the modern atmosphere had CO_2 levels anything like those reconstructed for the Early Palaeozoic all carbonates would dissolve in the seawater. As it is, even the comparatively small increase of CO_2 in today's atmosphere caused by burning fossil fuels is sufficient to influence seawater pH and the dissolution of calcium carbonate, causing fears for the destruction of modern coral reef ecosystems within the next few hundred years (Munnecke et al. 2010). But invertebrate fossils in calcite sea deposits ***are usually dominated by forms with thick calcite shells***, so much so that we can all list a few examples; remember trilobites and ammonites?

Trilobites are so abundant in fossil beds that they might be thought of as dominating their ecosystems, but trilobites were only a minor part of the total arthropod diversity of those times. Their profusion in the fossil record being due to their heavy armour reinforced by calcium carbonate fossilising far more readily than the chitinous exoskeletons of other arthropods. Trilobites first appeared in the fossil record 521 Mya, flourished throughout the lower Paleozoic before declining during the Devonian until finally disappearing in the mass extinction at the end of the Permian about 252 Mya. Ammonite fossils tell of a broadly similar 300-million-year success. The earliest ammonites appeared during the Devonian (419.2 to 358.9 Mya), and their last species vanished in the Cretaceous–Paleogene extinction event about 66 Mya, when the dinosaurs, among so many other striking organisms in the seas and on land, also left the scene. And Fig. 6.3 shows that coccolithophes have also flourished and diversified in our oceans since their first appearance in the fossil record about 250 Mya, though they have the distinct advantage of surviving to the present day.

The relevance of this historical diversion is that because the fossil record shows that the distant ancestors of today's marine calcifiers had the physiological tools to cope with both acidified oceans and great excesses of atmospheric CO_2 and still create vast remains of shells made from crystalline $CaCO_3$ we might reasonably expect today's marine calcifiers to be similarly equipped. We have shown in Chap. 1 that foraminifera actively pumps protons out of the site of calcification which is therefore surrounded by a low (acidic) external pH ***of their own making*** (Kawahata et al. 2019). This physiology might help explain why in today's oceans ***both*** calcification by coccolithophores and primary production of these and other algae are significantly ***increased*** by elevated CO_2 partial pressures and acidified seawater, even though this might be counterintuitive to some more pessimistic simulation models (Krumhardt et al. 2019).

We could start engineering our present atmosphere from an established, and impressive, platform because coccolithophore calcification ***already accounts for about a third of the total marine*** $CaCO_3$ ***production of today's oceans***. Evidence from the deep ocean indicates that over the past 220 years there has been a 40% increase in coccolith mass in the deep-sea sediments (Iglesias-Rodriguez et al. 2008). Clearly, the coccolithophores have already reacted to the anthropogenic rise in atmospheric CO_2 partial pressures by doing what they have done before: detoxifying their environment. The difference this time is that they are providing **us** with the service of detoxifying atmospheric CO_2.

6.4 Why Coccolithophores Could Be Good for Us and Our Planet

Overall, coccolithophores exhibit characteristics that, if the fossil record is anything to go by, enable them to be a more effective and more dramatic engineer of this planet and its atmosphere than *Homo sapiens*. Of course, there is a difference in the timescale. *H. sapiens* makes things (good and bad) happen in a few years or decades; coccolithophore climate engineering occupies tens of millions of years. One way of looking at this is that these algae have a much longer track record in the climate engineering field than us. Maybe we should get together to do something about today's atmosphere.

Much of the more recent literature on coccoliths has expressed concern about the effects of climate change and ocean acidification. Fox et al. (2020), for example, compared historic plankton collections made in 1872 to 1876 with those made in 2009 to 2016 to quantify the effect of acidification on planktonic calcifying organisms. A small proportion of the readily available literature even hints at the possibility of using coccolithophore algae for industrial purposes, and in some cases to sequester atmospheric CO_2.

Jakob et al. (2018) describe the successful development of a batch culture process (see below) suitable for the production of coccoliths of *Emiliania huxleyi* for industrial process developments in the many areas of industry (the cement industry, and potable water filtration, among many others) that depend on chemical reactions at the calcite-water-interface (Heberling et al. 2014). Currently, most industrial calcite ($CaCO_3$) is made by crushing mined limestone.

Ultrafine synthetic calcite is produced by bubbling CO_2 into 'lime milk' (an aqueous solution of calcium hydroxide), when nanometer scale calcite particles precipitate. Jakob et al. (2017) demonstrate that coccoliths of *Emiliania huxleyi* are likely to be of value as industrial calcite particles. Skeffington and Scheffel (2018) go further by illustrating how coccoliths might be used as component parts of nanodevices (Figs. 6.4 and 6.5). As they are formed under genetic control, coccolith structure is highly specific and reproducible in any one species; such reproducibility cannot be achieved by chemical/physical syntheses. The coccoliths can be obtained culture in high yields of around $5 \text{ g l}^{-1} \text{ d}^{-1}$.

Coccolith calcite can be modified by the incorporation of metal ions or adsorption of enzymes to the surface; feeding the microalgae with terbium (III) salts drives incorporation of this element into the coccoliths of *Emiliania huxleyi* and they become photoluminescent. Skeffington and Scheffel (2018) speculate that genetic modification of coccolithophores may permit the production of coccoliths with customised architectures and surface properties.

For carbon sequestration, Jiao et al. (2010) detail the role of ocean-dwelling microorganisms in the generation of a pool of long-lived carbon, using a new concept they call the microbial carbon pump. They set out one hypothetical scenario as:

> "... the concurrent elevation of pCO_2 and ocean temperature could increase microbial activity, channelling a greater fraction of the fixed carbon into recalcitrant dissolved organic matter ..."

Perrin et al. (2016) used light- and nutrient-limited batch photobioreactors to simulate conditions in the lower photic zone and study the

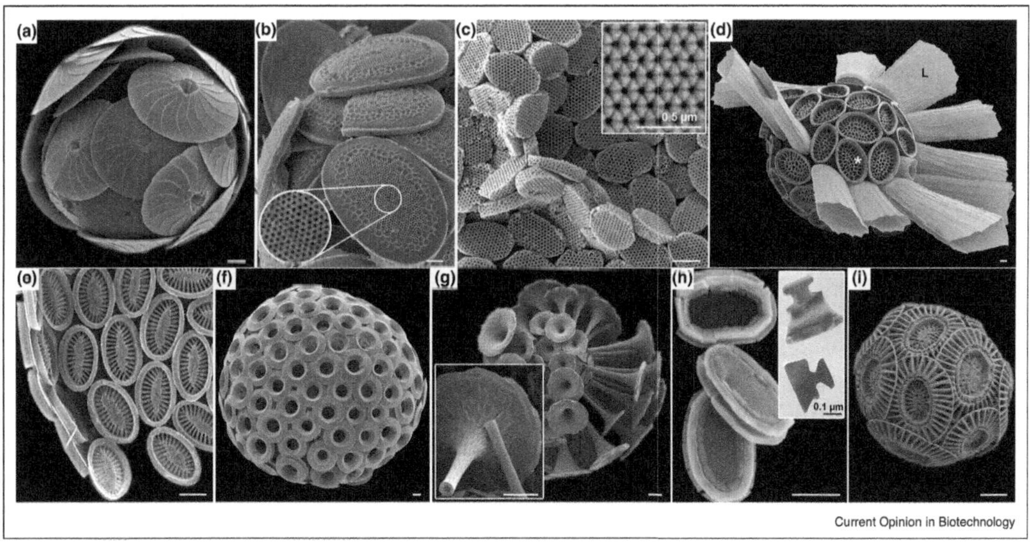

Fig. 6.4 Coccoliths for nanodevices. Electron micrographs showing the coccoliths of **a** Calcidiscus leptoporus subsp. leptoporus, **b** Pontosphaera japonica, **c** Calyptrolithophora papilifera, **d** Scyphosphaera porosa, **e** Michaelsarsia elegans, **f** Umbilicosphaera sibogae, **g** Discosphaera tubifera, **h** Pleurochrysis carterae and **i** Emiliania huxleyi. Inset in c shows the hexagonal array packing of the simple-shaped crystallites in these holococcoliths. S. porosa (d) produces dimorphic coccospheres with vase-like 'lapodoliths' (L) and oval casserole-like body coccoliths (*). Inset in g shows the narrow end of the trumpet-like spine which is hollow. Inset in h shows high-magnification image of the complex-shaped calcite crystals of which these coccoliths are composed. Scale bars: 1 μm. Images a–g by Jeremy Young, University College London, London, UK. From Skeffington and Scheffel (2018) under a Creative Commons license (CC BY-NC-ND 4.0)

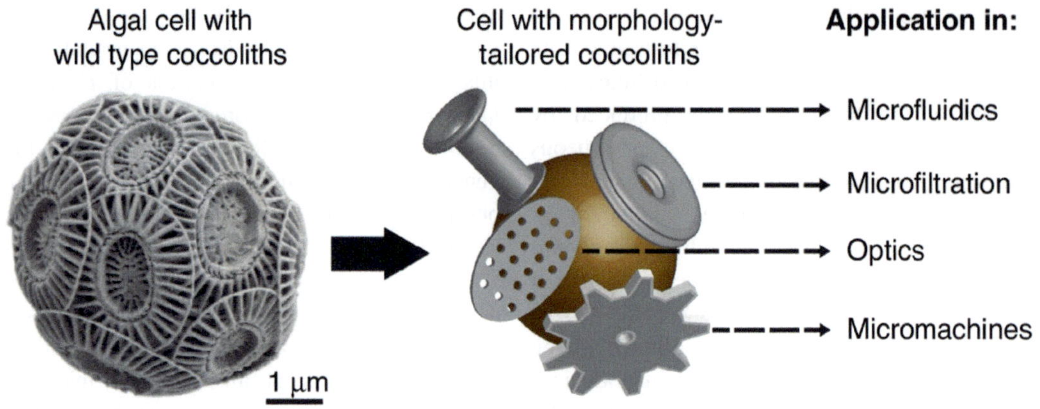

Fig. 6.5 Cartoon diagram showing the potential applications of coccoliths that might be produced by genetic modification of coccolithophores aimed at creating coccoliths with customised architectures and surface properties. From Skeffington and Scheffel (2018) under a Creative Commons license (CC BY-NC-ND 4.0)

physiological of *Emiliania huxleyi* in the oligotrophic gyres of the South Pacific. They were able to reproduce the in situ conditions of light and nutrient (nitrate and phosphate) limitation, showing that *E. huxleyi* growth in that zone is probably limited by the availability of light and nitrate.

Another example dealing specifically with prospects for carbon sequestration is the project ***AlgaCO2*** (entitled: *Industrial cultivation of microalgae as a green strategy for atmospheric CO_2 sequestration*) a research programme of the Portuguese Marine And Environmental Sciences Centre (MARE) funded by *Fundação para a Ciência e a Tecnologia* (FCT), which is the Portuguese national funding agency for science, research and technology [https://www.mare-centre.pt/en/proj/algaco2].

Vicente et al. (2019), who are researchers in this *AlgaCO2* Programme, state that because of microalgae:

> "… remove CO_2 from the atmosphere through photosynthesis, their efficient industrial production may represent a sustainable technology for carbon sequestration …"

They identify coccolithophores as useful candidates among the phytoplankton because their calcite coccoliths have many potential uses in nanotechnology (Jakob et al. 2017, 2018; Skeffington and Scheffel 2018; Santomauro et al. 2020) (Figs. 6.4 and 6.5). Industrial cultivation of coccolithophores, on which all of these potential uses depend, will be explored in the next section.

6.5 Coccolithophore Cultivation

Any of the interventions just mentioned are likely to require in vitro cultivation of coccolithophores as starter cultures/inocula. The laboratory culture of marine planktonic organisms is not challenging and was first reviewed by Allen and Nelson (1910), and there are many insights that are still worth reading in this publication despite its more than 100-year age. Guillard 1975 published a collection of:

> "… relatively simple and reliable methods for the culture of marine phytoplankton species useful for feeding marine invertebrates…" and applied the methods for production "… of sterile cultures of considerable density in volumes up to 18 liters , in commercially available 5 gallon borosilicate glass carboys" (Siegelman and Guillard 1971).

Keller et al. (1987) described a seawater-based medium which was found suitable for a wide range of phytoplankton, giving its important aspects as the addition of selenium, the inclusion of both nitrate and ammonium, an increased level of chelation and a moderate level of pH buffering.

A more recent review of laboratory culture specifically of coccolithophores was published by Probert and Houdan (2004). They point out that laboratory culture experiments have focussed on two easily cultured species, *Emiliania huxleyi* and *Pleurochrysis carterae*, and there is a lack of comparative data for culture of other coccolithophore species, especially those from oligotrophic oceanic habitats. Nevertheless, they suggest ways of culturing these species, such as reducing concentrations of macro- and micronutrients in culture media, and the possible use of organic nutrients in place of mineral nutrients. More general guidance can be found in Lavens and Sorgeloos (1996), Helm et al. (2004), Rincon et al. (2017) and Jerney and Spilling (2018).

6.6 Large-Scale Cultivation of Coccolithophores

Vicente et al. (2019) isolated cocolithophores from Portuguese coastal waters (strains of *Emiliania huxleyi* and *Coccolithus braarudi*) for cultivation under laboratory conditions, using batch cultures and a standard laboratory medium ('Guillard's F/2' (Guillard and Ryther 1962; Guillard 1975); a commercially available *Marine Water Enrichment Solution*) and an industrial medium to mimic conditions of an industrial unit.

There is a considerable industry devoted to large-scale production of microalgae, and there is no shortage of knowhow because the commercial farming of microalgae goes back to the middle of the twentieth century (Borowitzka 2013). Kurano and Miyachi (2004) pointed out that microalgal photosynthesis is efficient enough to fix CO_2 from the atmosphere ***and*** from industrially discharged gases, representing a possible future alternative for CO_2 reduction in both [and see https://www.powermag.com/breakthrough-carbon-capturing-algae-project/].

According to Borowitzka & Moheimani (2013):

> "… Microalgae are currently probably the most studied potential source of ***biofuels***, and in the US alone there are some 30+ companies working in the area and total investment in R&D is in excess of several billion $US worldwide … One of the key attractions of microalgae is the high lipid content of some species …, and the production … of biodiesel from these …"

Biodiesel is prepared from algal lipids by esterifying free fatty acids or transesterifying triacylglycerol fatty acids by reacting them with an alcohol, usually methanol or ethanol. Compared to oilseed crops of terrestrial plants, autotrophic microalgae are capable of achieving very high conversion efficiencies of solar energy into biomass and oil.

But as well as their high lipid and/or sugar content, another feature of microalgae that makes them attractive sources of renewable biofuels ('green diesel') is that they can be grown using saline water on land that is not suitable for agriculture (Borowitzka 2010; Torrey 2010; Davis et al. 2011; Ullah et al. 2014). Davis et al. (2011) claim that:

> "… It is well-established that microalgal-derived biofuels have the potential to make a significant contribution to the US fuel market…"

And, of course, every litre of biodiesel that is used saves the use of a litre of fossil fuel; merely ***cycling present day CO_2*** through the atmosphere rather than ***adding long fossilised CO_2*** to our present day atmosphere.

Another example of a successful microalgal industry is the product known as **Spirulina**, which is actually the dried biomass of a photosynthetic bacterium (cyanobacterium) *Arthrospira platensis*, which was once classified in the genus *Spirulina*, which is maintained as the common name for the commercial product. Spirulina is protein-rich and used widely as a niche healthfood.

Dried Spirulina typically contains 5% water, 24% carbohydrates, 8% fat and about 60% protein, together with numerous vitamins and minerals. It has the distinction of being advocated by both NASA and ESA (the European Space Agency) for cultivation on long-term space missions as food for astronauts travelling to Mars.

As it requires less land and water than farm animals to produce, and has a much lesser impact on the carbon balance of the environment than farm animals and their demands for feed, this nutritious protein and energy food is gaining wide use on Earth to supplement human diets and as an alternative feed for animals in agriculture and aquaculture (Sachdeva et al. 2004; Tuomisto 2010; Alexander et al. 2017).

Spirulina is also rich in bioactive substances that have bio-modulatory and immunomodulatory activities (Khan et al. 2005), and although the evidence is not yet sufficient to endorse Spirulina supplement as treatment for any human disorder, some of those bioactive substances are potent antioxidant and antiproliferative agents and have been shown to decrease the proliferation of experimental pancreatic cancer (Koníčková et al. 2014).

As a potential 'superfood' there is plenty of advice on the Internet for 'home grown' cultivation of Spirulina. An example is the YouTube video *Smart Microfarms—Algae Growing Systems for Home & Backyard* [https://youtu.be/XZW0NpvxTH8]; and compare this with the commercial production process which is shown at this URL: http://www.aurospirul.com/production-process.html. Much more basic methods of cultivating home-grown algae aimed at aquarists is shown at: https://www.wikihow.com/Grow-Algae and https://bitesizebio.com/27998/open-closed-two-ways-grow-algae/; and you can buy your Spirulina starter cultures from ***Amazon*** [https://www.amazon.co.uk/HealthAlgae-Spirulina-platensis-living-culture/dp/B07F93L1C7].

On the commercial scale there are a number of production technologies in use and under development that fall into two basic types:

- Open ponds (raceways).
- Closed, illuminated, 'fermentation' tanks, usually called ***photobioreactor*** (PBR) systems.

The common feature of all technologies is maximisation of algal growth for the production of the desired industrial product(s) (fuel, chemicals/pharmaceuticals, biomass or, in our case, $CaCO_3$). Therefore, apart from the intended end-product, the most suitable approach to employ depends on location, available work force and, of course, economics and finance [http://biomassmagazine.com/articles/3618/open-ponds-versus-closed-bioreactors].

Davis et al. (2011) established the baseline economics for using microalgae to produce biofuels using either open pond cultivation or closed photobioreactor (PBR) systems. They found that to achieve a 10% return on investment the cost of production of diesel (including hydrogenation of algal oils to produce a 'green diesel' blend stock) was \$9.84 gallon^{-1} for open ponds and \$20.53 gallon^{-1} of diesel. Which makes the open pond approach the one to use, providing the location and other negative factors permit.

Open-culture systems rely on natural light for illumination and are inexpensive to instal and run. They may be based on natural small ponds, lakes or lagoons; or be entirely artificial ponds, containers or tanks. The most popular open pond system is the artificial raceway pond in which nutrients, algae and water flow along a circular path, the circulation being maintained by a paddlewheel (Fig. 6.6). Ponds vary in size from 0.5 m^2 to 100 m^2; a single paddlewheel can provide sufficient mixing for a 5 ha (50,000 m^2) cultivation area.

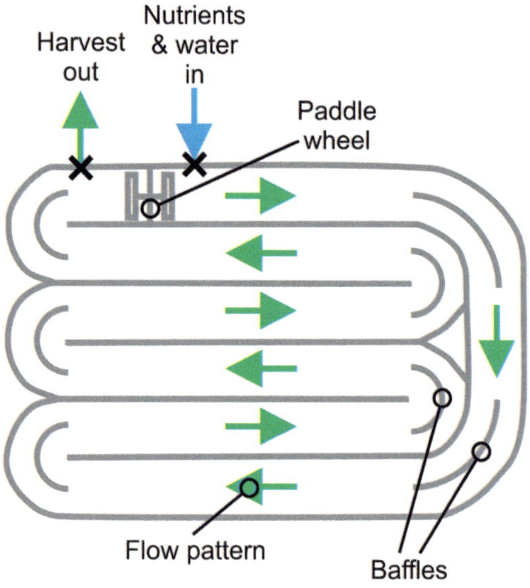

Fig. 6.6 Diagrammatic plan view of a raceway pond and its water circulation (redrawn after Xu et al. 2009)

Being outdoor facilities, open-culture systems suffer from several outdoor-related problems:

- As the cultures are usually not axenic, photosynthetic contaminants (algal, cyanobacterial) can out-compete the desired species.
- Contaminating algal predators (and that grouping might include flocks of water birds!) can graze the culture, causing significant crop losses.
- Weather conditions can cause evaporative losses if too warm, while rainfall dilutes the growth medium and reduces light intensity; all of which make proper control of nutrients, light intensity and CO_2 levels challenging.
- Uncontrolled changes in water temperatures can inhibit high production; for this reason, geographical locations where temperatures range higher than 15 °C are favoured for open-culture systems.

Closed photobioreactors, in contrast, allow control of all these features. But in their case the technical challenge starts with making the bioreactors axenic, and then proceeds to providing mechanisms to control temperature, nutrients, gas exchange and adequate mixing, just like any other fermenter-style bioreactor. By their very nature, closed reactors allow better and more immediate control of culture conditions than open systems.

Unfortunately they are also usually more expensive to instal, but the biggest design challenge is to provide illumination for the photosynthetic microorganisms, generally using fluorescent or LED lighting rigs. Photobioreactors used for the cultivation of microalgae vary in their architecture and include flat-plate, horizontal, inclined or vertical and serpentine tubular airlift photobioreactors (Fig. 6.7), and biofilm reactors, where the algal cells are immobilised onto surfaces within the reactor (Qureshi et al. 2005; Ugwu et al. 2008; Xu et al. 2009; Ketheesan and Nirmalakhandan 2012; Narala et al. 2016; Acién et al. 2017; Rincon et al. 2017).

Xu et al. (2009) published comparisons of the characteristics of open and closed systems, shown in Table 6.2. They state that "… In view of potential applications, development of a more controllable, economical, and efficient closed culturing system is needed. Further developments still depend on continued research in the design of photobioreactors and break-throughs in microalgal culturing technologies…".

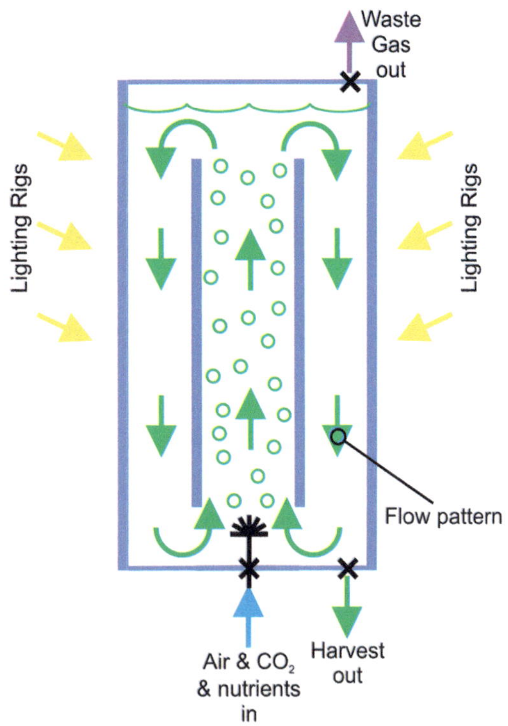

Fig. 6.7 Schematic of an airlift photobioreactor (redrawn after Xu et al. 2009)

Narala et al. (2016) developed a ***two-stage hybrid cultivation system*** using a marine species of the green microalga *Tetraselmis* (*Chlorophyta*), which may have future use in biofuel production. The hybrid system gave significantly higher yields of algal lipids than either single stage system; it combined an initial exponential biomass production in air lift photobioreactors, with a second high lipid induction phase in nutrient depleted open raceway ponds. Nutrients were added only to the closed photobioreactors, greatly improving biomass yields; while the open raceway ponds had turnovers of only a few days, which reduced crop losses due to microalgal grazers.

Table 6.2 A comparison of the levels of risk of open and closed systems for microalgae

Characteristic	Open systems	Closed systems
Contamination risk	High	Low
CO_2 losses	High	Low
Evaporative losses	High	Low
Light use efficiency	Poor	Excellent
Area/volume ratio	Low	High
Area required	High	Low
Process control	Difficult	Easy
Biomass productivities	Low	High
Investment costs	Low	High
Operation costs	Low	High
Harvesting costs	High	Relatively low
Scale-up	Easy	Difficult

From Xu et al. 2009.

There remain a few particular problems related to the biology of microalgae. One is the matter of the fragility of microalgal cells which can have adverse effects on production in closed photobioreactors. Cell damage results in a reduced growth rate, but the cause is the hydrodynamic stress resulting from the vigorous pumping and mixing needed to ensure the turbulent flow of the culture, which is necessary to optimise the light regime. Factors influencing hydrodynamic stress (bioreactor geometry, type of pump, and morphology and physiology of algal cells) are discussed by Gudin and Chaumont (1991).

Another problem, applying specifically to the diploid phase of the life cycle of coccolithophores, particularly *Emiliania huxleyi*, is their susceptibility to a lytic infection caused by giant DNA-containing viruses, known as *E. huxleyi* viruses or EhVs. EhVs infect the coccosphere and induce programmed cell death of the diploid algal cell (Vardi et al. 2012).

Long-term cultivation of the coccolithophore *Pleurochrysis carterae* in outdoor raceway ponds has been reported by Moheimani and Borowitzka (2006). The experiments on this calcifying marine haptophyte alga were carried out because of the likely impact of their blooms in nature on the carbon cycle. The coccolithophore was grown.

> "…semi-continuously in paddlewheel-driven outdoor raceway ponds over a period of 13 months in Perth, Western Australia."

The biomass yield achieved was 0.19 g l^{-1} d^{-1} (dry weight), of which cell lipid amounted to 33% and $CaCO_3$ to 10%. Overall, the total productivity of *P. carterae* biomass averaged an annual total of 60 tonnes ha^{-1} y^{-1}, representing 21.9 tonnes ha^{-1} y^{-1} total lipid and 5.5 tonnes ha^{-1} y^{-1} calcium carbonate. On a molar mass basis, carbon represents 12% of the mass of calcium carbonate; consequently, each hectare (10,000 m^2) of raceway pond devoted to the cultivation of the coccolithophore *Pleurochrysis carterae* removes 0.66 tonnes of carbon from the atmosphere each year.

The *Intergovernmental Panel on Climate Change* Special Report of 2018 (IPCC 2018) suggested that an increase of 1 billion hectares of forest will be necessary to limit global warming to 1.5 °C by 2050. There seems to be a wide acceptance that we should contemplate that remedy, even though it is becoming increasingly clear that planting trees for carbon capture is, at best, only a ***temporary*** sequestration. But before that endeavour becomes a done-deal, should we not at least consider creating 1 billion hectares of coccolithophore bloom in the open ocean and/or raceway ponds in appropriate locations? That amount of coccolithophore cultivation could

permanently remove 0.66 billion tonnes of carbon from the atmosphere each year, which is equivalent to about 7% of our annual global carbon emissions from fossil fuel.

The area of the Pacific Ocean is 16 billion hectares; there's probably enough space there to generate coccolithophore algal blooms that would make a serious dent in the atmosphere's carbon load. If anybody could be bothered to make it happen.

6.7 What We Might Do

In the bullet points below, I suggest the actions that could be taken to exploit fully the potential for cultivation of coccolithophore algae on such large scales that (**a**) their carbon sequestration will contribute to detoxifying our present day atmosphere; (**b**) they will provide today's $CaCO_3$ that will free the cement industry from its use of long-fossilised CO_2; (**c**) their cellular biomass will provide lipids and biofuels to replace fossil fuel usage, as well as other bioactive substances with potential pharmaceutical uses; and (**d**) they will provide tailor-made coccoliths for developments in the nanotechnology industries. All of these features are currently known as **ecosystem services**, which are defined as "… the benefits provided by ecosystems that contribute to making human life both possible and worth living …" [http://uknea.unep-wcmc.org/Home/tabid/38/Default.aspx]. It may be unlikely that any one installation could provide all of these services. For example, a large terrace of raceway ponds intended to supply crystalline $CaCO_3$ to the cement industry is unlikely to be able to provide the nanoscale engineering needed by nanotechnology industries. Nevertheless, all of these services are available from a human-enhanced coccolithophore ecosystem; we just need to find the human get-up-and-go to ***DO*** it.

There is one other ecosystem service that ***very*** large-scale deployment of coccolithophores in the oceanic high seas could provide, which is to contribute to cloud brightening in the expectation that this would produce an atmospheric cooling effect by reflecting solar radiation back into space before it reaches the Earth's surface which it would otherwise warm and re-radiate infrared that greenhouse gases then trap in the atmosphere. This has been suggested as a geoengineering technique that might use sub-micrometre sea water particles injected into the atmosphere in sufficient number to enhance cloud droplet formation (Latham et al. 2012), although "… altering the Earth's radiative energy budget …" may not be an entirely reliable step (Lawrence et al. 2018). However, I remind you of the paragraph above describing how algal blooms produce the volatile gas dimethyl sulfide (DMS), which itself promotes cloud formation (Keller 1989; Alcolombri et al. 2015). So, here is a potential ecosystem service (**e**): promote the formation of clouds that reflect solar radiation, which cools the ocean by altering the radiative energy budget, consequently reduces coccolithophore activity, thereby reducing levels of DMS in a classic, self-regulating feedback loop.

The actions I suggest we should take to exploit the potential of coccolithophore algal cultivation and deployment are as follows.

- **A tropical agriculture solution**, using ponds, raceways or terraces (Schwab et al. 1996; Baryła and Pierzgalski 2008) on tropical (desert?) coastlines continuously filled with ocean water (pumped by solar-energy), and continuously trickling downhill into successively lower terraces during the day and left to settle at night. The sludge of insoluble crystals of calcium carbonate being dredged from the lowest terraces, providing a renewable feedstock of quicklime for cement production in place of the fossiliferous limestone that is currently used. Our way of life uses a lot of cement and cement production is the source of about 8% of the world's CO_2 emissions (Lehne and Preston 2018).
- **A biotechnology solution**, cultivating coccolithophore algae in large industrial LED-illuminated fermenters (photobioreactors) operating in continuous-culture mode (powered by renewable energy sources), which, again, could yield a continuous harvest of insoluble crystals of calcium carbonate,

providing a renewable feedstock for cement production to replace the fossil limestone that is currently used to make quicklime. Additionally, the closer control and greater sensitivity of closed photobioreactors would allow arrays of such devices to be used for the development, using gene-editing technology, and production of 'designer coccoliths' for use in nanodevices.

- **A high seas solution**, by creating an artificial upwelling of nutrient-rich waters using a Perpetual Salt Fountain as discussed in Chap. 4, above "... This can be made to work where you have a warm and salty water mass above a colder and fresher one. The technique is to insert a vertical duct between these two layers, and then pump it out until the pipe is filled with deep water. You can then stop pumping. The upflow from the lower layer will last perpetually, without any other external energy expenditure ... it is the density difference caused by the salinity difference that drives the upward flow..." (see Fig. 10 in Chap. 4). This could be done either using the sea mount installations described in Chap. 4 or by the use of floating processing plants or 'factory ships' [https://en.wikipedia.org/wiki/Factory_ship] (Heilweck and Moore 2021). These would be similar to those used to process oceanic capture fishery catches but these would instead have 'factories' capable of cultivating ***both*** suitable species of coccolithophores ***and*** suitable species of bivalve molluscs in aquaculture nurseries onboard ship during their open ocean cruises. The ships would also be able to produce biodegradable floatation devices already spawned with fixed juvenile bivalve molluscs that could be released into the ocean currents and ocean gyres well away from shipping lanes and commerce routes. The ships would also be equipped to create Perpetual Salt Fountains (Chap. 4) to bring deep water nutrient streams closer to the surface into which coccolithophore algae, cultivated in photobioreactors onboard, could be released to create and maintain blooms of coccolithophores in the oceanic high seas.

References

Acién FG, Molina E, Reis A, Torzillo G, Zittelli GC, Sepúlveda C, Masojídek J (2017). Photobioreactors for the production of microalgae. In Gonzalez-Fernandez C, Muñoz R (eds) Microalgae-based biofuels and bioproducts, Woodhead Publishing Series in Energy, pp 1–44. ISBN 9780081010235. https://doi.org/10.1016/B978-0-08-101023-5.00001-7

Alcolombri U, Ben-Dor S, Feldmesser E, Levin Y, Tawfik DS, Vardi A (2015) Identification of the algal dimethyl sulfide-releasing enzyme: a missing link in the marine sulfur cycle. Science 348:1466–1469. https://doi.org/10.1126/science.aab1586

Alexander P, Brown C, Arneth A, Dias C, Finnigan J, Moran D, Rounsevell MDA (2017) Could consumption of insects, cultured meat or imitation meat reduce global agricultural land use? Glob Food Sec 15:22–32. https://doi.org/10.1016/j.gfs.2017.04.001

Allen EJ, Nelson EW (1910) On the artificial culture of marine plankton organisms. J Mar Biol Assoc UK 8:421–474. https://doi.org/10.1017/S0025315400073690

Balch WM (2018) The ecology, biogeochemistry, and optical properties of coccolithophores. Ann Rev Mar Sci 10:71–98. https://doi.org/10.1146/annurev-marine-121916-063319

Balch WM, Bowler BC, Drapeau DT, Lubelczyk LC, Lyczkowski E, Mitchell C, Wyeth A (2019) Coccolithophore distributions of the North and South Atlantic Ocean. Deep Sea Research Part I: Oceanographic Research Papers, 151. Article 103066. https://doi.org/10.1016/j.dsr.2019.06.012

Baryła A, Pierzgalski E (2008) Ridged terraces - functions, construction and use. J Environ Eng Landsc Manag 16:1–6. https://doi.org/10.3846/1648-6897.2008.16.104-109

Bentaleb I, Grimalt JO, Vidussi F, Marty JC, Martin V, Denis M, Hatté C, Fontugne M (1999) The C_{37} alkenone record of seawater temperature during seasonal thermocline stratification. Marine Chem 64:301–313. https://doi.org/10.1016/S0304-4203(98)00079-6

Blanco-Ameijeiras S, Lebrato M, Stoll HM, Iglesias-Rodriguez D, Müller MN, Méndez-Vicente A, Oschlies A (2016) Phenotypic variability in the coccolithophore Emiliania huxleyi. PLoS One 11: Article e0157697. https://doi.org/10.1371/journal.pone.0157697

Boeckel B, Baumann KH, Henrich R, Kinkel H (2006) Coccolith distribution patterns in South Atlantic and Southern Ocean surface sediments in relation to environmental gradients. Deep-Sea Research Part I: Oceanographic Research Papers, 53: 1073–1099. https://doi.org/10.1016/j.dsr.2005.11.006

Borowitzka MA (2010) Algae oils for biofuels: chemistry, physiology, and production. In: Cohen Z, Ratledge C (eds) Single Cell Oils (Second Edition). Urbana, IL, USA: American Oil Chemists Society (AOCS) Press, an imprint of Elsevier Inc., pp 271–289. ISBN

9781893997738. https://doi.org/10.1016/B978-1-893997-73-8.50017-7

Borowitzka MA (2013) Energy from microalgae: a short history. In: Borowitzka MA, Moheimani NR (eds) Algae for biofuels and energy. developments in applied phycology 5, pp 1–16. Dordrecht: Springer Science +Business Media. pp 288. ISBN: 9789400754782. https://doi.org/10.1007/978-94-007-5479-9_1

Borowitzka MA, Moheimani NR (eds) (2013) Algae for biofuels and energy. Developments in applied phycology 5, Springer Science+Business Media, Dordrecht. pp 288. ISBN: 9789400754782. https://doi.org/10.1007/978-94-007-5479-9

Bown PR, Lees JA, Young JR (2004) Calcareous nannoplankton evolution and diversity through time. In: Thierstein HR, Young JR (eds) Coccolithophores: From Molecular Processes to Global Impact. Springer, Berlin, Heidelberg, pp. 481–508. ISBN: 9783540219286. https://doi.org/10.1007/978-3-662-06278-4_18

Buitenhuis ET, Pangerc T, Franklin DJ, Le Quéré C, Malin G (2008) Growth rates of six coccolithphorid strains as a function of temperature. Limnol Oceanogr 53:1181–1185. https://doi.org/10.4319/lo.2008.53.3.1181

Chang FH, Northcote L (2016) Species composition of extant coccolithophores including twenty six new records from the southwest Pacific near New Zealand. Marine Biodiversity Records, 9. Article 75. https://doi.org/10.1186/s41200-016-0077-7

Chimileski S, Kolter R (2017) Life at the edge of sight: a photographic exploration of the microbial world. Harvard University Press, Cambridge, MA, USA. 250 pp. ISBN: 9780674975910. https://www.hup.harvard.edu/catalog.php?isbn=9780674975910

Connor SA (2008) Can seashells save the world? Science News in: The Independent, of Wednesday 23 April 2008. https://www.independent.co.uk/news/science/can-seashells-save-the-world-813915.html

Davis R, Aden A, Pienkos PT (2011) Techno-economic analysis of autotrophic microalgae for fuel production. Appl Energy 88:3524–3531. https://doi.org/10.1016/j.apenergy.2011.04.018

de Vargas C, Aubrey MP, Probert I, Young J (2007) Origin and evolution of coccolithophores: from coastal hunters to oceanic farmers. In: Falkowski PG, Knoll AH (eds) Evolution of primary producers in the sea, Chap. 12, pp 251–285. ISBN 9780123705181. Boston: Elsevier. Chapter DOI: https://doi.org/10.1016/B978-012370518-1/50013-8. Book DOI: https://doi.org/10.1016/B978-0-12-370518-1.X5000-0

Egge JK, Aksnes DL (1992) Silicate as regulating nutrient in phytoplankton competition. Marine Ecology Progress Series, 83:281–289. https://doi.org/10.3354/meps083281

Eikrem W, Medlin LK, Henderiks J, Rokitta S, Rost B, Probert I, Throndsen J, Edvardsen B (2017) Haptophyta. In: Archibald J, Simpson A, Slamovits C (eds) Handbook of the Protists. Springer, Cham, Switzerland, pp 893–953. ISBN: 9783319281476. https://doi.org/10.1007/978-3-319-28149-0_38

Fitzer SC, Chung P, Maccherozzi F, Dhesi SS, Kamenos NA, Phoenix VR, Cusack M (2016) Biomineral shell formation under ocean acidification: a shift from order to chaos. Sci Rep 6: Article number 21076. https://doi.org/10.1038/srep21076

Foissner W (2005) Biogeography and dispersal of microorganisms: a review emphasizing protists. Acta Protozool 45:111–136. http://www.wfoissner.at/data_prot/Foissner_2006_111_136.pdf

Fox L, Stukins S, Hill T, Miller CG (2020) Quantifying the effect of anthropogenic climate change on calcifying plankton. Sci Rep 10. Article 1620. https://doi.org/10.1038/s41598-020-58501-w

Geisen M, Young JR, Probert I, Sáez AG, Baumann KH, Sprengel C, Bollmann J, Cros L, de Vargas C, Medlin LK (2004) Species level variation in coccolithophores. In: Thierstein HR, Young JR (eds) Coccolithophores: from molecular processes to global impact. Springer, Berlin, Heidelberg, pp 327–366. ISBN: 9783540219286. https://doi.org/10.1007/978-3-662-06278-4_13

Gudin C, Chaumont D (1991) Cell fragility—the key problem of microalgae mass production in closed photobioreactors. Biores Technol 38:145–151. https://doi.org/10.1016/0960-8524(91)90146-B

Guillard RRL (1975) Culture of phytoplankton for feeding marine invertebrates. In: Smith WL, Chanley MH (eds) Culture of marine invertebrate animals. Plenum Press, New York, USA, pp. 26–60 ISBN: 9781461587163 (2013 softcover reprint of the original). https://doi.org/10.1007/978-1-4615-8714-9_3

Guillard RRL, Ryther JH (1962) Studies of marine planktonic diatoms. I. *Cyclotella nana* Hustedt and *Detonula confervacea* Cleve. Can J Microbiol 8:229–239. https://doi.org/10.1139/m62-029

Heberling F, Bosbach D, Eckhardt JD, Fischer U, 13 others (2014) Reactivity of the calcite–water-interface, from molecular scale processes to geochemical engineering. Appl Geochem 45:158–190. https://doi.org/10.1016/j.apgeochem.2014.03.006

Heilweck M, Moore D (2021) Saving the planet with appropriate biotechnology: 3. The High Seas Solution/Salvando el planeta con biotecnología apropiada: 3. La solución de alta mar. Mexican J Biotechnol 6:92–128. https://doi.org/10.29267/mxjb.2021.6.1.92

Helm MM, Bourne N, Lovatelli A (2004) Hatchery culture of bivalves. A practical manual. Part 3: Hatchery operation: culture of algae. FAO Fisheries Technical Paper. No. 471, pp 203. Rome: Food and Agriculture Organization of the UN. ISBN 9251052247. http://www.fao.org/3/y5720e/y5720e08.htm

Houdan A, Bonnard A, Fresnel J, Fouchard S, Billard C, Probert I (2004) Toxicity of coastal coccolithophores (Prymnesiophyceae, Haptophyta). J Plankton Res 26:875–883. https://doi.org/10.1093/plankt/fbh079

Iglesias-Rodriguez MD, Halloran PR, Rickaby REM, Hall IR, Colmenero-Hidalgo E, Gittins JR, Green DRH, Tyrrell T, Gibbs SJ, von Dassow P, Rehm E, Armbrust EV, Boessenkool KP (2008) Phytoplankton calcification in a high-CO_2 world. Science 320:336–340. https://doi.org/10.1126/science.1154122

IPCC (2018) Special report on the impacts of global warming of 1.5 °C above pre-industrial levels and related global greenhouse gas emission pathways. https://www.ipcc.ch/sr15/

Irie T, Bessho K, Findlay HS, Calosi P (2010) Increasing costs due to ocean acidification drives phytoplankton to be more heavily calcified: optimal growth strategy of coccolithophores. PLoS ONE 5. Article e13436. https://doi.org/10.1371/journal.pone.0013436

Jakob I, Chairopolus MA, Vučak M, Posten C, Teipel U (2017) Biogenic calcite particles from microalgae—coccoliths as a potential raw material. Eng Life Sci 17:605–612. https://doi.org/10.1002/elsc.201600183

Jakob I, Weggenmann F, Posten C (2018) Cultivation of *Emiliania huxleyi* for coccolith production. Algal Res 31:47–59. https://doi.org/10.1016/j.algal.2018.01.013

Jerney J, Spilling K (2018) Large scale cultivation of microalgae: open and closed systems. In: Spilling K (ed) Biofuels from Algae. Methods in Molecular Biology, 1980: 1–8. https://doi.org/10.1007/7651_2018_130

Jiao N, Herndl G, Hansell D, Benner R, Kattner G, Wilhelm SW, Kirchman DL, Weinbauer MG, Luo T, Chen F, Azam F (2010) Microbial production of recalcitrant dissolved organic matter: long-term carbon storage in the global ocean. Nat Rev Microbiol 8:593–599. https://doi.org/10.1038/nrmicro2386

Jordan RW, Chamberlain AHL (1997) Biodiversity among haptophyte algae. Biodivers Conserv 6:131–152. https://doi.org/10.1023/A:1018383817777

Kawahata H, Fujita K, Iguchi A, Inoue M, Iwasaki S, 8 others (2019) Perspective on the response of marine calcifiers to global warming and ocean acidification — behavior of corals and foraminifera in a high CO_2 world "hot house". Progr Earth and Planet Sci 6. article 5 (37 pp). https://doi.org/10.1186/s40645-018-0239-9

Kazmierczak J, Kempe S, Kremer B (2013) Calcium in the early evolution of living systems: a biohistorical approach. Curr Org Chem 17:1738–1750. https://doi.org/10.2174/13852728113179990081

Keller MD (1989) Dimethyl sulfide production and marine phytoplankton: the importance of species composition and cell size. Biol Oceanogr 6:375–382. https://doi.org/10.1080/01965581.1988.10749540

Keller MD, Selvin RC, Claus W, Guillard RRL (1987) Media for the culture of oceanic ultraphytoplankton. J Phycol 23:633–638. https://doi.org/10.1111/j.1529-8817.1987.tb04217.x

Ketheesan B, Nirmalakhandan N (2012) Feasibility of microalgal cultivation in a pilot-scaleairlift-driven raceway reactor. Biores Technol 108:196–202. https://doi.org/10.1016/j.biortech.2011.12.146

Khan Z, Bhadouria P, Bisen PS (2005) Nutritional and therapeutic potential of Spirulina. Curr Pharm Biotechnol 6:373–379. https://doi.org/10.2174/1389201057774370607

Kinkel H, Baumann K-H, Čepek M (2000) Coccolithophores in the equatorial Atlantic Ocean: response to seasonal and Late Quaternary surface water variability. Mar Micropaleontol 39:87–112. https://doi.org/10.1016/S0377-8398(00)00016-5

Koníčková R, Vaňková K, Vaníková J, Vánová K, Muchová L, 10 others (2014) Anti-cancer effects of blue-green alga *Spirulina platensis*, a natural source of bilirubin-like tetrapyrrolic compounds. Annals Hepatol 13:273-283. https://doi.org/10.1016/S1665-2681(19)30891-9

Krumhardt KM, Lovenduski NS, Long MC, Levy M, Lindsay K, Moore JK, Nissen C (2019) Coccolithophore growth and calcification in an acidified ocean: insights from community earth system model simulations. J Adv Model Earth Syst 11:1418–1437. https://doi.org/10.1029/2018MS001483

Kurano N, Miyachi S (2004) Microalgal studies for the 21st century. Hydrobiologia 512:27–32. https://doi.org/10.1023/B:HYDR.0000020362.58480.ed

Langer G, Taylor AR, Walker CE, Meyer EM, Ben Joseph O, Gal A, Harper GM, Probert I, Brownlee C, Wheeler GL (2021) Role of silicon in the development of complex crystal shapes in coccolithophores. New Phytol 231:1845–1857. https://doi.org/10.1111/nph.17230

Latham J, Bower K, Choularton T, Coe H, Connolly P, Cooper G, 19 others (2012) Marine cloud brightening. Philos Trans Royal Soc London A, Math, Phys Eng Sci 370:4217–4262. https://doi.org/10.1098/rsta.2012.0086

Lavens P, Sorgeloos P (1996).Manual on the production and use of live food for aquaculture. FAO Fisheries Technical Paper, No. 361. Rome: FAO. 306 pp. ISBN 9251039348. http://www.fao.org/3/w3732e/w3732e01.htm. https://www.researchgate.net/publication/285237285.

Lawrence MG, Schäfer S, Muri H, Scott V, Oschlies A, Vaughan NE, Boucher O, Schmidt H, Haywood J, Scheffran J (2018) Evaluating climate geoengineering proposals in the context of the Paris agreement temperature goals. Nature Commun 9: article number 3734. https://doi.org/10.1038/s41467-018-05938-3

Lehne J, Preston F (2018) Making concrete change: innovation in low-carbon cement and concrete. A Chatham house report. London: The Royal Institute of International Affairs. ISBN: 9781784132729. Open source URL: https://www.chathamhouse.org/2018/06/making-concrete-change-innovation-low-carbon-cement-and-concrete

Lloyd GT, Smith AB, Young JR (2011) Quantifying the deep-sea rock and fossil record bias using coccolithophores. Geol Soc London, Spec Publ 358:167–177. https://doi.org/10.1144/sp358.11

Mackinder L, Wheeler G, Schroeder D, Riebesell U, Brownlee C (2010) Molecular mechanisms underlying

calcification in coccolithophores. Geomicrobiol J 27:585–595. https://doi.org/10.1080/01490451003703014
Mejia R (2011) Will ion channels help coccolithophores adapt to ocean acidification? PLoS Biol 9. article e1001087. https://doi.org/10.1371/journal.pbio.1001087
Menschel E, González HE, Giesecke R (2016) Coastal-oceanic distribution gradient of coccolithophores and their role in the carbonate flux of the upwelling system off Concepción, Chile (36°S). J Plankton Res 38:798–817. https://doi.org/10.1093/plankt/fbw037SN0142-7873
Mock T (2021) Silicon drives the evolution of complex crystal morphology in calcifying algae. New Phytol 231:1663–1666. https://doi.org/10.1111/nph.17507
Moheimani NR, Borowitzka MA (2006) The long-term culture of the coccolithophore *Pleurochrysis carterae* (Haptophyta) in outdoor raceway ponds. J Appl Phycol 18:703–712. https://doi.org/10.1007/s10811-006-9075-1
Moheimani NR, Webb JP, Borowitzka MA (2012) Bioremediation and other potential applications of coccolithophorid algae: a review. Algal Res 1:120–133. https://doi.org/10.1016/j.algal.2012.06.002
Monteiro FM, Bach LT, Brownlee C, Bown P, Rickaby REM, Poulton AJ, Tyrrell T, Beaufort L, Dutkiewicz S, Gibbs S, Gutowska MA, Lee R, Riebesell U, Young J, Ridgwell A (2016) Why marine phytoplankton calcify. Sci Adv 2. article e1501822. https://doi.org/10.1126/sciadv.1501822
Moore TS, Dowell MD, Franz BA (2012) Detection of coccolithophore blooms in ocean color satellite imagery: a generalized approach for use with multiple sensors. Remote Sens Environ 117:249–263. https://doi.org/10.1016/j.rse.2011.10.001
Morrissey JF, Sumich JL, Pinkard-Meier DR (2016) Introduction to the Biology of Marine Life, 11th, revised. Jones and Bartlett Publishers, Inc., Burlington, MA, USA, p 450 9781284090505
Müller MN (2019) On the genesis and function of coccolithophore calcification. Front Mar Sci. 6: article 49. https://doi.org/10.3389/fmars.2019.00049
Müller MN, Barcelos e Ramos J, Schulz KG, Riebesell U, Kaźmierczak J, Gallo F, Mackinder L, Li Y, Nesterenko PN, Trull TW, Hallegraeff GM (2015) Phytoplankton calcification as an effective mechanism to alleviate cellular calcium poisoning. Biogeosciences 12:6493–6501. https://doi.org/10.5194/bg-12-6493-2015
Munnecke A, Calner M, Harper DAT, Servais T (2010) Ordovician and Silurian sea–water chemistry, sea level, and climate: a synopsis. Palaeogeogr Palaeoclimatol Palaeoecol 296:389–413. https://doi.org/10.1016/j.palaeo.2010.08.001
Narala RR, Garg S, Sharma KK, Thomas-Hall SR, Deme M, Li Y, Schenk PM (2016) Comparison of microalgae cultivation in photobioreactor, open raceway pond, and a two-stage hybrid system. Front Energy Res 4. article 29. https://doi.org/10.3389/fenrg.2016.00029
Nomura M, Atsuji K, Hirose K, Shiba K, Yanase R, Nakayama T, Ishida K-I, Inaba K (2019) Microtubule stabilizer reveals requirement of Ca^{2+}-dependent conformational changes of microtubules for rapid coiling of haptonema in haptophyte algae. Biology Open 8. article bio036590. https://doi.org/10.1242/bio.036590
Okada H, Susumu H (1973) The distribution of oceanic coccolithophores in the Pacific. Deep-Sea Research and Oceanographic Abstracts 20:355–374. https://doi.org/10.1016/0011-7471(73)90059-4
Perrin L, Probert I, Langer G, Aloisi G (2016) Growth of the coccolithophore *Emiliania huxleyi* in light- and nutrient-limited batch reactors: relevance for the BIOSOPE deep ecological niche of coccolithophores. Biogeosciences 13:5983–6001. https://doi.org/10.5194/bg-13-5983-2016
Probert I, Houdan A (2004) The laboratory culture of coccolithophores. In: Thierstein HR, Young JR (eds) Coccolithophores: from molecular processes to global impact. Springer, Berlin, Heidelberg, pp 217–249. ISBN: 9783540219286. https://doi.org/10.1007/978-3-662-06278-4_9
Qureshi N, Annous BA, Ezeji TC, Karcher P, Maddox IS (2005) Biofilm reactors for industrial bioconversion processes: employing potential of enhanced reaction rates. Microb Cell Fact 4:24. https://doi.org/10.1186/1475-2859-4-24
Raven JA, Crawfurd K (2012) Environmental controls on coccolithophore calcification. Mar Ecol Prog Ser 470:137–166. https://doi.org/10.3354/meps09993
Read BA, Kegel J, Klute MJ, Kuo A, Lefebvre SC, 34 others (2013) Pan genome of the phytoplankton *Emiliania* underpins its global distribution. Nature 499:209–213. https://doi.org/10.1038/nature12221
Rigual-Hernández AS, Trull TW, Flores JA, Nodder SD, Eriksen R, nine others (2020) Full annual monitoring of Subantarctic *Emiliania huxleyi* populations reveals highly calcified morphotypes in high-CO_2 winter conditions. Sci Rep 10. article number 2594. https://doi.org/10.1038/s41598-020-59375-8
Rincon SM, Romero HM, Aframehr WM, Beyenal H (2017) Biomass production in *Chlorella vulgaris* biofilm cultivated under mixotrophic growth conditions. Algal Res 26:153–160. https://doi.org/10.1016/j.algal.2017.07.014
Saavedra-Pellitero M, Baumann KH, Flores JA, Gersonde R (2014) Biogeographic distribution of living coccolithophores in the Pacific sector of the Southern Ocean. Mar Micropaleontol 109:1–20. https://doi.org/10.1016/j.marmicro.2014.03.003
Sachdeva R, Kaur R, Sangha JK (2004) Effect of supplementation of *Spirulina* on the haematological profile and intellectual status of school girls (7–9 years). J Hum Ecol 15:105–108. https://doi.org/10.1080/09709274.2004.11905674
Sáez AG, Probert I, Young JR, Edvardsen B, Eikrem W, Medlin LK (2004) A review of the phylogeny of the Haptophyta. In: Thierstein HR, Young JR (eds) Coccolithophores: from molecular processes to global impact. Springer, Berlin, Heidelberg, pp 251–269.

ISBN: 9783540219286. https://doi.org/10.1007/978-3-662-06278-4_10

Santomauro G, Stiefel M, Jeurgens LPH, Bill J (2020) In vivo shaping of inorganic functional devices using microalgae. Adv Biosyst 4. article 1900301. https://doi.org/10.1002/adbi.201900301

Schwab GO, Fangmeier DD, Elliot WIE (1996) Soil and water management systems, 4th edn. John Wiley & Sons Inc., Hoboken, NJ, USA 9780471109730. 384 Pp

Self-Trail JM, Powars DS, Watkins DK, Wandless GA (2012) Calcareous nannofossil assemblage changes across the Paleocene-Eocene Thermal Maximum: evidence from a shelf setting. Mar Micropaleontol 92–93:61–80. https://doi.org/10.1016/j.marmicro.2012.05.003

Siegelman HW, Guillard RRL (1971) Large-scale culture of algae. Methods Enzymol 23:110–115. https://doi.org/10.1016/S0076-6879(71)23085-8

Simkiss K (1977) Biomineralization and detoxification. Calcif Tissue Res 24:199–200. https://doi.org/10.1007/BF02223316

Skeffington AW, Scheffel A (2018) Exploiting algal mineralization for nanotechnology: bringing coccoliths to the fore. Curr Opin Biotechnol 49:57–63. https://doi.org/10.1016/j.copbio.2017.07.013

Smith HEK, Tyrrell T, Charalampopoulou A, Dumousseaud C, Legge OJ, 9 others (2012) Predominance of heavily calcified coccolithophores at low $CaCO_3$ saturation during winter in the Bay of Biscay. Proceedings of the National Academy of Sciences of the United States of America 109:8845–8849. https://doi.org/10.1073/pnas.1117508109

Taylor AR, Brownlee C, Wheeler G (2017) Coccolithophore cell biology: chalking up progress. Ann Rev Mar Sci 9:283–310. https://doi.org/10.1146/annurev-marine-122414-034032

Thierstein HR, Young JR (eds) (2004) Coccolithophores: from molecular processes to global impact. ISBN: 9783540219286. https://doi.org/10.1007/978-3-662-06278-4

Torrey M (2010) Survey of commercial developments of microalgae as biodiesel feedstock. In: Cohen Z, Ratledge C (eds) Single cell oils (Second Edition). American Oil Chemists Society (AOCS) Press, Urbana, IL, USA, pp 243–270. An imprint of Elsevier Inc. ISBN 9781893997738. https://doi.org/10.1016/B978-1-893997-73-8.50016-5

Tuomisto HL (2010) Food security and protein supply—cultured meat a solution? Aspects Appl Biol 102:99–104. http://static.elmercurio.cl/Documentos/Campo/2011/08/02/20110802122710.pdf

Ugwu CU, Aoyagi H, Uchiyama H (2008) Photobioreactors for mass cultivation of algae. Biores Technol 99:4021–4028. https://doi.org/10.1016/j.biortech.2007.01.046

Ullah K, Ahmad M, Sofia, Sharma VK, Lu P, Harvey A, Zafar M, Sultana S, Anyanwu CN (2014) Algal biomass as a global source of transport fuels: overview and development perspectives. Progr Natural Sci Mater Int 24:339. https://doi.org/10.1016/j.pnsc.2014.06.008

Vardi A, Haramaty L, Van Mooy BAS, Fredricks HF, Kimmance SA, Larsen A, Bidle KD (2012) Host-virus dynamics and subcellular controls of cell fate in a natural coccolithophore population. Proc Natl Acad Sci USA 109:19327–19332. https://doi.org/10.1073/pnas.1208895109

Vicente B, Cachão M, David H, Silva JL, Tenreiro A, Amorim A (2019) Coccolithophores, a green technology for mitigation of carbon emissions. In: Poster at IMMR'18, International Meeting on Marine Research 2018. https://doi.org/10.3389/conf.FMARS.2018.06.00034

Walker CE, Taylor AR, Langer G, Durak GM, Heath S, Probert I, Tyrrell T, Brownlee C, Wheeler GL (2018) The requirement for calcification differs between ecologically important coccolithophore species. New Phytol 220:147–162. https://doi.org/10.1111/nph.15272

Winter A, Jordan RW, Roth PH (2006) Biogeography of living coccolithophores in ocean waters. In: Winter A, Siesser WG (eds) Coccolithophores. Cambridge University Press, Cambridge, UK, 256 pp. ISBN: 9780521031691. https://www.cambridge.org/

Winter A, Siesser WG (2006) Coccolithophores. Cambridge University Press, Cambridge, UK, 256 pp. ISBN: 9780521031691. https://www.cambridge.org/gb/academic/

Xu L, Weathers PJ, Xiong X-R, Liu C-Z (2009) Microalgal bioreactors: challenges and opportunities. Eng Life Sci 9:178–189. https://doi.org/10.1002/elsc.200800111

Young JR, Andruleit H, Probert I (2009) Coccolith function and morphogenesis: insights from appendage-bearing coccolithophores of the family syracosphaeraceae (haptophyta). J Phycol 45:213–226. https://doi.org/10.1111/j.1529-8817.2008.00643.x

Yunev OA, Carstensen J, Moncheva S, Khaliulin A, Ærtebjerg G, Nixon S (2007) Nutrient and phytoplankton trends on the western Black Sea shelf in response to cultural eutrophication and climate changes. Estuar Coast Shelf Sci 74:63–76. https://doi.org/10.1016/j.ecss.2007.03.030

Zahnle K, Schaefer L, Fegley B (2010) Earth's earliest atmospheres. Cold Spring Harbor Perspectives in Biology 2010, 2: article a004895. https://doi.org/10.1101/cshperspect.a004895

Comparing Industrial and Biotechnological Solutions for Carbon Capture and Storage

7

Peter Petros and David Moore

7.1 In this Chapter...

We deal with the current artificial/industrial Carbon Dioxide Capture, Utilisation and Storage (CCUS) solutions and show their power and potential in curtailing greenhouse gas (GHG) emissions. Key evaluation models of sustainability for current carbon capture and storage (CCS) infrastructure are used to explain what problems could arise and potential ways to avoid the likely risks through drastic changes in fundamental attitudes. The shortfalls of each industrial solution are also presented in the context that all activities should be carried out with due regard for long-term human and environmental well-being, rather than economic growth alone.

Overall, we discuss solutions for atmospheric carbon reduction; the carbon market; industrial/artificial carbon dioxide capture, utilisation and storage systems; carbon emissions reduction targets. We make comparisons between 'soft' nature-based biotechnological solutions, including coastal blue carbon and the ultimate blue carbon, which is the ocean's calcifiers. Following a discussion of sustainability assessment of CCUS methods, we conclude that changing the paradigm of shellfish farming from '***shellfish as food***' to '***shellfish for carbon sequestration***' places the value of the exercise of shellfish cultivation onto the production of shell, taking the food value of the animal protein as one of the several ecosystem services that bivalve molluscs and calcifying microalgae (specifically, coccolithophores) supply. We calculate that this paradigm shift makes mussel farming, and by default other bivalve molluscs and microalgal farming enterprises, viable alternatives to all the CCUS industrial technologies in use today.

7.2 Solutions for Atmospheric Carbon Reduction

The current global industrial trend towards adoption of carbon capture and storage (CCS) technologies will be examined in this chapter, focusing on key CSS technologies, such as flue gas CCS injection facilities in fossil fuel and other heavy industry plants (others include steel, concrete and fertiliser production). Current climate policies and industry trends are directing and incentivising the increase of industrial CCS as central technology for reaching climate change targets. While CCS is essential in meeting the emissions targets, as already stated by the IPCC in 2005, complications have arisen in putting all our eggs in that basket. To date, the developed carbon emissions market along with major heavy industry players have integrated and adopted a major CCS solution that allows for a 'business as usual' approach.

> ... Talking up carbon capture is good for fossil fuel companies - it makes the next few decades look profitable for them. Companies from ExxonMobil to Shell to Occidental Petroleum have all boasted about investments in carbon capture while continuing to double down on their core business model of finding and digging up as much oil and gas as possible. (Aronoff 2020).

D. Moore et al., *Aquaculture: Ocean Blue Carbon Meets UN-SDGS*,
Sustainable Development Goals Series, https://doi.org/10.1007/978-3-030-94846-7_7

What is lacking in this approach, namely environmental ecosystem services and circular economic value, is in fact guaranteed by certain biotechnological CCS solutions available to us. To ensure our humanity's future, these biotechnological solutions are vital. They offer sustainable engineering solutions, environmental ecosystem services, guaranteed life cycle extension and circular economic value economy in addition to carbon sequestration potential. Lenton (2014) concluded that "...Ultimately, CDR [carbon dioxide removal] could be used to bring atmospheric CO_2 concentration down to whatever is considered a safe level. CDR may also be used to counter-balance some 'essential' or 'unavoidable' fossil fuel CO_2 emissions, without increasing the [atmospheric] CO_2 concentration. However, most CDR technologies are more expensive than most conventional emissions reduction options, and hence are unlikely to be used..." (Lenton 2014). Unfortunately, the words '**calcifier**', '**mollusc**' or '**bivalve**' do not appear in this chapter; although some quite exotic potential methods of CDR can be found in the book, including artificial trees, stratospheric aerosols, and solar radiation management like brightening clouds and space-based geoengineering.

7.3 The Carbon Market

The importance of carbon sequestration will be increasingly significant as we proceed further into the twenty-first century. Not only is carbon sequestration an environmental and atmospheric issue, it is now considered an economic market, whereby carbon credits are offered by legislators and a carbon market continues to be expanded and refined. Nations currently have a monetary value assigned to the quantity of carbon directly emitted into the atmosphere. By doing so, we have created the greenhouse gas (GHG) emissions market or emission trading systems (ETSs). As such, we have 'put a price on carbon' and from this point on will call it simply '**the carbon market**'.

Described as a unique environmental commodity, the carbon market was created out of the Kyoto Protocol. This international treaty extends the 1992 United Nations Framework Convention on Climate Change (UNFCCC), committing nations to reduce greenhouse gas emissions, based on the scientific consensus that global warming is occurring and is most likely caused by human-made CO_2 emissions. The Kyoto Protocol, completed in December 1997, required industrialised countries to reduce their total greenhouse gas emissions to 5.2% below 1990 levels (Jacobson 2001). As listed in Annex A of the Protocol, developed countries must limit all GHG emissions, which are carbon dioxide (CO_2), methane (CH_4), nitrous oxide (N_2O), hydrofluorocarbons [HFCs], perfluorocarbons [PFCs], and sulfur hexafluoride (SF6) (Fig. 7.1).

The carbon market deals with a specific **Environmental Commodity**. Environmental Commodities are commodities that take the form

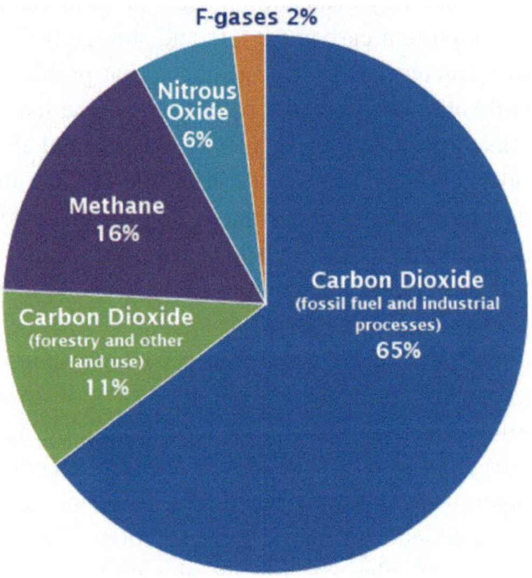

Fig. 7.1 Global greenhouse gas emissions by type of gas. 65% of carbon dioxide emissions derives from fossil fuel use and industrial processes and 11% of carbon dioxide is emitted by deforestation, decay of biomass, etc. Methane represents 16% of the total and nitrous oxide 6%. 2% of the total is from fluorinated gases (hydrofluorocarbons [HFCs], perfluorocarbons [PFCs], and sulfur hexafluoride [SF_6]). Image from the ***United States Environmental Protection Agency*** website (https://www.epa.gov/ghgemissions/global-greenhouse-gas-emissions-data), data from the Contribution of Working Group III to the Fifth Assessment Report of the Intergovernmental Panel on Climate Change (IPCC 2014)

of non-tangible energy credits, the value of which derives from the need for cleaner forms of energy. The market formed as a result of governmental efforts to deal with GHG emissions by tax reductions or other financial incentives and was first implemented by regulatory policies from government bodies. Many industries produce GHGs in the manufacturing of their products and as governments across the world place strict limits on the rights of individuals or institutions to pollute by generating GHGs, those rights become scarce, valuable and tradeable (Pines 2020). Without such limitation by governmental regulation the right to pollute would have no economic value as production and supply could be unlimited theoretically.

This is a point worth remembering: ultimately, regulatory policy has the power to assign value and create economic markets, no matter what the value-assigned object might be (a service, a chemical, object or organism, an environment or a pollutant). The markets or ETSs that trade Environmental Commodities emerged as a way to buy and sell the right to pollute. The question that needs to be asked is whether the future of humanity on this planet would be better served by markets based on **Global Health** rather than **Global Pollution**?

Many would agree that after more than two decades since adoption of the Kyoto Protocol, ETSs and the 16 compliance carbon markets in operation across the world have failed in their primary objective of ensuring significant reductions in GHG emissions to curtail anthropogenic-inputs and mitigate rising atmospheric GHG input. Indeed, Pearse and Böhm (2014) argue that:

> …carbon markets do not have a role to play in a policy scenario that requires radical emissions reductions in order to avoid dangerous greenhouse gas concentrations. We put forward 10 reasons why carbon markets should not be the preferred climate policy choice… (Pearse & Böhm 2014).

More clearly as of late, has been the misguided allocation of carbon credits and carbon offsets in the name of business, rather than in the name of climate change; meaning, in short, that the rich and powerful win more than poorer nations. Carbon credits are being used increasingly to finance nature-based solutions but are of varying quality, with some being of doubtful permanence and/or having little regard for social and ecological factors (Girardin et al. 2021). These authors recognise that:

> … Nature-based solutions need both public and private finance; in particular, governments need to reward ecosystem stewardship while taxing polluters and ramping up regulation to ensure that companies meet strict social and environmental safeguards. (Girardin et al. 2021).

While Cziesielski et al. (2021) comment that (the emphasis is ours):

> … Sustainable ocean management is, in its essence, a political process that requires coordination across governments as well as relevant stakeholders, including scientists, local communities, and industries. This model of ***inclusive governance*** will be central to ensuring an equitable and just development of the blue economy…

Though we share this opinion, we do not wish to develop this political narrative here; suffice to say in summary that the current rules and regulations built by policy-makers have created a flawed carbon market in order to solve the climate change crisis, albeit with good initial intentions. So, what is the alternative?

In short, we consider that **redefining market value** is the key. An ideal, possibly utopian, scenario might be one where the market focuses primarily on ***improving and sustaining global environmental health*** and secondly on GHG emissions reductions (although the latter is a significantly-weighted factor).

Global health fundamentally relies on:

- raising environmental awareness,
- continuous educated decision-making,
- sympathetic planning protocols,
- timely action,
- full implementation,
- extensive monitoring,
- conservation of environmental systems.

Whereas GHG emissions and carbon trading, by definition, can be produced, reduced, moved around, traded and sequestered, global health cannot and should not be passed around. The policies would ideally settle on any management body or agency holding responsibility for their local environment and the global environmental impact of their businesses.

If value ***is*** assigned to global health, then global markets must be regulated with rules that uphold the natural capital values that the Earth's natural ecosystems offer as services (also known as ***ecosystem services***). Such a move would fundamentally shift us towards planning and implementing a true ***circular economy*** with our planet and a healthy and harmonious relationship from market to industrial and commercial ventures to communities. We will return to this theme towards the end of this Chapter.

7.4 Industrial Carbon Dioxide Capture, Utilisation and Storage (CCUS)

Industrial or artificial, carbon capture and storage is usually considered essential to meeting climate goals. However, what is not discussed very often are the ***potential negative implications*** of widespread adoption of certain artificial **carbon capture and utilisation** (CCU) and **carbon capture and storage** (CCS) solutions (under the overall acronym **CCUS**). Technology being developed now, which is likely to be constructed over the next few years, with the expectation of operating for at least 10 years to become economically viable, will place enormous unforeseen burdens on all aspects of the activities into the short-term. This is particularly worrisome given the very short (decadal) timeframes which are implicit in the climate models describing future GHG emissions to the atmosphere and consequential climate change used by the IPCC and other expert bodies that describe the climatic paths we may already be heading into due to historic rates of GHG emissions.

The implications emerge more clearly when we understand how the carbon market works and who are the current big players. It is also important to remember that money is the key hurdle for change and in this case, **where** the money is channelled and what it is directed towards. Carbon dioxide capture, utilisation and storage is, in many ways, a twenty-first-century technological marvel as a climate solution. A major reason for CCUS being so readily embraced is its mitigation potential of ***significantly large amounts*** of CO_2 from point sources.

As a brief background of its inception, the IPCC 2005 meeting on climate change first brought CCS into global attention in a weighty expert reviewed special report on *Carbon Dioxide Capture and Storage* (IPCC 2005), which outlined the technology, the costs, the benefits, the complications and the potential for playing a significant role in climate change mitigation.

In 2011, six years after CCS was first presented in that IPCC special report, the UN Framework Convention on Climate Change agreed upon CCS as a ***Clean Development Mechanism*** (CDM), which under Article 12 of the Kyoto Protocol, allows such projects to.

> ... earn saleable certified emission reduction (CER) credits, each equivalent to one tonne of CO_2, which can be counted towards meeting Kyoto targets.

Generally speaking, CCUS has a key role in achieving the goals of the Paris Climate Agreement targets and are deemed as ***vital emissions reduction technologies*** by both the IPCC and International Energy Agency (IEA). The global CCS programme, between 2010 and 2020, expressed in terms of annual capacities for carbon sequestration from 2010 to 2020 is illustrated in Fig. 7.2 and the global distribution of key CCS projects in 2019 is shown in Fig. 7.3.

An important question that is raised as the cost of CCUS roll-outs increases is simply this: ***is it really worth it*?** The answers given to that question are certainly not a unanimous 'yes' because recent innovations in biotechnological solutions could provide better alternatives, such as improved energy efficiency, renewable energy, or ***bio***technological innovations.

Before we go further with that proposition, we should establish exactly what CCS is. According to the IPCC 2005 Special report on Carbon Capture and Storage (IPCC 2005), CCS is a

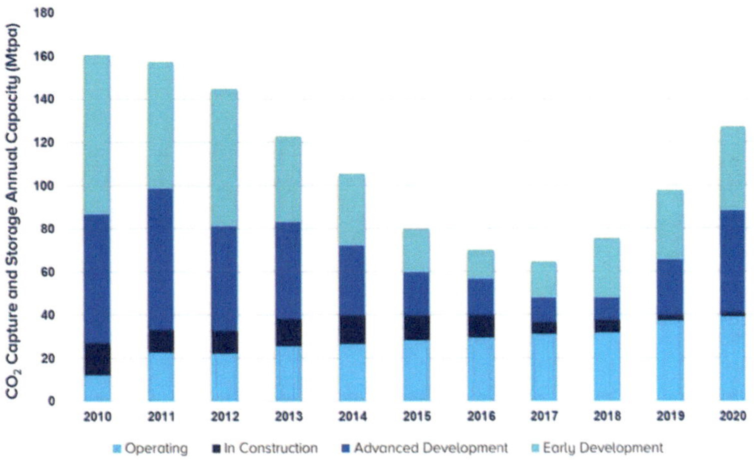

Fig. 7.2 Progress of carbon capture and storage (CCS) programmes in terms of annual capacities for carbon sequestration around the world from 2010 to 2020. *Source* Marshall et al. (2020)

process consisting of the separation of CO_2 from industrial and energy-related sources, transportation to a specified storage location and long-term storage and isolation from the atmosphere (Fig. 7.4).

This is currently considered to be the primary tool for mitigation and stabilisation of atmospheric greenhouse gas concentrations. The utilisation aspect of GHG emissions, or CCU, has more recently been developed as a ***better practice*** as compared to CCS due to the utilisation of the emissions as a secondary resource rather than solely storing them. CCU is therefore more closely suited to a circular economy, but more on that later.

The capture of CO_2 and other GHG emissions via CCUS can be applied to large point sources, where the emissions can be compressed and transported for storage in geological formations, in the ocean, in bedrock as mineral carbonates or for use in further industrial processes (IPCC 2005). According to Zevenhoven and Fagerlund (2010), CCS involves injecting CO_2 into host rocks or employing an ex situ application step, reacting huge volumes of CO_2 as carbonate minerals, and storing these in geological formations. The initial steps involve capturing the CO_2 emissions, followed by transportation and injection. Each step can involve variations in physical and chemical processes, each major CCS project utilising different solutions of varying efficiencies. The end results are nonetheless similar; CO_2 either in liquified or mineralised form which is now available for either utilisation or direct storage in geological underground pockets. A more recent review (Hills et al. 2020) discusses mineralisation in geologically derived minerals and industrial wastes, emphasising the manufacture of products with value. The authors suggest that this sort of CCUS technology can manage significant quantities of CO_2.

Leakage and escape of injected CO_2 have been a topic of major concern over the last two decades and many of these concerns have been allayed by pilot experimental studies by expert geologist teams. Possible escape routes for geologically injected sequestered CO_2 are shown in Fig. 7.5. Larkin et al. (2019) listed 29 potential hazards in a risk assessment of CCS injection and storage activities, suggesting that for 0–50 year, 51–499 year and > 500 year time periods, the likelihood of the occurrence of ***major leakage*** from CCS storage resulting in "… measurable negative effects on human health or the environment …" is approximately 1 in 10^3.

The paper also makes note of the enormously wide uncertainties involved with CCS leakage potential, such as uncertainties in worldwide saline aquifer storage capacity (0.1–76,000 Gt), uncertainties of ultimate CO_2 sequestration capacity in solution, as a per cent of a deep saline aquifer (0.2–76%), and uncertainties in the distances affected by salt precipitation (1–175 m). This last is when salt crystals form during CCS

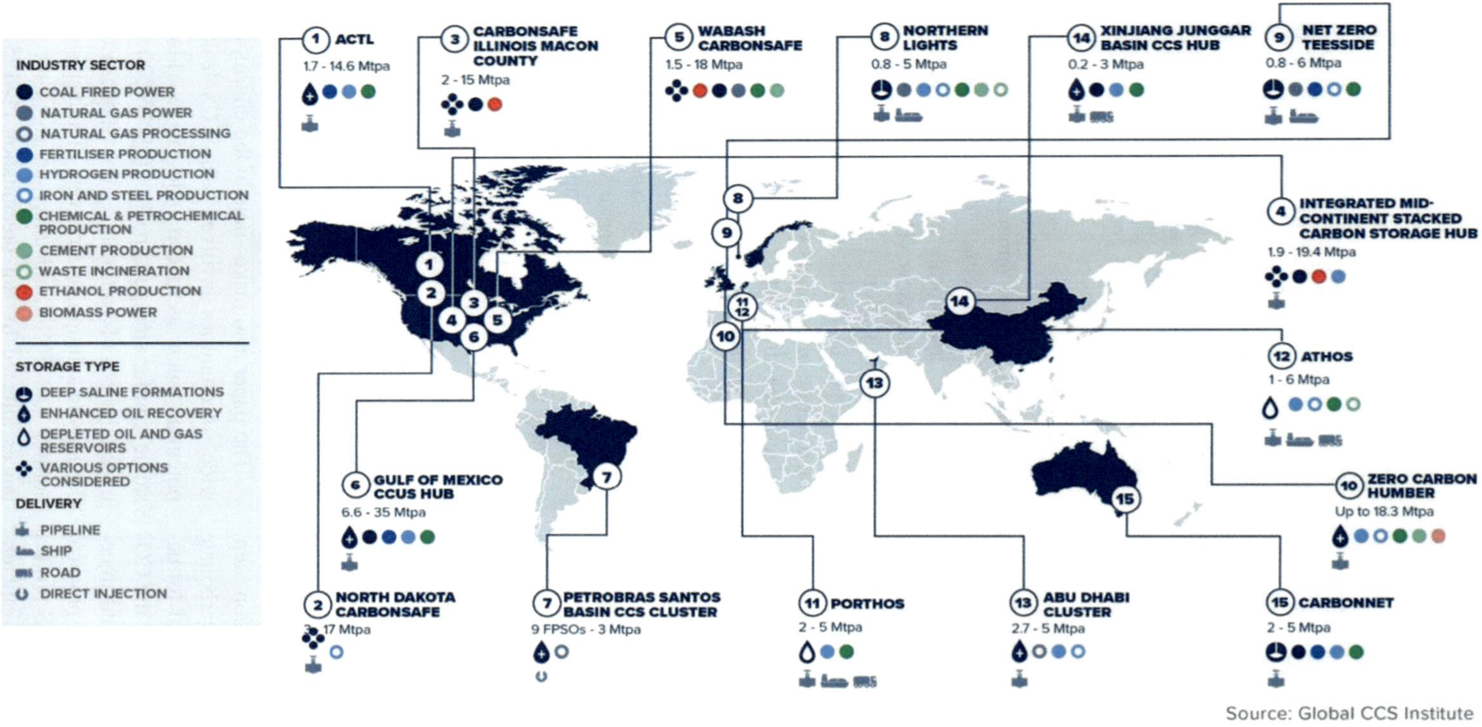

Fig. 7.3 Global distribution of key CCS projects in 2019. *Source* Marshall et al. (2020)

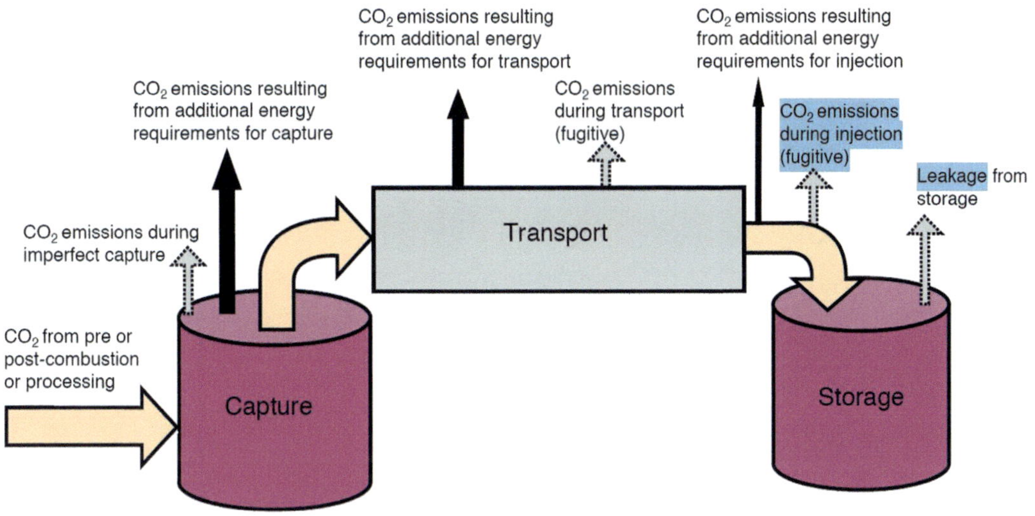

Fig. 7.4 Simplified flow diagram of possible CO_2 emission sources during carbon capture and storage. From the Special Report prepared by Working Group III of the Intergovernmental Panel on Climate Change (IPCC 2005)

and inhibit the well's pores, thereby reducing CO_2 holding capacity and consequently decreasing storage capacity and increasing possibilities for well permeability and leakage (Ho & Tsai 2020). Most CCS projects that have been successful to date are site-specific, either pilot or small-to-medium-scale and have yet to reach annual expected injection capacities. Put simply, there is not enough historical data on long-term, wide-ranging, and large-scale CCS to really gauge the impact of potential hazards to be comfortable about global-scale CCS implementation.

Confidence in the technology continues to be expressed, however. Miocic et al. (2019) calculated leakage rates from a 420,000-year-old naturally occurring, but faulted, CO_2 reservoir in Arizona, USA. Surface travertine ($CaCO_3$) deposits provide evidence of vertical CO_2 leakage which can be dated by uranium–thorium decay. The data show that leakage varies along faults and that individual seeps have lifespans of up to 200,000 years. Time-averaged leakage equated to a linear rate of less than 0.01% y^{-1}. Friedmann et al. (2020) estimate that 85 Gt of CO_2 must be captured and stored from coal-fired power generation alone between 2030 and 2050:

> … Most gas power plants operate for about 30 years, while coal-fired generation plants operate for 40–50 years, and this newly installed capacity will remain in operation through to 2060 without premature closure - CO_2 emissions from the global coal fleet are expected to approach 10 Gt CO_2 in 2030 and exceed 7 Gt CO_2 in 2050 (Cui et al. 2019). If they operate, around 90 percent of those emissions must be captured and stored in 2030, and effectively all emissions must be captured and stored in 2050 to achieve net-zero. If power production from the global coal fleet is only half what has been assumed in this simple illustrative analysis, approximately 85 Gt of CO_2 must be captured and stored from coal-fired power generation alone between 2030 and 2050 to be consistent with a 1.5 °C climate outcome. (Friedmann et al. 2020).

If that 85 Gt reservoir leaks back into the atmosphere at a rate of about 0.01% y^{-1}, the reservoir's total content of sequestered CO_2 will be returned to the atmosphere in 10,000 years. In comparison with the human lifetime, 10,000 years is an unimaginable length of time, but it is totally insignificant compared with the length of time that atmospheric CO_2 has remained sequestered in, for example, coccolithophore limestone layers laid down in the Triassic Period. Due to the sheer size and capacities anticipated for CCS storage sinks, assuming the current global trend for fossil fuel use with CCS

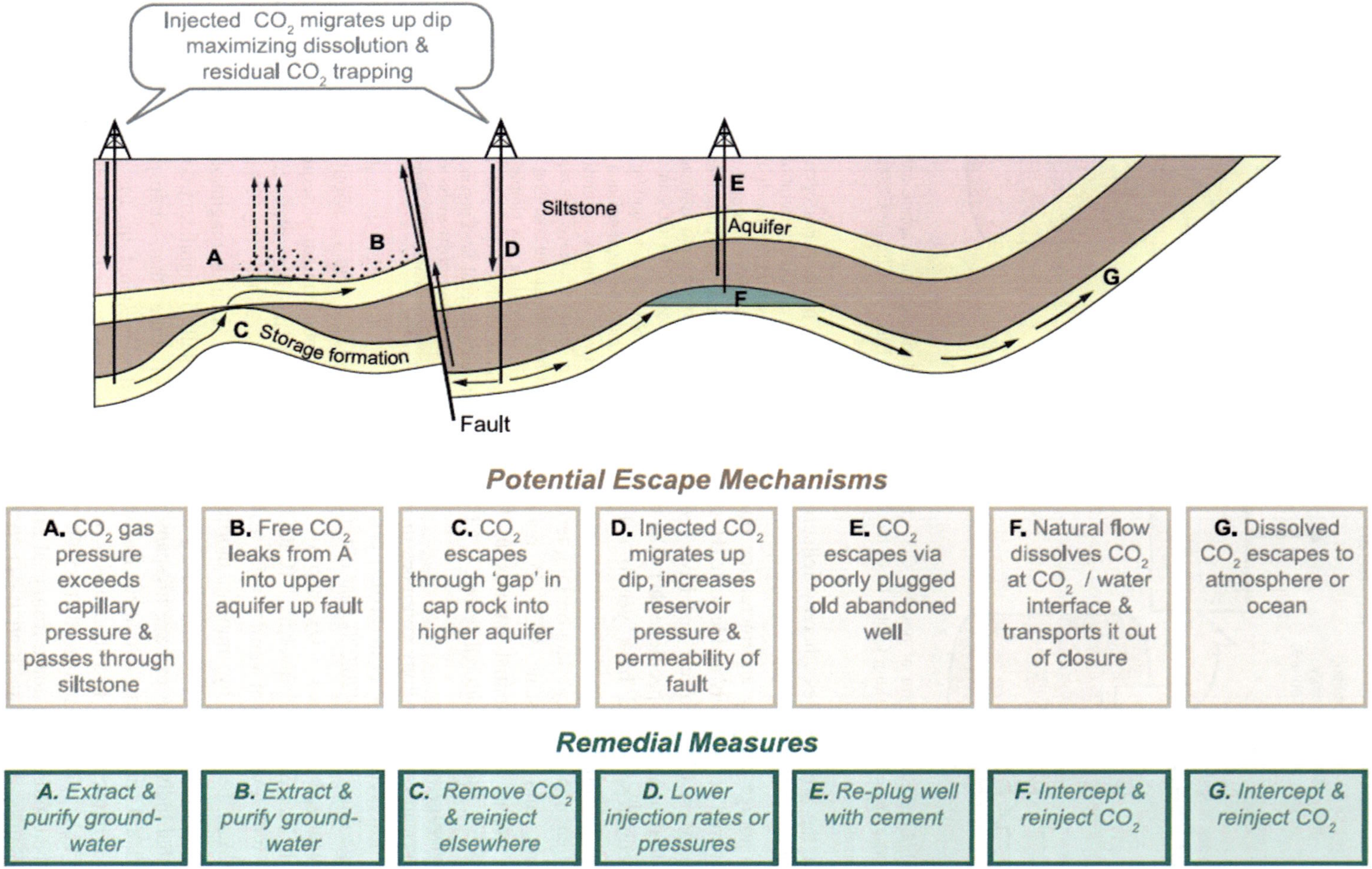

Fig. 7.5 Potential leakage routes and remediation techniques for CO2 injected into saline formations. The remediation technique used would depend on the leakage route identified in a reservoir as shown in the bottom set of panels. From the Special Report prepared by Working Group III of the Intergovernmental Panel on Climate Change (IPCC 2005)

continues, even tiny error margins could result in thousands of tonnes of CO_2 leaking back into terrestrial and coastal ecosystems with potential for environmental damage along the same lines as contaminating leachates from historic landfills or mines implemented by our engineering forefathers.

While the economic and energy-system risks due to potential CCS leakage are arguably modelled with confidence (Liu et al. 2016; Deng et al. 2017), it is our environmental ecosystems that are calling for more attention. Industrial CCS has small risks, but huge consequences for our environment. The key question is 'what if?' Once the gas is in storage, there is no going back, and the environmental risks can only be managed after complications arise. CCS technology is arguably the most significant and powerful carbon sequestration tool we have that can serve as a point-source, 'brute-force' carbon sink solution. Although relatively few sites, globally, are suitable for CCS (because the geological characteristics must be perfect), several sites have been found and classified as having the giga tonnage (Gt) CO_2-storage potential required to meet Paris Agreement climate goals (Fig. 7.3, above).

The *Global CCS Institute* (https://www.globalccsinstitute.com/) is the leading organisation and knowledge-base on CCS projects for industry as well as research and development. According to this Institute's website, current CCS projects either in operation or under procurement or construction (Fig. 7.3) have been estimated to sequester CO_2 at rates from 100,000 to 30 million tonnes per annum, per CCS project site. Operational lifetimes are expected to be at least 25 years. As an example, the *CarbonNet Project* located in South Gippsland, Victoria, Australia is working towards establishing a commercial scale CCS network with storage at the project's Pelican site in Bass Strait, off the South East coast of Australia's 'Ninety Mile Beach'.

The site is projected to sequester up to 5 million t of CO_2 annually (it is site 15 in Fig. 7.3). This is a significant quantity of CO_2 gas. On a molar mass basis, carbon represents 27.29% of the mass of CO_2. Consequently, that 5 million t of CO_2 corresponds to **1,364,500 t of carbon** removed from the atmosphere ***annually*** by the individual **Pelican Site CCS facility**. The key consideration here is that these large point-source quantities of CO_2 are, for the most part, found in heavy industrial plant sites. Artificial CCUS solutions include but are not limited to CO_2 injection or subsurface mineralisation, CO_2 flooding and enhanced oil recovery (EOR), deep sea storage (such as deep water pressurised storage conveyed by pipe), which are the major solutions. Less impactful, but equally innovative are: Direct Air capture and storage (DAC), Dry Ice Emissions capture (e.g., DecarbonIce™) or capturing CO_2 from hydrogen production (e.g., CryoCap™).

Although sceptics have raised significant concern for the environmental risks involved with CCS projects, the science has (so far) proved its safety and efficacy, albeit, at very small scales beyond pilot field trials alone. As a result of stricter government policies towards fossil fuel use and of heavy GHG emissions in general, the major CO_2 emitters (namely fossil fuel companies) have sought to invest into CCS as a **business solution** to become carbon neutral. In turn, the highest quantifiable CCU/CCS technologies are capitalising on a new market demand created by government policy, where major heavy industries and GHG emitters are needing to protect themselves ***and their banks*** against possible future sanctions.

As discussed in Chap. 1, the 2019 report of the US National Academies of Sciences, Engineering, and Medicine entitled *Negative Emissions Technologies and Reliable Sequestration: A Research Agenda* (NASEM 2019) describes negative emissions technologies, or NETs as optimal carbon sequestration solutions. NETs are technologies that remove and sequester CO_2 from the atmosphere with the intention of mitigating climate change, with a biotechnological component.

NETs have previously received less attention than industrial technologies aimed at reducing the level of future CO_2 emissions by reducing fossil fuel consumption, though this requires massive deployment of low-carbon technologies and agricultural land use change between now and the target date of 2050. One key point here is that CCUS is more useful for achieving zero or carbon–neutral operations, not negative, especially when the CCUS-facilitated plant is not processing biological or waste resources (also known as 'BECCS', Bioenergy with CCS).

According to the Global Carbon Project, about 37 billion tonnes of CO_2 gas was emitted globally by heavy industries in 2019 [https://www.globalcarbonproject.org/]. The number of heavy emitting plants is rising, particularly in Asia as mentioned earlier. To date, there are more than 5,000 large industrial plants globally that produce CO_2 emissions above 1 million tonnes per year. Again, due to recent industrial development in Asia and lacking regulatory action or initiative, this number continues to grow at significant capacity. Interestingly, the number of CCS plants under development between 2010 and 2017 reduced significantly, followed by a recent resurgence in the development of the technology (Fig. 7.2).

To date, close to 40 CO_2 injection facilities have been brought into operation (mostly in the USA) and many more are in development (Fig. 7.3, above). This activity is monitored by the Center for Climate and Energy Solutions (https://www.c2es.org/), an independent, non-partisan, nonprofit organisation which is "... working to forge practical solutions to climate change ..." (and view https://www.c2es.org/content/carbon-capture/).

Facilities already in operation are implemented as an add-on or retrofit to heavy industrial plants; particularly in the oil and gas industries and fossil fuel energy generators, but also cement, steel, and fertiliser producers, though the technologies are generally applicable to any CO_2 emitting facility. The CCS system captures CO_2 produced directly from the industrial plant's output flue gases and pumps it underground into deep saline pockets under cap rock.

Although injection into sedimentary basins has been commonly conducted for enhancing oil recovery from certain wells (Enhanced Oil Recovery is one of the business goals of CSS; Fig. 7.6), it has been proved that basaltic cap rock pockets provide much more safety and encapsulation for mineralised CCS storage into stone (with pioneer work laid out via pilot studies in Iceland; see https://www.carbfix.com/).

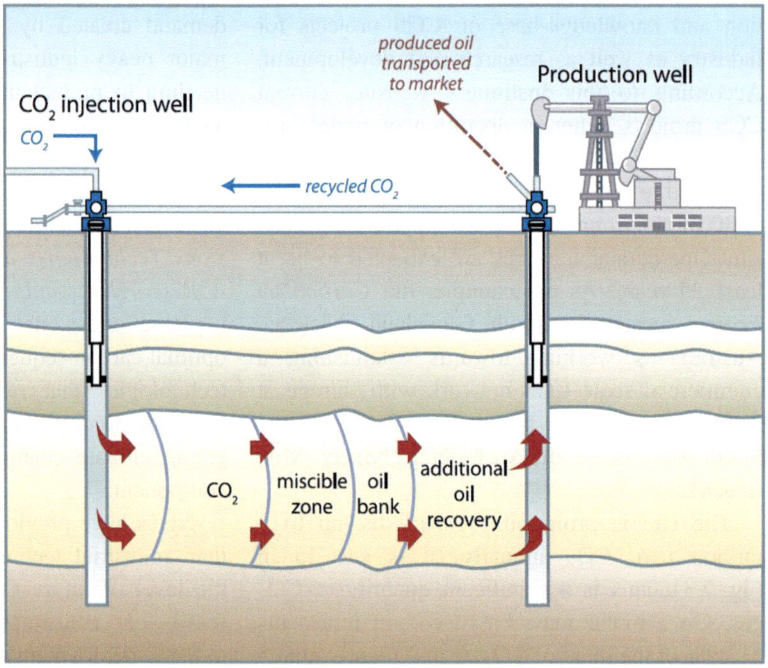

Fig. 7.6 Enhanced Oil Recovery (EOR) by CO_2 injection with some storage of retained CO_2. The CO_2 that is produced with the oil is separated and reinjected back into the formation; recycling CO_2 this way decreases the amount of CO_2 that must be purchased and avoids emissions to the atmosphere. From the Special Report prepared by Working Group III of the Intergovernmental Panel on Climate Change (IPCC 2005)

Figure 7.3 displays the main CCS projects as of 2019, as listed by the *Center for Climate and Energy Solutions* (URL: https://www.c2es.org/content/carbon-capture/). Many of the projects shown in Fig. 7.3 are pioneering new approaches and/or new technologies, a few examples will illustrate the range of these technological innovations:

- **The Northern Lights project** is part of the Norwegian full-scale CCS project, which includes capture of CO_2 from industrial capture sources in the Oslo-fjord region (cement and waste-to-energy industries). The process uses CO_2 mixtures with amine-gases and cryogenic separation and distillation to separate and liquify CO_2 gas. Amine gas treatment, also known as amine scrubbing, is widely used to remove hydrogen sulfide (H_2S) and carbon dioxide (CO_2) from gases in refineries, petrochemical plants, natural gas processing plants and other chemical industries. The process uses aqueous solutions of various alkylamines, most commonly diethanolamine (DEA), monoethanolamine (MEA) and methyldiethanolamine (MDEA). The gas mixtures have advantageous physical properties under pressure that permit gas liquefaction and cryogenic distillation to purify and liquify the CO_2 (Mandal et al. 2001; Xu et al. 2014). Liquid CO_2 is shipped from the capture sites to an onshore terminal on the Norwegian west coast. From there, the liquified CO_2 will be transported by pipeline to an offshore permanent storage location 2700 m below seabed of the North Sea. The facility is capable of sequestering 5 Mt y^{-1}.
- **The CarbonNet Project/CO2CRC** in Australia is capable of up to 5 million tonne/year and utilises metal organic framework (MOF) material to capture CO_2. Metal organic frameworks resemble a sponge, filled with magnetic nanoparticles that adsorb carbon dioxide gas. Otherwise known as magnetic induction swing adsorption (MISA), the advantage of the process is that it requires one-third of the energy input (used mainly to regenerate the capture media) compared to any other reported CO_2 capture method (Sadiq et al. 2020).
- **CarbFix** in Iceland is a project that commenced in 2006 and has since developed innovative geological carbon storage by capturing and rapidly storing CO_2 as a mineral formed in reactive, porous, basaltic subsurface. The project has also explored mineral fluid interactions to predict the fate and impact of CO_2 injected into the subsurface. The process involves first dissolving the CO_2 gas into water and then injecting it into the subsurface. "… This had two advantages: firstly, CO_2-charged water is denser than pure water, so it tends to sink. Secondly, the acidic CO_2-charged water promotes reactions in the subsurface, specifically the dissolution of basalt, which in turn leads to the fixation of carbon as stable mineral phases … Once it is made into a mineral the carbon is immobile over geologic time frames, representing a safe, long time solution for CO_2 storage…" The process was field-tested at the CarbFix pilot site in Hellisheidi, Iceland, where the original injection was shown to fix over 90% of the injected 170 tonnes of pure CO_2 as stable carbonate minerals in less than 18 months. Economic studies show costs in the order of "… 30–40 US$ per tonne, which is no more expensive than other less safe alternatives…" (quotations above taken from the ***Carbfix.com*** website at https://www.carbfix.com/co2-react-2013-2017). Hellisheidi has achieved costs less than \$US25 t^{-1} and as of January 2020 "… over 50,000 tonnes have been injected into reactive basalts … for permanent storage". Here, the CO_2 is captured in a scrubbing tower with annual capacity of about 12,000 tonnes of CO_2 and 6,000 tonnes of H_2S, about 30% and 75% of the plant's emissions respectively." (https://www.carbfix.com/faq).

Possibly the most exotic carbon storage plan is that which intends to convert captured CO_2 to methane (CH_4) and use that to make diamonds [https://skydiamond.com/].

The costs of CCS adoption were discussed in the Special Report *Carbon Dioxide Capture and*

Storage prepared by Working Group III of the Intergovernmental Panel on Climate Change (IPCC 2005). According to Kheshgi et al. (2012) the publication of this report "… raised the profile of CCS, particularly among the expert community dealing with international climate policy (Meadowcroft & Langhelle 2011).

The expert community now commonly sees CCS as a major option for reducing global emissions of CO_2. The technology plays a major role in long-term scenarios where there is significant reduction in greenhouse gas emissions (Clarke et al. 2009; IEA 2010). For CCS to play such a major role, the separation, transport and storage would have to handle large volumes of CO_2 and involve huge investments in facilities and infrastructure …".

We illustrate costs of CCS adoption in Table 7.1, for which we have recalculated the cost ranges given in the original 2005 publication using the Consumer Price Index inflation calculator of the US Bureau of Labor Statistics as featured on ***Ian Webster*'s** website (https://www.in2013dollars.com/us/inflation/).

Despite the economic advantages of CCUS apparent from Table 7.1, the technologies face a number of practical and economic barriers that must be overcome before they can be deployed on a sufficiently large scale, and over a sufficiently long time interval, to make serious inroads into the atmosphere's accumulated fossil-CO_2 burden. The main economic and environmental hurdles in sight are:

- the significantly large capital investment and hard infrastructure required for implementation, operation and maintenance; and
- the extremely energy-intensive process required for carbon utilisation (CU) or sequestration (CS).

Table 7.1 Cost ranges for the components of a CCS system as applied to a given type of power plant or industrial source

CCS system components	Cost range	Remarks
Capture from a coal or gas-fired power plant	21–104 US$ per t CO_2 net captured US$ per t CO_2	Net costs of captured CO_2, compared to the same plant without capture
Capture from hydrogen and ammonia production or gas processing	7–76 US$ per t CO_2 net captured	Applies to high-purity sources requiring simple drying and compression
Capture from other industrial sources	35–159 US$ per t CO_2 net captured	Range reflects use of a number of different technologies and fuels
Transportation	1.4–11 US$ per t CO_2 transported	Per 250 km pipeline or shipping for mass flow rates of 5 (high end) to 40 (low end) Mt CO_2 yr^{-1}
Geological storage[a]	0.7–11 US$ per t CO_2 net injected	Excluding potential revenues from EOR or ECBM
Geological storage: monitoring and verification	0.14–0.4 US$ per t CO_2 injected	This covers pre-injection, injection, and post-injection monitoring, and depends on the regulatory requirements
Ocean storage	7–41 US$ per t CO_2 net injected	Including offshore transportation of 100–500 km, excluding monitoring and verification
Mineral carbonation	69–138 US$ per t CO_2 net mineralised	Range for the best case studied. Includes additional energy use for carbonation

All numbers are representative of the costs for large-scale, new installations, with natural gas prices assumed to be 3.9–6 US$ GJ^{-1} and coal prices 1.4–2 US$ GJ^{-1}. Monitoring costs are also reflected. [a]Over the long term there may be additional costs for remediation and liabilities. Data Source: The Special Report prepared by Working Group III of the Intergovernmental Panel on Climate Change (IPCC 2005); all costs recalculated for inflation using the factor $1 in 2004 is equivalent in purchasing power to about $1.38 in 2021

Those two points identify the most important disincentive to CSS implementation: its cost. This was foreshadowed in IPCC's special report on CCS, which stated that fossil fuel-based power plants equipped with CCS for mineralised subsurface injection, will require 60–180% ***more energy*** (more energy = more cost) than a power plant without CCS (IPCC 2005).

Table 7.2 shows the total costs of CCS and electricity generation for three power systems with pipeline transport and two geological storage options. Again, the data is sourced from the Special Report *Carbon Dioxide Capture and Storage* prepared by Working Group III of the Intergovernmental Panel on Climate Change (IPCC 2005), with costs adjusted for inflation as in Table 7.1.

Overall, the situation is well summarised by this quotation from the Wikipedia article on *Carbon Capture and Storage [CSS]*:

> The increased energy required for the carbon capturing process is also called an energy penalty. It has been estimated that about 60% of the energy penalty originates from the capture process itself, 30% comes from compression of CO_2, while the remaining 10% comes from electricity requirements for necessary pumps and fans. CCS technology is expected to use between 10% and 40% of the energy produced by a power station. CCS would increase the fuel requirement of a plant with CCS by about 15% for a gas-fired plant. The cost of this extra fuel, as well as storage and other system costs, are estimated to increase the costs of energy from a power plant with CCS by 30%–60%, depending on the specific circumstances. (source: https://en.wikipedia.org/wiki/Carbon_capture_and_storage).

The early recognition of this energy penalty may well be the reason for the relatively late uptake of CSS technology by the power generation industries, as compared with gas processing industries (Fig. 7.7). Though, of course, the scale of the infrastructure required by power generation facilities and the long lead times required for its design and implementation must also have contributed to the marked difference evident in Fig. 7.7 between the operation of CCS applications in these two types of industry. We have assembled a summary of cost estimates of CCUS technologies and their CO_2 removal rates in Table 7.3.

Table 7.2 The costs of CO_2 capture, transport and geological storage for new power plants using bituminous coal or natural gas

Power plant performance and cost parameters[a]	Pulverised coal power plant	Natural gas combined cycle power plant	Integrated coal gasification combined cycle power plant
Reference plant without CCS			
Cost of electricity (US$ per kWh)	0.062–0.075	0.045 0.073	0.060–0.089
Power plant with capture			
Increased fuel requirement (%)	24–40	11–22	14–25
CO_2 captured (kg per kWh)	0.82–0.97	0.36–0.41	0.67–0.94
CO_2 avoided (kg per kWh)	0.62–0.70	0.30–0.32	0.59–0.73
% CO_2 avoided	81–88	83–88	81–91
Power plant with capture and geological storage[b]			
% increase in cost of electricity	43–91	37–85	21–78
Power plant with capture and enhanced oil recovery[c]			
% increase in cost of electricity	12–57	19–63	(−10)–46

All changes are relative to a similar (reference) plant without CCS. Data sourced from Table TS.10 in IPCC (2005); see Table TS.3 in that report for the assumptions underlying quoted cost ranges. Costs recalculated for inflation using the factor $1 in 2002 is equivalent in purchasing power to about $1.45 in 2021

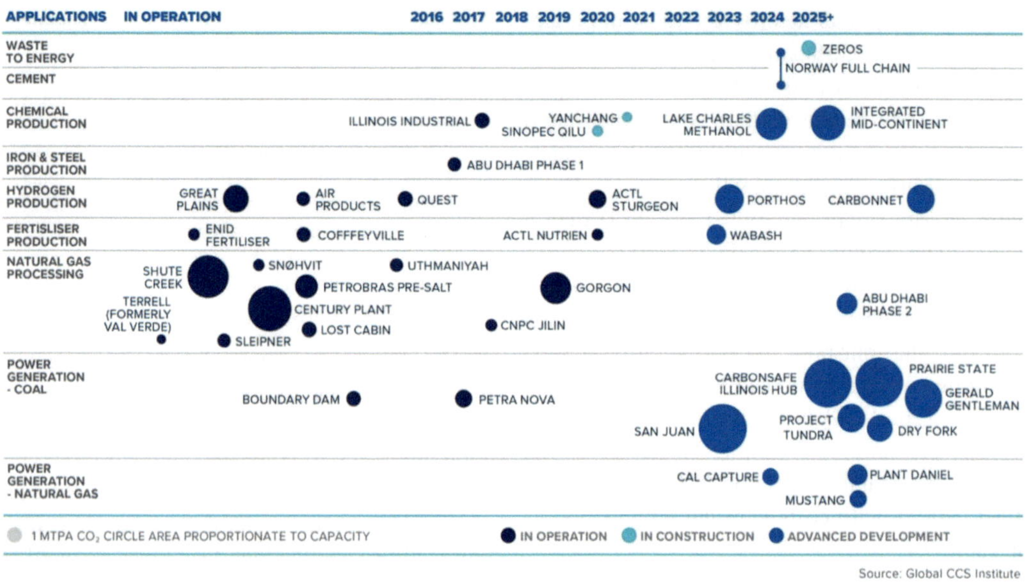

Fig. 7.7 CCS projects around the world since the 2005 IPCC Special Report *Carbon Dioxide Capture and Storage. Source* Marshall et al. (2020)

Table 7.3 Summary of CCUS solutions including cost estimates, CO_2 removal rate estimates and UN Sustainable Development Goals (UN SDGs) addressed

Solution	Estimated global potential removal rate of CO_2 (current) (Gt y^{-1} CO_2)	Estimated cost of implementation at scale (US$ t^{-1} CO_2)	Number of UN SDGs addressed (/ 17)
Terrestrial afforestation	2.5–9 (higher values directly impact food security)[x]	15–50[x]	10[a]–13[b,h]
Blue carbon afforestation	0.13–0.84 (only based on post-1980 coastal wetland recovery)[x, h]	10[x]	12[e,f,h]
Enhanced Weathering (TEW)	2[g]	75-250 [g]	9 [h]
Ocean Fertilisation (Macronutrient only)	3.7[i]	≥ 20[i]	2[j]
Agricultural & Other Soil Management (e.g., biochar)	0–3[x,h]	0–50[x]	12 [h]
Bioenergy with carbon capture and sequestration (BECCS)	3.5–5.2 (assumes only waste biomass as feedstock)[x] 10–15 (assumes waste biomass ***and*** dedicated energy crop feedstocks)[x]	Electricity: 70[x] Fuels: 37–132[x]	7–9 [h]
Direct Air Capture	< 0.01[k]	90–600 (current demonstrated cost of DAC)[x]	< 8 [l]
CCUS	15 [m]	25–210[n]	CCUS: 4[o]–6[c]

Sources [x] NASEM, (2019). [a] The State of the World's Forests 2018 (FAO, 2018). [b] De Jong et al. (2018)
[c] Aker Carbon Capture Presentation: https://www.akersolutions.com/globalassets/investors/presentations/aker-carbon-capture-company-presentation-aug-6-2020.pdf. [d] CCM Technologies: http://ccmtechnologies.co.uk/. [e] Kuwae and Hori (2019). [f] United Nations Development Programme - Thailand (UNDP Thailand, 2019). [g] Beerling et al. (2020). [h] Smith et al. (2019). [i] Jones (2014). [j] Secretariat of the Convention on Biological Diversity (2009). [k] Budinis (2020). [l] Beuttler et al. (2019) [note that all authors are employed by Climeworks AG, which is one of the main proponents of direct air capture]. [m] IOGP (2019). [n] Irlam (2017). [o] Zapantis (2017). View the 17 Sustainable Development Goals of the United Nations at this URL: https://sdgs.un.org/goals. View the 17 Sustainable Development Goals of the United Nations at this URL: https://sdgs.un.org/goals

The UN Sustainable Development Goals (SDGs [https://sdgs.un.org/goals]) are shown in the final column of Table 7.3 because in pursuance of the Paris Climate targets through climate change mitigation technologies (artificial or bio-based), we must consider both the opportunities and risks associated with such solutions that remove GHGs from the atmosphere. Such an approach is helpful in determining the true sustainability of solutions, because value factors such as land and water use, cultural and land heritage as well as biodiversity and nutrient stocks are given ***significant*** weighting.

Smith et al. (2019), also explored this for land-based solutions, by:

> … looking through the lens of the functions …" provided by each solution and "… their impact on ecosystem services [classified according to the new Intergovernmental Science-Policy Platform on Biodiversity and Ecosystem Services (IPBES) classification known as Nature's Contributions to People (NCPs) … (Smith et al. 2019).

Especially for solutions that help conserve or improve natural ecosystem services, the valued benefits usually go far beyond what project engineering or financial models would normally include. Meaning, ***we should be placing even higher-than-usual value on natural capital and global environmental health improvement indicators on current and future decision-making (Fig. 7.8)***.

There are some other issues that seem to be held in the background of the CCS arena, though common in the business world. These result in some ambiguity in regards to ***how*** climate is to be managed, raising the questions: where do the controlling influence and interest lie, and who are the major stakeholders? These are robust questions that need to be asked, especially in a situation where CCUS is most wholeheartedly backed by the major fossil fuel-based enterprises themselves.

A quick analysis of the **Global CCS Institute**'s current (December 2020) 88 members (https://www.globalccsinstitute.com/membership/our-members/), at least 48 out of 88 members rely on or have direct business interests in fossil fuel use. A further 17 members currently rely on fossil fuel industries either indirectly or partially, leaving ***only 22 of the 88 members with no immediate evidence of business reliance or connection to fossil fuel use***. However, it is important to keep in mind that these members might also have significant shareholders or be subsidiaries of upper tier companies who do have vested interests in continued fossil fuel use. Here, we looked only as far as each company's web page or Wikipedia descriptions where available.

The ***Global CCS Institute*** recognises the IPCC's latest targets in a September 2020 report (Friedmann et al. 2020) these certain actions are:

- A 50% reduction of CO_2 emissions is needed to achieve net-zero climate goals by 2030.
- A rapid implementation of climate mitigating infrastructure is needed urgently, including the expansion of CO_2 pipelines from the current 8,000 km to 43,000 km by 2030.
- Urgent development and implementation of clear climate policies to optimise financial and regulatory risk mitigation for CCS infrastructure.

The report also offers this advice:

> Due to the urgency of the climate crisis, time is of the essence. There are no important technical barriers to scale-up. The costs are well within the conventional boundaries of global energy investments and the policy options well understood. The next ten years will prove decisive – if the governments of the world are to meet their climate goals, these key policies must enter into force with deliberate speed (Friedmann et al. 2020).

Indeed, 43,000 km of CO_2 pipeline is a lot of hard infrastructure. So, let us assume that by 2030 we achieve a reduction in fossil fuel usage and then ask ourselves: will that not make some of these pipelines redundant? We must not ignore the fact that retrofitting conventional fossil fuel plants with CCS serves not only to assist in climate change mitigation, but also to create redundant hard infrastructure for future generations, not to mention the enormous continual efforts required to monitor and manage the thousands of highly concentrated CO_2 sinks that come with this direction.

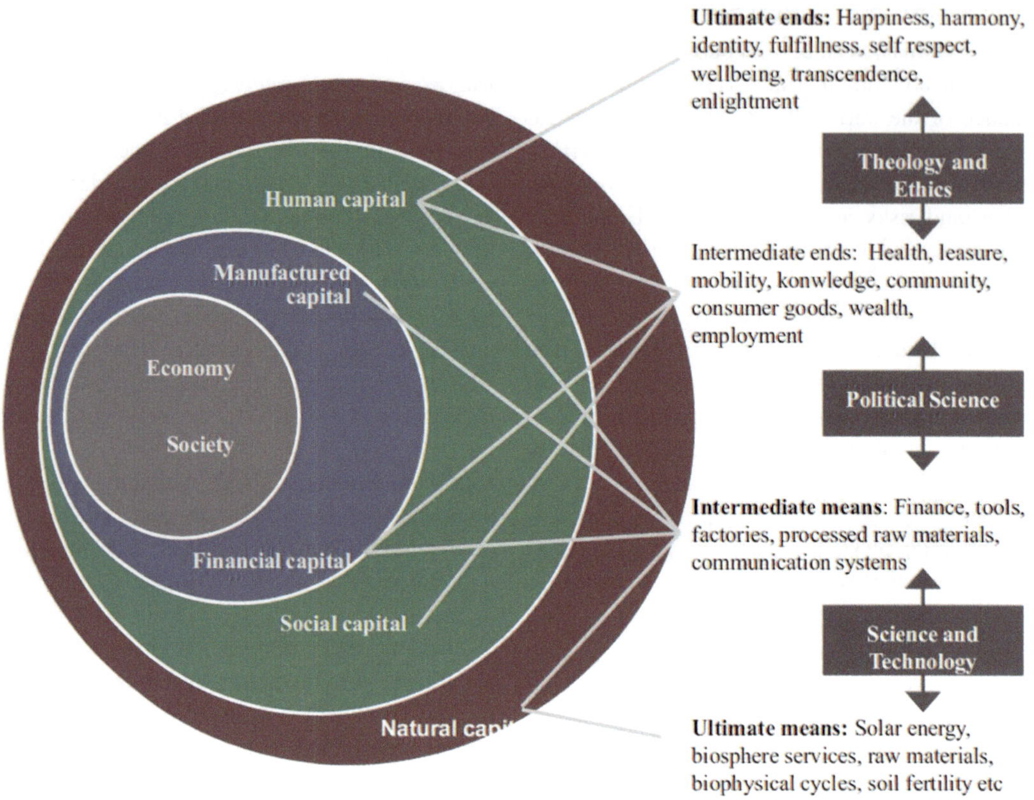

Fig. 7.8 Capital categories supporting wellbeing shown as embedded circles (left) with concepts of human needs for wellbeing listed at right. Redrawn after Maack and Davidsdottir (2015)

Of course, some plants *may* be able to convert to biomass-use instead of total decommissioning, but the costs of conversion will usually outweigh the construction of a whole new plant, particularly given the likelihood of more cost-effective and optimised designs, construction/manufacturing materials and technological services that will be available decades from now. The scenario can be seen as similar to mine tailings ponds; we are now seeing more and more closed mining sites requiring growing amounts of risk management, primarily environmental.

On another important note, Krüger (2017) published an interesting piece on the conflicts over CCS in international climate governance, namely postulating two theses:

- That the future of climate governance is contingent on decisions about the **continued use of fossil fuels**.
- That CCS-conflicts have an unpredictable influence that could lead to implications and cracks within the paradigm of ecological modernisation and thus could **politicise international climate policy**.

Krüger (2017) discusses the consequences of allowing private business interests to determine the direction of humanity's future. The problem, however, is one of necessity. On the one hand, CCUS is a power-house technology that could play a central role in deciding where humanity ends up by the end of the twenty-first century. On the other hand, because it is desired most by fossil fuel-reliant enterprises to safeguard their own business, CCUS is tainted with contention. It may be the magical release from our worst nightmares; or it could be the Poisoned Apple which will send us into the Sleeping Death of our times.

Artificial CCS solutions are researched, developed, and engineered to address specifically the question of 'how can we prevent GHG emissions entering our atmosphere?' However, if these artificial CCS solutions are continuously implemented, unchecked rapidly and widely, they could result in serious implications and even more problems for our future generations of scientists and engineers.

As we see it, the problem is that CCUS has attracted market-trading, but without the optimal regulatory framework and market rules that would alleviate mistrust, misguidance, and corruption. The carbon trading schemes that have been opened in many nations to date have yielded both positive and negative results in relation to the problem posed by climate change. As the initial goal of carbon sequestration is to reduce atmospheric CO_2 levels, the primary goal of a carbon market or carbon trading scheme is to sequester the most carbon. As a result, industries and corporations have started to look at technologies that will sequester the most carbon, and that aligns with their future business plans. These are the methods of carbon sequestration best supported by fossil fuel companies and are therefore not the ideal solutions for our environment and its ecosystems. It is the technology that secures the industry's business plan and market position heading forward into the future, rather than the technology that is best for planet Earth.

As we all know, increasing carbon emissions, atmospheric GHG levels and global warming result from a complex system of biogeochemical processes affected by many anthropogenic practices. Because of this, rather than a ***carbon trading market***, it would make more sense to introduce a **global environmental health market** that offers traders and participating industries and businesses, alongside the carbon credits, trading credits that could be equally important contributors to our attempts to avert global warming. For example, **biodiversity credits**, **ecosystem service credits**, and **biomimicry-of-technology credits**.

That's not what we have. Instead of introducing an ***environmental with carbon market***, we only have a carbon market. What is concerning about current practices is that removal of carbon from the atmosphere is the only environmental concern and those other global environmental health concerns are not at the forefront of any aspect of the carbon trading market. The value is placed on the removal of carbon from the atmosphere at almost any cost. Consequently, the money (what little is left of it after successive traders have taken their top slice) therefore, goes to carbon credits, not ***environmental*** credits (listen to The Climate Question 2021a podcast).

We would rather see a market, that consists of rules and regulations based on a global environmental health market focused on altering the root anthropogenic causes that have resulted not only in global warming, but in active destruction of ecosystems by over-exploitation, global loss of biodiversity, and anthropogenic species extinctions at rates not seen since the darkest days of the planet's geological history.

The carbon market is already established, with the ebb and flow of supply and demand circulating. But it is important, as we make more serious attempts to ameliorate the damage our industrial activities have already done to the atmosphere, that rather than concentrating solely on the symptomatic results of unsustainable anthropogenically-raised GHG emissions, we do not forget those broader anthropogenic mistakes that should be change-incentivised towards restoring and maintaining the natural circular economies of healthy environmental ecosystems. Between the additional energy required for industrial CCS, the CO_2 emissions during the process and the leakage during storage (which certainly increases with the years), it seems that twice as much oil and gas would have to be extracted to store the CO_2 emitted simply by the current use of these fossil fuels. Widespread use of CSS would be like being blindfolded on the edge of a precipice and taking a big step forward!

7.5 Carbon Emissions Reduction Targets

As already mentioned in earlier chapters, key climate-focused actions are required in order to avoid climate catastrophe. As we progress into the third decade of the twenty-first century, climate records proved that 2011–2020 was the warmest decade on record, with the warmest six years all being since 2015 (WMO 2020), while the ***Copernicus Climate Change Service*** satellite data showed that 2020 was statistically at a dead heat with 2016 as the world's warmest year on record. Copernicus data comes from a constellation of Sentinel satellites that monitor the Earth from orbit, as well as measurements taken at ground level. Temperature data from the system shows that 2020 was 1.25 °C warmer globally than the average from 1850–1900, a time often described as the 'pre-industrial' period. (https://climate.copernicus.eu/). Furthermore, the *Carnegie Climate Governance Initiative* report (Mace et al. 2021) makes clear that it is ***no longer sufficient to reduce emissions alone***. Instead, CO_2 will also need to be ***removed from the atmosphere, on a scale never previously attempted***. But, while a number of reporting rules and accounting practices are already in place with direct applicability to the implementation of carbon dioxide removal options, many governance gaps remain. From their analysis of why private and public sectors must invest in protecting, preserving, and enhancing the blue natural capital of the Red Sea, Cziesielski et al. (2021) conclude that communication, participation, and transparency ***of all involved parties*** are required to successfully build a blue economy that thrives with its natural resources.

7.6 The Comparison with 'Soft' (Nature-Based) Carbon Sequestration

The 'hard' carbon sequestration solutions available to us include the following processes.

CCUS & mineralisation; in the latter part of this combined process, CO_2 from the atmosphere forms a chemical bond with reactive rocks, like mantle peridotite and basaltic lava, both at the surface (ex situ) where CO_2 in ambient air is mineralised on exposed rock, and in the subsurface (in situ) where concentrated CO_2 streams are injected into rocks to mineralise in the pores.

Direct air capture (DAC) uses chemical processes that capture CO_2 from ambient air and concentrate it, so that it can be injected into a storage reservoir or utilised in the value-chain of secondary industries.

Bioenergy with carbon capture and sequestration (BECCS). BECCS involves using plant biomass as an energy source, primarily to produce electricity by one of two methods ***combustion*** or ***conversion***. ***Combustion*** uses the biomass directly as a furnace fuel for conventional electricity generation or for other furnace-based industrial applications (cement, paper pulping, waste incineration, petrochemicals and steel and iron production). Emitted CO_2 is captured from the flue gas stream resulting from combustion. ***Conversion*** of biomass involves digestion or fermentation to produce gaseous or liquid fuels, respectively; the main one being bioethanol, which produces almost pure CO_2 during fermentation.

The subsequent combustion of the biofuel or gas (methane is generated by anaerobic digestion of biomass, including household food and garden wastes) also produces CO_2 which, ***if stored*** by the end user, results in overall lower emissions reduction by BECCS (if not stored the CO_2 is returned to the atmosphere by the end user). In 2019 there were five BECCS facilities around the world, collectively capturing approximately 1.5 million tonnes of CO_2 per year (Mt y^{-1}). BECCS is a way to avoid use of fossil fuels, in addition to its capture and storage aspects.

This energy production method recycles today's CO_2, which was extracted from the atmosphere by the biomass as it grew, back to the atmosphere; in contrast to fossil fuels, which make a net increase of ancient CO_2 to today's atmosphere. The biomass feedstock can be derived from a waste material (e.g., sugarcane wastes which are widely used for bioethanol) or dedicated energy crops (e.g. fast-growing tree species) planted purely as an energy production feedstock. At the present time, biomass feedstock supply for energy generation by burning is dominated by forest management schemes (Consoli 2019).

When combined with capture and sequestration of CO_2 the overall BECCS process can provide a ***net reduction of CO_2 in the atmosphere***. However, the irrigation needs of bioenergy crop plantations can constrain the potential of BECCS (Ai et al. 2021). Industry opinion of BECCS is essentially that it is the best solution to decarbonise emission-intensive industries. However, public perceptions of this technology are variable and seem to be linked to the regulatory policies by which its use is incentivised (Bellamy et al. 2019). Payments based on the amount of CO_2 removed from the atmosphere were approved but guarantees of higher prices for producers selling energy derived from BECCS were strongly opposed. It remains to be seen whether the recently (April 19, 2021) announced winners of the $20 M ***NRG COSIA Carbon XPRIZE***, a prize that set out to convert CO_2 emissions into valuable products, can change these public perceptions, at least as far as production of traditional concrete is concerned. Concrete being "*the world's most abundant human-made material* …[accounting]… *for seven of all global CO_2 emissions*" [view: https://www.xprize.org/prizes/], both $7.5 M grand prize winners developed technologies focused on decarbonising concrete and converted the most CO_2 into products with the highest value, while minimising their overall CO_2 footprint, land use, water use, and energy use.

- **CarbonCure Technologies** [https://www.carboncure.com/] produce concrete with a reduced water and carbon footprint without sacrifice to the material's reliability by injecting a precise dosage of CO_2 into a concrete plant's reclaimer system, which contains the water used to wash out concrete trucks and mixers. In tests under industrial conditions, 25 tonnes of CO_2 per day supplied by the flue gasses from an adjacent natural gas-fired power plant was converted to a permanently embedded mineral with strength-enhancing properties which can be incorporated into new concrete mixes. Overall, the technology reduces the material costs and increase profitability for concrete producers.
- The Los Angeles-based **UCLA CarbonBuilt** [https://www.carbonbuilt.com/] developed technology that reduces the carbon footprint of concrete by more than 50% while reducing raw material costs and increasing profitability. The CarbonBuilt concrete formulation significantly decreases the need for ordinary Portland cement by direct injection of CO_2 from flue gas streams during the curing process of concrete mixtures. In this process, also, the CO_2 is mineralised and permanently stored.

Additionally, the ***NRG COSIA Carbon XPRIZE*** awarded ***X-Factor awards*** to two finalists that created other valuable products from waste CO_2:

- **Carbon Upcycling-NLT** [https://carbonupcycling.com/] produces nanoparticles with applications in various industries, particularly concrete, construction and plastics.
- **Carbon Corp** [http://carboncorp.org/] transforms CO_2 into carbon nanotubes, with applications such as lightweight, ultra-strong and cost-effective replacements for metals; stronger cement-composite building materials; and expanding applications in industrial catalysis, batteries, and nanoelectronics.

Enhanced weathering. Enhanced weathering ***or accelerated weathering*** refers to geoengineering approaches intended to remove CO_2 from the atmosphere by using specific natural or artificially created minerals which absorb CO_2 and transform it into other substances through chemical reactions occurring in water (https://en.wikipedia.org/wiki/Enhanced_weathering).

Ocean fertilisation has also been suggested as a CO_2 removal technique involving dumping iron filings or other nutrients (e.g., urea) into seawater to ***stimulate phytoplankton growth*** in areas that have low photosynthetic production. The idea is that the new phytoplankton will absorb atmospheric CO_2 and, when the phytoplankton die, the carbon is expected to be sequestered 'as they sink to the ocean floor'. Over the last 30 years there have been at least 13 ocean iron fertilisation experiments. However, scientific studies have shown that the amount of carbon exported to the deep sea is either very low or undetectable because ***much of the carbon is released again* via *the food chain*** (https://www.geoengineeringmonitor.org/2018/05/ocean-fertilization/).

The section below briefly outlines the nature-based (or 'soft') alternative solutions. We will look at each solution holistically and from a sustainable infrastructure point of view, including consideration of all capital value offered by each solution to society. Following the outlining of each solution, a comparison of the value capital offered by each will be presented.

7.7 'Soft' Carbon Sequestration Solutions (Nature-Based)

Soft carbon sequestration solutions include all the nature-based negative emissions technologies (NB-NETs). NB-NETs differ from 'hard' solutions mainly in terms of natural capital. The 'hard' solutions (CCUS and direct air capture in particular) lack natural capital, primarily biomimicry-of-technology functionality, and ecosystem services. These aspects are provided by the 'soft' NB-NETs. As described elsewhere (Moore et al. 2021a), these NB-NETs have low to medium costs (US\$100 t^{-1} CO_2 or less) and offer substantial potential for safe scale-up from current deployment.

Griscom et al. (2017) provide a succinct overview of ***natural climate solutions*** (NCSs), which encompass 'soft' carbon sequestration potential. According to the study, NCSs can provide over one-third of the cost-effective climate mitigation needed between now and 2030 to satisfy the IPCC's 'below 2 °C model'. However, this can only be achieved via aggressive fossil fuel emissions reductions, which if achieved can allow NCSs to offer a powerful set of solutions for Paris Climate Agreement nations. As an added natural capital benefit, 'soft' solutions help improve soil health and productivity, clean air and water and help restore and maintain biodiversity and healthy nutrient flow. They showed that most NCSs, when implemented effectively, offer additional benefits such as water filtration, flood risk reduction, improved soil health, improved habitat biodiversity, and enhanced climate resilience, and they concluded that:

> … existing knowledge … provides a robust basis for immediate global action to improve ecosystem stewardship as a major solution to climate change… (Griscom et al. 2017).

Another valuable source of detailed information is the 2019 report of the US National Academies of Sciences, Engineering, and Medicine entitled ***Negative Emissions Technologies and Reliable Sequestration: A Research Agenda*** (NASEM 2019). The *Committee on Developing a Research Agenda for Carbon Dioxide Removal and Reliable Sequestration*, which produced this report, was created to recommend a detailed research development plan for what are known as **negative emissions technologies** (NETs), which are technologies that remove and sequester CO_2 from the atmosphere with the intention of mitigating climate change. NETs have received much less attention than the 'hard' technologies, but this report concludes that:

> … If the goals for climate and economic growth are to be achieved, negative emissions technologies will likely need to play a large role in mitigating climate change by removing ~10 Gt y^{-1} CO_2 globally by mid-century and ~20 Gt y^{-1} CO_2 globally by the end of this century.

Deploying NETs may be less expensive and less disruptive than reducing some emissions, such as a substantial portion of agricultural and land use emissions and some transportation emissions. NETs are envisaged by this Committee to:

- use biological processes to increase carbon stocks in soils, forests, and wetlands,
- produce energy from biomass, while capturing and storing the resulting CO_2 emissions,

- use chemical processes to capture CO_2 directly from the air and then sequester it in geologic reservoirs,
- enhance geologic processes that capture CO_2 from the atmosphere and permanently bind it with rocks (quoted from NASEM 2019).

The summary of this report lists several conclusions that outline the main thrust of the research agenda it goes on to develop. Their Conclusion 2 lists some negative emissions technologies described as ready for large-scale deployment:

- afforestation/reforestation,
- changes in forest management,
- uptake and storage by agricultural soils.

All of these involve land use and management practices such as planting trees, changes in management of existing forests, or changes in agricultural practices that enhance carbon storage in agricultural soils. This is possibly the most conventional aspect because photosynthetic carbon capture by trees and other photosynthetic organisms is widely considered to be an effective strategy to limit the rise of CO_2 concentrations in the atmosphere by sequestering carbon in the plant body. The Intergovernmental Panel on Climate Change Special Report of 2018 (Masson-Delmotte et al. 2019) suggested that an increase of 1 billion hectares of forest will be necessary to limit global warming to 1.5 °C by 2050.

The authors of this book like trees (and other plants) and we are in favour of planting more of them, ***but they should be planted for their intrinsic ecosystem value***, because there are negative aspects to relying on them so heavily as a way to sequester carbon from the atmosphere on the long term basis required for full and lasting benefit (Moore et al. 2021a; and see Chap. 2).

The ***Trillion Tree Initiative*** is a World Economic Forum initiative, designed to support the UN Decade on Ecosystem Restoration 2021–2030, led by the United Nations Environment Programme (UNEP) and the Food and Agriculture Organization of the United Nations (FAO) (https://www.1t.org/) and the parallel programme ***Trillion Trees***, which is a joint venture between BirdLife International, Wildlife Conservation Society (WCS) and the World Wide Fund for Nature (WWF) (https://trilliontrees.org/) sometimes seem to be the only nature-centric solutions catching the attention of mainstream media.

Such reforestation practices incorporating large-scale tree planting could reduce the atmospheric carbon pool by about 25% by capturing more than 200 Gt of carbon (Bastin et al. 2019). Thus, aligning with IPCC 2018 climate targets to limit global warming to 1.5 degrees above pre-industrial levels before 2050 (Masson-Delmotte et al. 2019).

However, while tree planting in general is usually considered by the mainstream public as one of the only natural solutions to counter climate change, such large-scale restoration efforts should be carefully considered to avoid negative impacts. Large-scale forest restoration projects in China (Hua et al. 2018) have revealed that while monoculture tree planting can assist in carbon sequestration goals, they do not provide the same ecosystem services as native forests do, which are more valuable and should be further protected by policy. Indeed, similar concerns about adverse impacts on carbon sequestration being caused by 'the wrong trees in the wrong places' have been expressed by studies of ecosystems as far apart as Chile (Heilmayr et al. 2020) and China (Hong et al. 2020).

For decades, trees have been an inspiration and a powerful symbol of change, a symbol of sustainability, representing healthy growth both within us as individuals and all around us in our environment. Trees represent life. The phrase "just plant trees" has the power of the local hippy, the nature-lover, the "greeny", nested within its meaning.

At times it is a symbol of rebellion and a simple response when faced with our greatest challenge in the present modern day, which must surely be climate change. "Just plant trees" contains within it a love for mother nature and a respect for our planet and our humanity, but unfortunately, it funnels our knowledge and action and conveys it through just that: "trees".

Unfortunately, recent research suggests the conclusion that mass tree planting will ***harm the environment*** if not planned properly. Importantly, forests are only effective CO_2 sinks while they remain alive. Seasonally shed leaves, petals, ripe fruit, and dead wood are digested and respired to CO_2 ***in the same year*** the CO_2 was fixed from the atmosphere (see Fig. 7.6 in Chap. 2). And when the tree dies there are legions of animals, bacteria and, especially, fungi (see Fig. 7.7 in Chap. 2, above) just waiting for the chance to digest the forest's biomass and convert it back to atmospheric CO_2 as quickly as possible. As we say in Chap. 2:

> … 'That's life'. Of course, sustainably managed forests can be harvested to provide wood fuels as environmentally benign alternative to fossil fuels (but still returning their CO_2 to the atmosphere), or timber for buildings and furniture. There are about 60 or so indoor wood decay fungi from which you need to protect your timber buildings and furniture, including dry rot, wet rot, cellar rot, and oak rot. The longevity of the carbon pools represented by wood products derived from harvested timber depends upon their use: lifetimes may range from less than one year for fuelwood, to several decades or centuries for lumber; but still, timber is only ever ***a temporary remedy*** for the atmosphere.

Brandão et al. (2013) indicate that even if the carbon storage is temporary, any carbon removal and storage from the atmosphere has the potential to mitigate climate change. However, there is firm evidence that current projections of global forest carbon sink persistence are too optimistic because the ***increased growth rates*** of trees caused by increased levels of CO_2 in the atmosphere ***may shorten the lifespan of forest trees*** (Brienen et al. 2020):

> … Faster growth has a direct and negative effect on tree lifespan, independent of the environmental mechanisms driving growth rate variation. Growth increases, as recently documented across high latitude and tropical forests, are thus expected to reduce tree lifespans…" and that "… recent increases in forest carbon stocks may be **transient** due to lagged increases in mortality … (quoted from Brienen et al., 2020).

So, current plans for tree planting on a massive scale are not the panaceas that many believe. Putting such plans into effect could do more harm than good (Friggens et al. 2020; Heilmayr et al. 2020; Hong et al. 2020; Natural Capital Committee 2020; and listen to The Climate Question 2021b podcast).

In addition, our current forests are suffering from the effects of the climate changes that have already occurred: forested areas are dying due to newly emerged, virulent and invasive, pests and diseases as well as drought, often amplified by more devastating wildfires (Demeude and Gadault 2020). These threats to forest ecosystems are worldwide. We cannot rely on forests to mitigate the effects of climate change while they are dying because of it!

Despite all these negatives there remains some hope that better management of forests and their carbon stocks can help improve overall terrestrial carbon cycle management providing knowledge of the role of fungi and soil microbes in carbon cycling is implemented into ***sustainable forest management*** practices (Soudzilovskaia et al. 2019; Domeignoz-Horta et al. 2020).

There is more to terrestrial plant cover than just trees, of course, but the limitation that plants only store carbon while they are alive applies to all photosynthetic organisms (including aquatic ones); wherever the plant dies, its stored carbon is returned to the atmosphere through the respiration of the animals, fungi and bacteria that digest its biomass.

In addition, there is a large amount of carbon stored in soils, and that includes peatlands and permafrost. Peatlands cover an area of about 3.7 million km^2 in the northern hemisphere, about half this being permanently frozen permafrost. These northern peatlands are estimated to store around 415 billion metric tonnes of carbon, which is equivalent to over 45 years of current global CO_2 emissions. Unfortunately, this is not a permanent sequestration. Global warming will cause the northern ***peatlands to become a major source of greenhouse gas emissions*** into the atmosphere (methane, carbon dioxide and nitrous oxide) (Hugelius et al. 2020).

And don't expect planting trees on peatland to help. Friggens et al. (2020) recorded a 58% reduction in soil organic carbon stocks 12 years after birch trees (*Betula pubescens*) had been planted in heather (*Calluna vulgaris*) moorland. This decline was not compensated by the gains in carbon represented in the growing trees. This was a continuation of a long term study of the effects of planting two native tree species which showed that 39 years after planting, the carbon sequestered into tree biomass did offset the carbon lost from the soil but, crucially, there was ***no overall increase*** in carbon sequestered by the ecosystem.

The UK's Office For National Statistics (ONS 2016) estimated that in 2007 UK soils contained approximately 4 million tonnes of carbon, of which 57% was the carbon stored in peat soils, but as the majority of UK peatlands are degraded (Natural England 2010), they are a highly significant ***source of greenhouse gas emissions***.

The aim of peatland restoration must be to reduce the extent of these emissions as a contribution to the 'net zero future' (Natural Capital Committee 2020): this report states that "The right tree in the right place for the right reason can bring a multitude of benefits…" but adds "the wrong trees in the wrong places can have adverse impacts on soil (including soil carbon), water flows, water quality, recreation, biodiversity and air quality."

In the UK, the Countryside Charity CPRE (originally the *Campaign to Protect Rural England*) has warned that emissions from UK peatland could cancel out all carbon reduction achieved through new and existing forests, in their August 2020 report entitled '*Net-zero virtually impossible without more ambition on peatlands*' (https://www.cpre.org.uk/).

It is also necessary to recognise that all soils incorporate carbon stocks that must be managed sensitively, especially when undertaking reforestation projects. Indeed, current carbon stocks are much larger in soils than in vegetation, particularly in non-forested ecosystems in middle and high latitudes (Table 7.4).

Bossio et al. (2020) stated that mitigating climate change requires clean energy and the removal of atmospheric carbon, commenting that "… building soil carbon is an appealing way to increase carbon sinks and reduce emissions owing to the associated benefits to agriculture." They quantify the role of soil carbon in natural (land-based) climate solutions showing that soil carbon represents 25% of the potential for nature-based solutions to the climate crisis with a total potential of 23.8 Gt of CO_2-equivalent ***per year***. 40% of which is protection of existing soil carbon and 60% is rebuilding depleted stocks. They point out that soil carbon comprises 9% of the mitigation potential of forests, 72% of that for wetlands and 47% for agriculture and grasslands. Finally, soil carbon is important to land-based efforts to prevent carbon emissions and remove atmospheric carbon dioxide and deliver ecosystem services ***in addition to*** climate mitigation.

Removing atmospheric carbon dioxide levels may be the primary objective, but to deliver additional ecosystem services in addition to this is a significant advantage of all natural biotechnological solutions. In particular, the potential role of biodiversity in helping society and nature face the linked challenges associated with biodiversity loss and climate change has received little attention but must be addressed if efforts to resolve our environmental crises are to be effective (Mori 2020).

What this means overall is that plans for terrestrial carbon sequestration are less promising because carbon storage by plants (a) is only ever temporary; (b) because large-scale reforestation may cause more problems than it solves; and (c) because disturbing the soil, as for example, is necessary for tree planting, can release carbon back to the atmosphere from the stabilised soil organic carbon pool in deeper horizons. Plant-rich environments have much to offer for both physical and mental well-being of humans, and biodiverse tree planting supports general biodiversity of woodlands and forest ecosystems. But tree planting, even on a monumental scale, will not contribute to solving the crisis of global warming.

But there is one further negative impact of any of these would-be cures of the climate crisis that involve growing plants on land, and this is that such activities are in direct competition for

Table 7.4 Global carbon stocks in vegetation and soil carbon pools down to a depth of 1 m

Biome	Area	Global Carbon Stocks (Gt C)		
	(× 10^9 ha)	Vegetation	Soil	Total
Tropical forests	1.76	212	216	428
Temperate forests	1.04	59	100	159
Boreal forests	1.37	88	471	559
Tropical savannas	2.25	66	264	330
Temperate grasslands	1.25	9	295	304
Deserts and semideserts	4.55	8	191	199
Tundra	0.95	6	121	127
Wetlands	0.35	15	225	240
Croplands	1.60	3	128	131
Total	15.12	466	2011	2477

Note There is considerable uncertainty in the numbers given, because of ambiguity of definitions of biomass, but the table still provides an overview of the magnitude of carbon stocks in terrestrial systems. Data from the 2000 IPCC Special Report: Land Use, Land-Use Change and Forestry (Watson et al 2000)

Note that the data of Table 7.4 are based on routine soil surveys for estimating the ***soil organic carbon*** (SOC) pool which account for a soil depth of only about ***one metre***. Deeper soil horizons, however, may have a high capacity to sequester significant amounts of SOC because the turnover time and chemical recalcitrance of soil organic matter increases with depth. In particular, the soil organic carbon (SOC) pool is ***the only terrestrial pool storing some carbon (C) for millennia,*** and it can be deliberately enhanced by agroforestry practices. Soil disturbance, especially, must be minimised and tree species with a high root biomass to above-ground biomass ratio and/or trees that have symbiotic nitrogen-fixing root nodules (to minimise fungal-recovery of nitrogen from otherwise stabilised soil organic matter) should be planted when carbon sequestration is the objective for the agroforestry system being established. The size of the Earth's soil organic carbon reservoir is estimated to be around 1,500 Gt C in the first metre, excluding permafrost areas (Hiederer and Köchy 2011). 58% of the chemically stabilised and 31% of the physically stabilised fractions of the soil organic carbon pool occurred in the subsoil horizons. The subsoil below the one m depth may have the potential to sequester between 760 and 1520 Gt C (Lorenz and Lal 2005, 2014; Lorenz et al. 2011)

cultivable land that might otherwise be used for growing food crops for human use. The situation we have ***today*** is that there is not enough land on Earth to support the diet recommended by authorities for the whole of the human population (Dockrill 2018; Rizvi et al. 2018). Consequently, if we wish, for the sake of carbon sequestration, to implement expansive plans for the restoration of peatlands and permafrost, and afforestation, and pasture rotation management, and wildlife biodiversity enhancement, we might have to set out parallel international plans to ***decide which members of the human population should be allowed to starve to death*** to make the necessary land available.

Or perhaps we should turn away from '**green carbon**' and look towards the 70% of the planet's surface that is covered in ocean for a cure for the climate crisis? '**Blue carbon**' to the rescue?

7.8 Coastal Blue Carbon

Coastal Blue Carbon, described as "… land use and management practices that increase the carbon stored in living plants or sediments in ***mangroves***, ***tidal marshlands***, ***seagrass beds***, and other ***tidal or salt-water wetlands*** are among the technologies considered by NASEM" (2019).

These approaches refer to coastal ecosystems instead of the open ocean and the report is at pains to point out that the committee's initial task statement (or 'job description') was to focus exclusively on ***near-shore coastal*** NETs despite the recognition that oceanic options for CO_2 removal and sequestration, which fall outside the scope of its task, could sequester an enormous amount of CO_2. Gattuso et al. (2021) conclude:

> … Ocean-based NETs are uncertain but potentially highly effective. They have high priority for research and development …

This is an attitude we wish to promote in this book. So much attention is given to afforestation in the conventional media that the potential of aquatic 'blue forests' and other prevalent marine biota to capture and sequester carbon in our coastal waters and the high seas is yet to be realised by the general public.

The blue carbon systems described by NASEM (2019) are usually categorised or labelled as ***shallow coastal ecosystems*** (SCEs). These include but are not limited to, mangrove forests, seagrasses, kelp and other aquatic biota that thrive in healthy blue carbon forests, including shellfish, algae and many other microbiota. Out of all the biological carbon captured in the world, over half is captured by marine living organisms and this is why it is called ***blue carbon*** (Nellemann et al. 2009; Pendleton et al. 2012). Moreover, compared to the average decadal time-scale for terrestrial systems to hold carbon before the aforementioned release back into the atmosphere (after their death), some blue carbon ecosystems could store the carbon for timescales of hundreds of millennia (Heilweck and Moore 2021; Moore 2021; Moore et al. 2021b).

Blue carbon science is relatively young but has revealed the importance of aquatic ecosystems in the carbon balance and ecosystem services (specific examples are the monetary value of mangroves and seagrasses in ecosystem services and the monetary value of the seafood industry) but it deserves significantly increased attention (Macreadie et al. 2019).

These coastal vegetation ecosystems (marshes, mangroves and seagrasses) have high rates of annual carbon sequestration as well as very large pools of previously-sequestered carbon, which is largely in their sediments and is in danger of being released to the atmosphere if these ecosystems are degraded (Pendleton et al. 2012; Nguyen et al. 2021).

Quite clearly, these systems deserve much more attention in the public eye, particularly because there seems to be solid experimental evidence that they are able to sequester more carbon than forest ecosystems (Table 7.5). Lee et al. (2020) tabulated annual carbon deposition estimates for a variety of ecosystems to show that European flat oyster beds (at a density of 75 oysters m^{-2}) in the Northern Hemisphere have the potential to deposit more carbon per square metre than terrestrial forests in the Northern

Table 7.5 Annual values of carbon deposition defined as sedimentary, carbonate, or sedimentary + carbonate per ecosystem

Ecosystem	Carbon store type	Carbon deposition per annum (g m^{-2})
Seagrass	Sedimentary	83
Saltmarsh	Sedimentary	210
Mangroves	Sedimentary	174
Maerl (coralline red algae)	Carbonate	74
Horse mussel (density 40 m^{-2})	Carbonate (+?sedimentary)	40 (+ about 360 organic matter deposition[a])
Oyster (density 75 m^{-2})	Sedimentary (+?carbonate)	50
Terrestrial forests[b]	Net sink	32

Notes + ? indicates data deficiency

[a]Data are available on organic content of sediment deposits rather than carbon deposition

[b]Net global sink/global forest cover (data from Pugh et al. 2019). Other data from Lee et al. (2020)

Hemisphere, through biodeposition to the seabed alone, and that oyster beds compare favourably with other shellfish habitats (Table 7.5).

Nellemann et al. (2009) state that while the.

> … contribution of forests in sequestering carbon is well known and is supported by relevant financial mechanisms. In contrast, the critical role of the oceans has been overlooked…" and go on to point out that oceans play a significant role in the global carbon cycle "… Not only do they represent the largest long-term sink for carbon, but they also store and redistribute CO_2. ***Some 93% of the Earth's CO_2 (40 Tt* [= 40 million Mt or 40 × 10^{12} t]*) is stored and cycled through the oceans…*** (the emphasis is ours).

Primavera et al. (2019) discuss the conservation and management of **mangroves**, the goods and services of these ecosystems, and factors causing mangrove loss and their restoration. Examples of large-scale mangrove reforestation can be seen in equatorial regions throughout the world and are monitored by the ***Mapping Ocean Wealth*** website (view https://oceanwealth.org/), from which we quote the following:

> … Global statistics on mangrove extents, gains and losses developed by our partners show the global extent of mangroves in 1996 was some 142,795 km^2, but in 2016 was some 136,714 km^2.
> In a first ever review of mangrove degradation, we have mapped some 1389 km^2 of degraded mangrove within the latest (2016) mangrove cover map.
> … an expert-derived model for 'restorability' has been developed based on key environmental components which influence the ease of restoration. Using this model, some 6,665 km^2 are considered highly restorable. Full restoration of the areas identified could enable:
>
> - Carbon sequestration in above-ground biomass amounting to 69 million tonnes of Carbon, equivalent of annual emissions from 25,000,000 US homes;
> - Soil carbon stocks of 296 million tonnes saved through a combination of avoided emissions and sequestration emissions equivalent to emissions from 117,000,000 US homes.
> - Addition of commercial fisheries species in mangrove waters totalling 23 trillion young-of-year finfish and 40 trillion crabs, shrimp and molluscs;
> - Coastal protection from annual flooding to hundreds of thousands of people… (all quoted from https://oceanwealth.org/applications/mangrove-restoration/).

There are examples of blue carbon restoration projects all over the globe, even in the coldest climates such as the arctic (see ***Nordic Blue Carbon Project***'s very informative website at https://nordicbluecarbon.no/).

Seagrasses (or **eelgrasses**) are submerged vascular flowering plants, found mostly along the coastline. The ***Ocean Health Index*** website estimates that globally they cover an area of 300,000 to 600,000 km^2 (http://www.oceanhealthindex.org/). Seagrasses have declined in area by about 29% since the beginning of the twentieth century, at an annual rate of about 1.5% and faster in recent years, due to change in land use, being replaced by mud and sandy marine 'soils' (Fourqurean et al. 2012; Asplund et al. 2021).

Healthy seagrass meadows store significant amounts of carbon. Röhr et al. (2018) sampled *Zostera marina* eelgrass meadows, spread across eight ocean margins and 36° of latitude, measuring organic carbon stocks in their sediments; this averaged 2,721 g C m^{-2}, which they extrapolated over the top 1 m of sediment to range between 23.1 and 351.7 Mg C ha^{-1} (equivalent to 23.1 to 351.7 tonnes C ha^{-1}). Using the lowest estimate of the seagrass meadow area globally these sedimentary carbon stocks extrapolate to between 693 Mt and 10.6 Gt of carbon currently sequestered in the sediment of the world's seagrass meadows.

Kelp forests. Seaweed farming to create kelp forests is another fashionable suggestion as a means to mitigate climate change. The crop is used for biofuel production, as an agricultural fertiliser for improving soil quality and substituting for synthetic fertiliser and is included in cattle feed to lower methane emissions from cattle. Kelp is large **brown algae,** in the Order *Laminariales*, which form prominent populations of 'underwater forests' in cool seas worldwide. There are 27 genera that vary in size, morphology, lifespan, and habitat. Although they are large, multicellular, photosynthetic and eukaryotic organisms, they are not plants; rather they are protists belonging to a group known as 'heterokonts' because when they produce motile cells (usually to reproduce) those cells have two flagella of different length and different morphology. This is a major group of eukaryotes ranging from the giant multicellular kelp to the unicellular diatoms, which are themselves a primary component of phytoplankton. Seaweed aquaculture has been described as the fastest-growing component of global food production.

Duarte et al. (2017) claim that the total global annual production of kelp was 27.3 million tonnes in 2014 and a growth rate of 8% y^{-1}, and seaweed aquaculture comprises 27% of total marine aquaculture production (although the value of the seaweed produced amounts to only 5% of the total value of aquaculture crops). The key features of seaweed farming that make it attractive include that the kelp forests provide habitat and several ecosystem services for very diverse coastal communities, which theoretically could range along 25% of the world's coastlines. Ecosystem services, apart from carbon sequestration, include climate change adaptation by damping wave energy and protecting shorelines, and by elevating pH and supplying oxygen to the waters, thereby locally reducing the effects of ocean acidification and de-oxygenation (Duarte et al. 2017).

Kelps exhibit a great diversity of growth forms and life strategies, with the largest fronds reaching lengths of more than 30 m with biomasses of 42 kg (Wernberg et al. 2019). There is controversy over the longevity of carbon sequestration by kelp forests (Hill et al. 2015; discussed in Duarte et al. 2017), and some are even described as 'perennial kelps' but this is a misnomer as the maximum lifespan of **fronds** has been calculated to be one year; it is the **holdfast** that is perennial (Tussenbroek 1989). Kelp forests face many threats and are quite dynamic and variable. As a result:

> … it seems almost certain that many kelp forests a few decades from now will differ substantially from what they are today… (Wernberg et al. 2019).

We wonder what happens to any sequestered carbon during this turnover.

Oceanic microalgae. Among the most important primary producers in our oceans are photosynthetic microalgae with chloroplasts similar to those derived from red algae in which chlorophyll is masked by the accessory carotenoid pigment fucoxanthin, giving them a brown or olive-green colour. These 'Haptophyte' algae account for about 40% of the total chlorophyll-a biomass in oceans, so they are a dominant marine primary producer in today's oceans. This has made them candidates for use in atmospheric carbon sequestration and there is a considerable literature dealing with biorefinery and other technologies applying to microalgae (Singh and Dhar 2019).

It is assumed, as with the kelps, that carbon fixation into their biomass makes them a carbon sink. For most haptophytes this is no more realistic than it is for any other primary producer; because these organisms are at the base of all food chains, all their biomass is converted into the biomass of organisms at higher levels in the food chain. And in that process the primary producer's biomass is metabolised and eventually respired as CO_2 that is returned to the atmosphere.

However, there is one group of haptophyte algae, called **coccolithophores**, that have played a central role in the global carbon cycle in the Earth's oceans for hundreds of millions of years. These organisms fix dissolved inorganic carbon, which all originates from the atmosphere, through both photosynthesis and **calcification**, because these single-celled algae surround themselves with microscopic plates, called **coccoliths**, made of limestone (calcite, $CaCO_3$). Coccolith $CaCO_3$ is indigestible and completely stable (until heated to over 1,000 °C).

> … A massive quantity of calcified cells has been sedimented throughout geological time, as seen in the White Cliffs of Dover; thus, coccolithophores contribute to sequester atmospheric CO_2 as limestone … (Tsuji and Yoshida 2017; and see references therein).

Now, ***that's an* effective *atmospheric carbon sink*** (Moore 2021)!

7.9 The Ultimate Blue Carbon: The Oceans' Calcifiers

Except for the coccolithophores, all of the blue carbon atmosphere mitigators mentioned so far suffer from the same disadvantages as the plant-based terrestrial mitigation projects we have already mentioned. Namely:

- Yes, the photosynthetic organisms fix atmospheric CO_2 into their biomass; but this is only **temporary** and remains in the biomass only as long as the organism is alive.
- Photosynthetic organisms, the **primary producers**, are at the base of all food chains (photosynthetically-fixed carbon is, ultimately, the **only** metabolic carbon available on the planet).
- When the organism dies its biomass is digested and the carbon in the biomass starts its journey through metabolism until it is respired as CO_2 and returned to the atmosphere.
- Any of the biomass that escapes being respired as CO_2 has a chance to be sequestered in the ocean sediment or, on land, in the deep soil organic carbon sink. But only as long as that sink remains undisturbed.
- Organisms at the base of food chains tend to be eaten fairly rapidly. So, the biomass-CO_2 that is returned to the atmosphere today may have only been fixed from yesterday's atmosphere.
- Longer lived primary producers, from the 1-year-old fronds of ('perennial' kelp; it's the holdfast that's perennial, not the frond that makes the kelp forest) to the thousand-year-old oak tree in a terrestrial woodland, will all die eventually, and their residual biomass will be digested and returned to the atmosphere as respired CO_2.

Finally, the coccolithophores lead us to ***the limestone elephant in the room***, the one that so few people talk about except to dismiss it from consideration, but which is the central thrust of the case presented in the book you are reading now:

- This is that the physiological chemistry of a few types of ocean creatures, the **calcifiers** of the coasts and open seas, (coccolithophore algae, corals, crustacea and molluscs) enables them to extract CO_2 from the atmosphere and sequester it permanently as crystalline $CaCO_3$.

Our ***case for the calcifiers*** is presented above in Chaps. 1–6 of this book, and in our recent publications (Heilweck and Moore 2021; Moore 2020, 2021; Moore et al.y2021a, b) so we will not repeat it here. We ***will*** reiterate that it is the certainty and permanence of the removal of CO_2 from the atmosphere that would make a biotechnology using calcifying organisms so attractive. Even NASEM (2019) notes that ***terrestrial*** options and the few coastal blue carbon options they consider are reversible if the carbon-sequestering practices are not maintained. "... Although temporary CO_2 storage will have some climate benefit, scientific and ***economic requirements to ensure the permanence of storage within ecosystems are substantial*** ..." (NASEM 2019). Storage of atmospheric carbon in calcifier shells ***IS*** permanent. Alonso et al. (2021) estimate that the CO_2 sequestration potential of bivalve aquaculture, using the current value of one metric tonne of CO2 in the carbon market is over 25 €, which would represent a value of around 125 to 175 million € y^{-1} to the European Union's bivalve aquaculture industry alone.

Solutions involving terrestrial land management are not permanent. Changes in policy could see afforested or reforested land cleared again and any return to intensive tillage would reverse any gains in soil carbon sequestration achieved by the afforestation. Restored coastal wetlands could be drained again for agricultural use, losing any advantage gained by the wetland restoration. Given the fact that there is insufficient agricultural land on Earth to grow food for the whole of the human population (Dockrill 2018; Rizvi et al. 2018) it may become impossible in the future to avoid returning restored forests, peatland or coastal wetlands to intensive agriculture just to safeguard basic food supply. If that is done, all the benefits to the atmosphere achieved by the restorations will be lost.

We have been asked how we overcome the issue of calcification being said to be a CO_2 emitting process and not a sink. Our usual response to this question is as follows. The calcifying reaction scheme shows that two bicarbonate ions (which ultimately were derived from the atmosphere) react with Ca ions and **one** of them is precipitated as $CaCO_3$, and the other released as CO_2. So, while it is true that "precipitation of calcium carbonate is a source of carbon dioxide (CO_2)" it is illogical to claim that returning one out of two carbons to the environment is a "major way by which CO_2 is returned to the atmosphere" as some have put it to us.

BUT, if we go one step further and ADD the consideration that there are a great many calcifying organisms in the oceans, which are all cycling through this reaction 24/7, then you can rightly claim that this is a major way by which CO_2 is returned to the atmosphere; PROVIDING you remember that the other one of those two carbons on the left of the reaction scheme is precipitated as $CaCO_3$ and ADMIT the matching claim that if this IS a major way by which CO_2 is RETURNED to the atmosphere then it is ALSO a major way by which carbon is REMOVED **permanently** FROM the atmosphere.

It might be time to **start taking Blue Carbon more seriously**, and not just on coastal sites, but over the whole of the High Seas as well, changing our attitudes and policies to recognise the enormous value that marine restoration projects represent to humanity (Gordon et al. 2020). Remember that these authors conclude that.

> … [marine] restoration projects could help maintain species survival and ecosystem services, ultimately providing humanity with the breathing space to stabilize the climate ….

To achieve this, we need to rebuild marine life and Duarte et al. (2020) argue that this.

> … represents a doable Grand Challenge for humanity, an ethical obligation and a smart economic objective to achieve a sustainable future …" and in their opinion "… substantial recovery of the abundance, structure and function of marine life could be achieved by 2050, if major pressures - including climate change - are mitigated …

The most influential report on climate change economics, policy and management is undoubtedly *The Stern Review* which was entitled *Economics of Climate Change*. Commissioned by the UK Government and released in October 2006, the report was published in January 2007 (Stern 2007). The main findings of this report were that:

- Climate change could have very serious impacts on growth and development.
- There is still time to avoid the worst impacts of climate change, if we take strong action now.
- The costs of stabilising the climate are significant but manageable; delay would be dangerous and much more costly.
- Action on climate change is required across all countries, and it need not cap the aspirations for growth of rich or poor countries.
- A range of options exists to cut emissions; strong, deliberate policy action is required to motivate their take-up.
- Climate change demands an international response, based on a shared understanding of long-term goals and agreement on frameworks for action such as ethics and equity.

The case for avoiding the dangerous risks of climate change by emphasising low-carbon economic development and growth is even stronger now than when the Stern Review was published (and see Stern 2015). Remember the climate records (not estimates or predictions, but ***records***) show that 2011–2020 is the warmest decade on record, with the warmest six years ever recorded all being since 2015 (WMO 2020). The implication being that the impacts of climate change are happening ever more quickly than previously expected.

This makes the action even more urgent, but action on climate change in any direction needs the application of insights from economic development and public policy and rigorous analysis of issues such as discounting, modelling the risks of unmanaged climate change, climate policy targets and estimates of the costs of mitigation. And significant obstacles remain in obtaining the international cooperation required.

The more recent Dasgupta Review (*The Economics of Biodiversity*; Dasgupta 2021) goes even further, and to give just a flavour of the findings of this authoritative 600-page review we list here its main headlines (the stress is ours):

- Our economies, livelihoods and well-being all depend on our most precious asset: Nature.
- We have collectively failed to engage with Nature sustainably, to the extent that our demands far exceed its capacity to supply us with the goods and services we all rely on.
- Our unsustainable engagement with Nature is endangering the prosperity of current and future generations.
- At the heart of the problem lies deep-rooted, widespread institutional failure.
- The solution starts with understanding and accepting a simple truth: our economies are embedded within Nature, not external to it.
- We need to change how we think, act and measure success.
  - Ensure that our demands on Nature do not exceed its supply, and that we increase Nature's supply relative to its current level.
  - Change our measures of economic success to guide us on a more sustainable path.
  - Transform our institutions and systems—in particular our finance and education systems—to enable these changes and sustain them for future generations.
- ***Transformative change is possible—we and our descendants deserve nothing less***.

7.10 Sustainability Assessment of CCS Methods

The global warming is a symptom of root-cause problems in our societies, representing a significant complexity of challenges that can all be linked together as threats to humanity's life support systems. Our primary concern as a generation must be to determine how we can use our talents and techniques to engineer a future that lessens the burdens that we pass on to future generations.

A comprehensive review of 27 life cycle assessment studies of environmental impacts of carbon capture and storage (CCS) and carbon capture and utilisation (CCU) technologies was reported by Cuéllar-Franca & Azapagic (2015). They point out that an advantage of CCU over CCS is that utilisation of CO_2 is normally a profitable activity as products can be sold. Also, CO_2 has the advantage over conventional petrochemical feedstocks, of being a low cost and non-toxic renewable resource. However, current global demand for chemicals does not have the capacity to sequester enough CO_2 emissions to contribute significantly to meeting global carbon reduction targets. while using CO_2 for fuel production only delays its emission rather than eliminating it as needed for mitigating climate change. They go on to state:

> … In addition, … there are other sustainability issues that must be considered before large-scale deployment of either CCS or CCU, notably environmental impacts. This is important to ensure that climate change is not mitigated at the expense of other environmental issues. It is also important that the impacts be assessed on a life cycle basis, to avoid shifting the environmental burdens from one life cycle stage to another. In an attempt to inform the debate in this field, this paper provides a comprehensive state-of-the-art review of different CCS and CCU technologies, analysing their life cycle environmental impacts based on the results of life cycle assessment (LCA) studies found in the literature … (Cuéllar-Franca and Azapagic 2015).

Various assessment models exist to compare and evaluate the sustainability of infrastructure systems. ***Life cycle assessment*** is just that, an analysis over a complete generational life cycle (birth-to-birth) that assesses environmental impacts associated with all the stages in the life of any manufactured product (or other processes) covering raw material extraction through materials processing, manufacture, distribution, and end use (with recycling/disposal where appropriate) (Wikipedia: https://en.wikipedia.org/wiki/Life-cycle_assessment).

The evaluation procedure is of significant importance in order to mitigate risks and manage uncertainties, while better adapting current infrastructure implementation with the future visions and plans for our societies. Proper ***sustainability*** evaluation is essential to enable better engineered futures, reducing waste of resources and reducing overburden for future societies. It is simply planning for a better future. Development and implementation of infrastructure systems that works towards the realisation of the United Nations' sustainable development goals is one way to map sustainability (UN 2016; https://sdgs.un.org/goals).

Maack and Davidsdottir (2015) formulated an approach to project appraisal different from the conventional concentration on Cost–Benefit assessment that deals with financial flows and rate of return on investments. Their approach to evaluation is based on the theory that ***five capital value types*** support long term ***well-being***, rather than economic growth alone; as they describe it: "… The theory states that humans depend on the size of stocks and flows from natural, manufactured, human, social and financial capital. We describe the five capitals to illustrate the value categories and outline an approach to evaluate all these in the context of energy development …" (Maack and Davidsdottir 2015) (Fig. 7.8).

They also comment that:

> … There seems to be a disciplinary gap between the European and North American schools of thought in assessing such values. The American thought is more rooted in economic theory and stresses supply, demand and efficiency. The European one rather leans towards accounting effectively the cost of all components in human lifestyle patterns using inventories in the spirit of LCA …
> … Our review reveals that assessing aspects of sustainable development is highly complicated. The methods that are offered to measure each aspect are evolving.... Still, the theoretical discourse must lead to a practical implementation frame. Otherwise further economic changes will lead to changes without progress towards sustainable development … (Maack and Davidsdottir 2015).

More recently, Müller et al. (2020) published comprehensive guidelines for application of life cycle assessment (LCA) specifically to carbon capture and utilisation (CCU) technologies, with the aim of improving comparability of LCA studies through clear methodological guidance. Improved comparability is expected to help strengthen knowledge-based decision-making so that funds and time can be allocated more efficiently towards climate change mitigation and emissions control.

The sustainability of a project can be assessed by the ***four-capital model*** of sustainable development evaluation (Ekins et al. 2008). The concept of ***capital*** in this model derives from economics; capital stocks (or assets) provide a flow of goods and services, which contribute to human well-being. In its narrowest interpretation capital can be used to mean manufactured goods, but the concept applies also to 'intangible' forms of capital, which may affect (and even account for the bulk of) the value of an activity. Four types of capital have been defined:

- **Manufactured capital**, the traditional production assets like machines, tools, buildings, and infrastructure.
- **Natural capital** includes obvious natural resources, such as water, energy, mineral reserves; but also, assets like biodiversity, endangered species, and ecosystem services (generally, assets with a bearing on human welfare).
- **Human capital** refers to the health, well-being, and productive potential of individual people, encompassing mental and physical health, education, motivation, and work skills. Assets contributing to a happy, healthy, and productive society.
- **Social capital**, again, related to human well-being, but on a societal level, such as neighbourhood associations, civic organisations, and co-operatives. Social networks that support an efficient, cohesive society and the political and legal structures that promote stability, democracy, governmental efficiency, and social justice.

Application of the model to an activity uses **indicators of sustainability** for the assessment, and there are two main approaches to constructing indicators:

- **The framework approach**, which sets out a range of indicators intended to cover the main issues and concerns related to sustainable development.
- **The aggregation approach**, which seeks to express changes in a common unit (normally money), so that they can be aggregated.
- An 'ideal' **indicator set** (aimed at evaluating the contribution of European Union structural funds to sustainable development) is listed in the appendix to Ekins et al. (2008).

A **three-pillar concept of sustainability**, the three pillars being social, economic, and environmental sustainability has been published by Purvis et al. (2019) who review and discuss historical sustainability literature, attempting to establish the origin of this three-pillar conception.

Assessing ***mariculture*** sustainability was formalised by Trujillo (2008) who developed a framework for evaluating sustainability of aquaculture production using a **Mariculture Sustainability Index** (MSI) with scores between 1 (poor) and 10 (very good). The MSI score is obtained as a combination of 13 indicators covering ecological, economic, and social aspects of the industry, and the original paper assessed sustainability in 64 countries over the 10 year period from 1994 to 2003 and involving 86 farmed species. Trujillo (2008) found the highest ranking countries for sustainable mariculture farm (a) native species, (b) of low trophic levels, (c) under non-intensive conditions, (d) for domestic consumption. The lowest ranking countries tend to farm (a) non-native species, (b) with high trophic levels, (c) under intensive conditions, (d) for export, often to countries ranking high for mariculture sustainability.

Mariculture assessment can be difficult because the required information is not always available about which species are cultivated, where they are cultivated, the methods used, local environmental impacts, sustainable yields expected for each species/location/method combination, etc., etc., but Campbell et al. (2016) have made a global analysis of mariculture production and its sustainability over the years 1950–2030; and Neori and Nobre (2012) correlated trophic level and economics in aquaculture. They demonstrated the overall ecological efficiency, sustainability and economics of culturing carnivorous fish are improved by growing them in an ecological balance with species from low trophic levels in integrated multi-trophic aquaculture.

Studies referenced so far deal with ***finfish aquaculture***, published studies of shellfish centre on considering only the ***sustainability of shellfish as food***. Filter-feeding bivalves (oysters, mussels, clams and scallops) are successfully farmed across the globe as a sustainable food source, and unlike all other aquaculture, and agriculture for that matter, commercially grown bivalves are the only sustainable form of human food that, when properly managed, has ***no negative impact on the environment*** [https://www.eco-business.com/opinion/sustainable-shellfish-aquaculture/]. Indeed, bivalve molluscs offer several ecosystem services that add value to their environment beyond their food value. These additional bivalve ecosystem services in the habitat restoration context have been listed (National Research Council 2010) as:

- Turbidity reduction by filtration.
- Biodeposition of organics containing plant nutrients.
- Induction of denitrification associated with organic deposition.
- Sequestration of carbon
- Provision of structural habitats (Reef structures) that promote diversity of fish, crustacea and other organisms.
- Habitat and shoreline stabilisation.

Jacquet et al. (2017), with the title '*Seafood in the future: bivalves are better*' add these advantages of ***bivalve farming*** to the above list:

- Bivalves don't require feeding.
- Bivalves build food security.
- Bivalve welfare is not as serious a concern as it is for terrestrial farm animals.

They point out that as human population expanded rapidly, terrestrial farmers domesticated sheep, goats, cows, and pigs, and chickens and these animals became part of a highly industrialised food system that destroys habitat, pollutes the environment, and is unsustainable. And go on to claim:

> … Aquaculture - the farming of aquatic animals and plants for food—is the fastest growing food production system in the world. But it is growing in the wrong way. We are farming carnivores, like salmon, that need us to catch additional fish to feed them, which is putting additional pressure on wild ecosystems. We are also completely ignoring welfare concerns.
> If done correctly, aquaculture could provide sustenance for our growing planet as well as reduce overfishing. But if we want to avoid repeating the same mistakes, we need to make changes now, including changing our diets generally to include more plants and fewer animals, and [particularly] eating more bivalves - oysters, mussels, and clams - instead of fish, shrimps, and octopus … (Jacquet et al. 2017).

In parallel to this study, Hilborn et al. (2018) examined the environmental cost of foods sourced from animals. They reviewed 148 assessments of food production from livestock, aquaculture, and capture fisheries, measuring four metrics of environmental impact (energy use, greenhouse-gas emissions, release of nutrients, and acidifying compounds), standardising these per unit of protein production. They found that the lowest impact forms of animal protein originated from species that feed naturally in the ocean and that can be harvested with low fuel requirements. Specifically, the lowest impact production methods were small pelagic fisheries and mollusc aquaculture, whereas the highest impact production methods were beef production and catfish aquaculture (Hilborn et al. 2018).

If aquaculture is to meet the growing demands for food around the world, its future will hinge on sustainable and ethical practices being used by the industry and a more consistent regulatory regime (Dumbauld et al. 2009). In terms of potential, Costello et al. (2020) have examined the main food-producing sectors of the ocean, wild fisheries, finfish mariculture and bivalve mariculture. to estimate 'sustainable supply curves' that account for ecological, economic, regulatory and technological constraints for an overall estimate of future seafood production. Finding:

> … that under our estimated demand shifts and supply scenarios (which account for policy reform and technology improvements), edible food from the sea could increase by 21–44 million tonnes by 2050, a 36–74% increase compared to current yields. This represents 12–25% of the estimated increase in all meat needed to feed 9.8 billion people by 2050… Costello et al. (2020).

7.11 Conclusions

There is no doubt that the concept, or paradigm, '***shellfish as food***' provides us with a food source that is widely accepted as healthy and nutritious meat, and a production industry that is productive, sustainable, ethical and environmentally friendly.

But that's not how ***we*** want this branch of aquaculture to be judged, because we want to **change the paradigm** to '***shellfish for carbon sequestration***'. Changing the paradigm means placing the **value** of the exercise of shellfish cultivation onto the **production of shell**, taking the food value of the animal protein as ***one of the several ecosystem services*** that bivalve molluscs supply (listed above).

Our claim is that cultivation of coccolithophores, corals, crustacea and molluscs on a massive scale would make a massive and continued ameliorative contribution to climate change on this planet; potentially achieving the UN's Sustainable Development Goal 14 (***to conserve and sustainably use the oceans, seas and marine resources for sustainable development***) (https://sdgs.un.org/goals).

That being the case the comparison that matters to us is not that between aquaculture and agriculture but ***the comparison between the aquaculture of calcifiers and industrial methods of carbon dioxide capture, utilisation and storage.*** Unfortunately, there is a dearth of data bearing on 'shellfish for carbon sequestration'; too little for an easy attempt at a formal life cycle assessment/sustainability assessment, but we can bring a few pertinent points to attention. Firstly, Turolla et al. (2020) have carried out a life cycle assessment of Manila Clam (*Ruditapes philippinarum*) farming in a lagoon in the Po River Delta and shown it to be a fully sustainable aquaculture practice. Indeed, they found that annual production of one tonne of fresh ready-to-sell clams sequestered in their shells 444.55 kg of CO_2, 1.54 kg of nitrogen and 0.31 kg of phosphorus per year.

This study brings home the fact that if you create an ***industrial*** carbon dioxide capture, utilisation and storage facility, that's what you get. Captured CO_2; nothing else. But secondly, if you create a ***bivalve mollusc farming enterprise***, then half the mass of the animals you cultivate is comprised of shell in which atmospheric CO_2 is captured and stored, permanently. But there's more. The other half of the animal's mass is meat that you can sell as a return on your initial investment. And while the animals were growing, they were performing all those other ecosystem services mentioned above (filtration, biodeposition, denitrification, reef building, enhanced biodiversity, shoreline stabilisation and wave management). How much value do you put on all that?

In terms of actual costs in monetary terms, Avdelas et al. (2020) provide a production cost (and farm gate sale price) for mussels produced by four different methods, averaged across eight EU countries and across the years 2010 to 2016 (see Table 7.2 in Chap. 2; Moore et al. 2021b). These authors showed that the overall average production cost of mussels in the EU over those years was 0.87 € kg^{-1} (for a farm gate price of 1.08 € kg^{-1}). Other useful data from the same source are:

- The total number of enterprises reporting was 2,720.
- The grand average value of assets per enterprise = approximately €700,000.
- The grand average turnover per enterprise = approximately €384,000.

From these data we can make these extrapolations:

- An average production cost of mussels of 0.87 € kg^{-1} is equivalent to 870 € t^{-1}.
- One tonne of fresh mussels has a farm gate value of 1,080 €.
- One tonne of fresh mussels is equivalent to 0.5 t of shell.
- The molar mass of CaCO3 = 100.0869 g; the molar mass of CO_2 = 44.01 g.
- Assuming the shell is made entirely of $CaCO_3$, 0.5 t of shell is equivalent to 0.5 t × 44/100 = 0.22 t CO_2.

This 0.22 t CO_2 cost 870 € to be converted to a permanent sink ***but*** was accompanied by highly nutritious mussel meat with a sales value of 1,080 €. And all this was achieved with a commercial cultivation process that has no negative impact on the environment when properly managed but offers several highly beneficial ecosystem services. ***We believe that this makes mussel farming, and by default other bivalve mollusc farming enterprises, viable alternatives to all the CCUS industrial technologies*** illustrated in Tables 7.2 and 7.3.

References

Ai Z, Hanasaki N, Heck V, Hasegawa T, Fujimori S (2021) Global bioenergy with carbon capture and storage potential is largely constrained by sustainable irrigation. Nat Sustain (2021). https://doi.org/10.1038/s41893-021-00740-4

Alonso AA, Álvarez-Salgado XA, Antelo LT (2021) Assessing the impact of bivalve aquaculture on the carbon circular economy. J Clean Prod 279:123873. https://doi.org/10.1016/j.jclepro.2020.123873.

Aronoff K (2020) Carbon capture is not a climate savior. New Repub Website. https://newrepublic.com/

Asplund ME, Dahl M, Ismail RO, Arias-Ortiz A, Deyanova D et al. (2021) Dynamics and fate of blue carbon in a mangrove–seagrass seascape: influence of landscape configuration and land-use change. Landsc Ecol (2021). https://doi.org/10.1007/s10980-021-01216-8

Avdelas L, Avdic-Mravlje E, Borges Marques AC, Cano S, Capelle JJ et al (2020) The decline of mussel aquaculture in the European Union: causes, economic impacts and opportunities. Rev Aquac; in press, online before inclusion in an issue. https://doi.org/10.1111/raq.12465

Bastin J-F, Finegold Y, Garcia C, Mollicone D, Rezende M, Routh D, Zohner CM, Crowthe TW (2019) The global tree restoration potential. Science 365:76–79. https://doi.org/10.1126/science.aax0848

Beerling DJ, Kantzas EP, Lomas MR, Wade P, Eufrasio RM et al (2020) Potential for largescale CO_2 removal via enhanced rock weathering with croplands. Nature 583:242–248. https://doi.org/10.1038/s41586-020-2448-9

Bellamy R, Lezaun J, Palmer J (2019) Perceptions of bioenergy with carbon capture and storage in different policy scenarios. Nat Commun 10:743. https://doi.org/10.1038/s41467-019-08592-5

Beuttler C, Charles L, Wurzbacher J (2019) The role of Direct Air Capture in mitigation of anthropogenic greenhouse gas emissions. Front Clim **1**:10. https://doi.org/10.3389/fclim.2019.00010

Bossio DA, Cook-Patton SC, Ellis PW, Fargione J, Sanderman J, Smith P, Wood S, Zomer RJ, von Unger M, Emmer IM, Griscom BW (2020) The role of soil carbon in natural climate solutions. Nat Sustain 3:391–398. https://doi.org/10.1038/s41893-020-0491-z

Brandão M, Levasseur A, Kirschbaum MUF, Weidema BP, Cowie AL, Jørgensen SV, Hauschild MZ, Pennington DW, Chomkhamsri K (2013) Key issues and options in accounting for carbon sequestration and temporary storage in life cycle assessment and carbon footprinting. Int J Life Cycle Assess 18:230–240. https://doi.org/10.1007/s11367-012-0451-6

Brienen RJW, Caldwell L, Duchesne L, Voelker S, Barichivich J, Baliva M, Ceccantini G, Di Filippo A, Helama S, Locosselli GM, Lopez L, Piovesan G, Schöngart J, Villalba R, Gloor E (2020) Forest carbon sink neutralized by pervasive growth-lifespan tradeoffs. Nat Commun 11:4241. https://doi.org/10.1038/s41467-020-17966-z

Budinis S (2020). Going carbon negative: what are the technology options? IEA Direct Air Capture Tracking report. Published by the International Energy Agency: Paris. https://www.iea.org/reports/direct-air-capture and https://www.iea.org/commentaries/going-carbon-negative-what-are-the-technology-options

Campbell B, Alder J, Trujillo P, Pauly D (2016) A global analysis of mariculture production and its sustainability, 1950–2030, pp 137–151. In: Pauly D, Zeller D (eds) Global atlas of marine fisheries: a critical appraisal of catches and ecosystem impacts. Washington, DC, USA: Island Press. https://tinyurl.com/3rjautnf

Clarke L, Böhringer C, Rutherford TF (eds) (2009) International, U.S., and E.U. climate change control scenarios: results from EMF 22. Energy Econ 31 (Supplement 2):S63–S305. https://www.sciencedirect.com/journal/energy-economics/vol/31/suppl/S2

Consoli C (2019) Bioenergy and carbon capture and storage. A perspective published by The Global CCS Institute. 14 pp. https://www.globalccsinstitute.com/wp-content/uploads/2019/03/BECCS-Perspective_FINAL_PDF.pdf

Costello C, Cao L, Gelcich S, Cisneros-Mata MÁ, Free CM et al (2020) The future of food from the sea. Nature 588:95–100. https://doi.org/10.1038/s41586-020-2616-y

Cuéllar-Franca RM, Azapagic A (2015) Carbon capture, storage and utilisation technologies: a critical analysis and comparison of their life cycle environmental impacts. *Journal of CO_2 Utilization*, **9**: 82–102. DOI: https://doi.org/10.1016/j.jcou.2014.12.001.

Cui RY, Hultman N, Edwards MR, He L, Sen A, Surana K, McJeon H, Iyer G, Patel P, Yu S, Nace T, Shearer C (2019) Quantifying operational lifetimes for coal power plants under the Paris goals. Nat Commun 10:article 4759. https://www.nature.com/articles/s41467-019-12618-3

Cziesielski MJ, Duarte Carlos M, Aalismail N, Al-Hafedh Y, Anton A et al (2021) Investing in blue natural capital to secure a future for the red sea ecosystems. Front Mar Sci 7:article 603722. https://doi.org/10.3389/fmars.2020.603722

Dasgupta P (2021) The economics of biodiversity: the dasgupta review. 606 pp. HM Treasury, London. ISBN:9781911680291. https://www.gov.uk/government/publications/final-report-the-economics-of-biodiversity-the-dasgupta-review

De Jong W, Pokorny B, Katila P, Galloway G, Pacheco P (2018) Community forestry and the sustainable development goals: a two way street. Forests 9:article 331. https://doi.org/10.3390/f9060331

Demeude H, Gadault T (2020) Le cri de détresse des forêts françaises confrontées aux crises écologiques. Website of The GoodPlanet Info magazine. https://www.goodplanet.info/2020/11/26/le-cri-de-detresse-des-forets-francaises-confrontees-aux-crises-ecologiques/

Deng H, Bielicki JM, Oppenheimer M, Fitts JP, Peters CA (2017) Leakage risks of geologic CO_2 storage and the impacts on the global energy system and climate change mitigation. Clim Change 144:151–163. https://doi.org/10.1007/s10584-017-2035-8

Dockrill P (2018) There is not enough land on Earth to support the diet recommended by authorities. Science Alert website 14 August 2018. https://www.sciencealert.com/not-enough-land-earth-support-diet-usda-recommends-guidelines-environment-ecosystem-health

Domeignoz-Horta LA, Pold G, Liu X-JA, Frey SD, Melillo JM, DeAngelis KM (2020) Microbial diversity drives carbon use efficiency in a model soil. Nat Commun 11:article 3684. https://doi.org/10.1038/s41467-020-17502-z

Duarte CM, Agusti S, Barbier E, Britten GL, Castilla JC, Gattuso JP, Fulweiler RW, Hughes TP, Knowlton N, Lovelock CE, Lotze HK, Predragovic M, Poloczanska E, Roberts C, Worm B (2020) Rebuilding marine life. Nature 580:39–51. https://doi.org/10.1038/s41586-020-2146-7

Duarte CM, Wu J, Xiao X, Bruhn A, Krause-Jensen D (2017) Can seaweed farming play a role in climate change mitigation and adaptation? Front Mar Sci 4: article 100. https://doi.org/10.3389/fmars.2017.00100

Dumbauld BR, Ruesink JL, Rumrill SS (2009) The ecological role of bivalve shellfish aquaculture in the estuarine environment: a review with application to oyster and clam culture in West Coast (USA) estuaries. Aquaculture 290:196–223. https://doi.org/10.1016/j.aquaculture.2009.02.033

Ekins P, Dresner S, Dahlström K (2008) The four-capital method of sustainable development evaluation. Eur Environ 18:63–80. https://doi.org/10.1002/eet.471

FAO (2018) The State of the World's Forests 2018 - Forest pathways to sustainable development. Published by the Food and Agriculture Organization of the United Nations: Rome under Licence: CC BY-NC-SA 3.0 IGO. 139 pp. http://www.fao.org/3/I9535EN/i9535en.pdf

Fourqurean JW, Duarte CM, Kennedy H, Marba N, Holmer M, Mateo AM (2012) Seagrass ecosystems as a globally significant carbon stock. Nat Geosci 5:505–509. https://doi.org/10.1038/NGEO1477

Friedmann SJ, Zapantis A, Page B, Consoli C, Fan Z, Havercroft I, Liu H, Ochu E, Raji N, Rassool D Sheerazi H, Townsend A (2020) Net-zero and geospheric return: actions today for 2030 and beyond. Published by the global CCS institute and the center on global energy policy of Columbia University, USA. 55 pp. https://drive.google.com/file/d/148lRxw0tg550bUCoy9IzOFTmO_a2DjZj/view

Friggens NL, Hester AJ, Mitchell RJ, Parker TC, Subke J-A, Wookey PA (2020) Tree planting in organic soils does not result in net carbon sequestration on decadal timescales. Glob Chang Biol; Online Version of Record before Inclusion in an Issue. https://doi.org/10.1111/gcb.15229

Gattuso JP, Williamson P, Duarte CM, Magnan AK (2021) The potential for ocean-based climate action: net negative emissions and beyond. Frontiers in Climate in press. https://www.frontiersin.org/articles/, https://doi.org/10.3389/fclim.2020.575716/abstract

Girardin CAJ, Jenkins S, Seddon N, Allen M, Lewis SL, Wheeler CE, Griscom BW, Malhi Y (2021) Nature-based solutions can help cool the planet - if we act now. Nature 593:191–194. https://doi.org/10.1038/d41586-021-01241-2

Gordon TAC, Radford AN, Simpson SD, Meekan MG (2020) Marine restoration projects are undervalued. Science 367:635–636. https://doi.org/10.1126/science.aba9141

Griscom BW, Adams J, Ellis PW, Houghton RA, Lomax G et al (2017) Natural climate solutions. Proc Natl Acad Sci USA 114:11645–11650. https://doi.org/10.1073/pnas.1710465114

Heilmayr R, Echeverría C, Lambin EF (2020) Impacts of Chilean forest subsidies on forest cover, carbon and biodiversity. Nat Sustain 2020; online https://doi.org/10.1038/s41893-020-0547-0

Heilweck M, Moore D (2021) Saving the planet with appropriate biotechnology: 3. The high seas solution/Salvando el planeta con biotecnología apropiada: 3. La solución de alta mar. Mex J Biotechnol 6:92–128. https://doi.org/10.29267/mxjb.2021.6.1.92

Hiederer R, Köchy M (2012) Global soil organic carbon estimates and the harmonized world soil database. EUR scientific and technical research series. 79 pp. Institute for Environment and Sustainability (Joint Research Centre) and European Soil Data Centre (ESDAC), ISBN: 9789279231087. https://doi.org/10.2788/13267

Hilborn R, Banobi J, Hall SJ, Pucylowski T, Walsworth TE (2018) The environmental cost of animal source foods. Front Ecol Environ 16:329–335. https://doi.org/10.1002/fee.1822

Hill R, Bellgrove A, Macreadie PI, Petrou K, Beardall J, Steven A, Ralph PJ (2015) Can macroalgae contribute to blue carbon? An Australian perspective. Limnol Oceanogr 60:1689–1706. https://doi.org/10.1002/lno.10128

Hills CD, Tripathi N, Carey PJ (2020) Mineralization technology for carbon capture, utilization, and storage. Front Energy Res 8:142. https://doi.org/10.3389/fenrg.2020.00142

Ho THM, Tsai PA (2020) Microfluidic salt precipitation: implications for geological CO_2 storage. Lab Chip 20:3806–3814. https://doi.org/10.1039/D0LC00238K

Hong S, Yin G, Piao S Dybzinski R, Cong N, Li X, Wang K, Peñuelas J, Zeng H, Chen A (2020) Divergent responses of soil organic carbon to afforestation. Nat Sustain 2020; online https://doi.org/10.1038/s41893-020-0557-y

Hua F, Wang L, Fisher B, Zheng X, Wang X, Yu DW, Tang Y, Zhu J, Wilcove DS (2018) Tree plantations displacing native forests: the nature and drivers of apparent forest recovery on former croplands in Southwestern China from 2000 to 2015. Biol Cons 222:113–124. https://doi.org/10.1016/j.biocon.2018.03.034

Hugelius G, Loisel J, Chadburn S, Jackson RB, Jones M, MacDonald G, Marushchak M, Olefeldt D Packalen M, Siewert MB, Treat C, Turetsky M, Voigt C, Yu Z (2020) Large stocks of peatland carbon and nitrogen are vulnerable to permafrost thaw. In: Proceedings of the national academy of sciences of the United States of America open access article 201916387. https://doi.org/10.1073/pnas.1916387117

IEA (2010) Energy technology perspectives 2010. Paris: OECD/IEA. https://www.iea.org/reports/energy-technology-perspectives-2010

IOGP (2019) The potential for CCS and CCU in Europe. Report to the thirty second meeting of the European Gas Regulatory Forum, June 2019. Coordinated by the International Association of Oil and Gas Producers (IOGP). 47 pp. https://ec.europa.eu/info/sites/default/files/iogp_-_report_-_ccs_ccu.pdf

IPCC (2005) Carbon dioxide capture and storage. In: Metz B, Davidson O, de Coninck H, Loos M, Meyer L (eds) Special report prepared by working group III of the intergovernmental panel on climate change. Cambridge University Press, Cambridge, UK, pp 431. ISBN: 0521685516. https://www.ipcc.ch/site/assets/uploads/2018/03/srccs_wholereport-1.pdf

IPCC (2014) Climate change 2014: mitigation of climate change. In: Edenhofer O, Pichs-Madruga R, Sokona Y, Farahani E, Kadner S, Seyboth K, Adler A, Baum I, Brunner S, Eickemeier P, Kriemann B, Savolainen J, Schlömer S, von Stechow C, Zwickel T, Minx JC (eds) The working group III contribution to the fifth assessment report (AR5) of the Intergovernmental panel on climate change. Cambridge University Press, Cambridge, UK and New York, USA. ISBN: 9781107654815. https://www.ipcc.ch/report/ar5/wg3/

Irlam L (2017) Global costs of carbon capture and storage. An update published by the Global CCS Institute. https://www.globalccsinstitute.com/resources/publications-reports-research/global-costs-of-carbon-capture-and-storage/

Jacobson HK (2001) Climate policy: international. In: Smelser NJ, Baltes PB (eds) 2011–2016 in international encyclopedia of the social & behavioral sciences. ISBN 9780080430768. https://doi.org/10.1016/B0-08-043076-7/04487-9

Jacquet J, Sebo J, Elder M (2017) Seafood in the future: bivalves are better. Solut J 8:27–32. https://www.thesolutionsjournal.com/article/seafood-future-bivalves-better

Jones ISF (2014) The cost of carbon management using ocean nourishment. Int J Clim Chang Strat Manag 6:391–400. https://doi.org/10.1108/IJCCSM-11-2012-0063

Kheshgi H, de Coninck H, Kessels J (2012) Carbon dioxide capture and storage: seven years after the IPCC special report. Mitig Adapt Strat Glob Change 17:563–567. https://doi.org/10.1007/s11027-012-9391-5

Krüger T (2017) Conflicts over carbon capture and storage in international climate governance. Energy Policy 100:58–67. https://doi.org/10.1016/j.enpol.2016.09.059. PDF download available at: http://hdl.handle.net/10419/215420

Kuwae T, Hori M (2019) The future of blue carbon: addressing global environmental issues. In: Kuwae T, Hori M (eds) Blue carbon in shallow coastal ecosystems: carbon dynamics, policy, and implementation. Springer Nature Singapore Pte Ltd., pp 347–373. https://doi.org/10.1007/978-981-13-1295-3_13

Larkin P, Gracie R, Shafiei A, Dusseault M, Sarkarfarshi M, Aspinall W, Krewski D (2019) Uncertainty in risk issues for carbon capture and geological storage: findings from a structured expert elicitation. Int J Risk Assess Manag 22:429–463. https://doi.org/10.1504/IJRAM.2019.103335

Lee HZL, Davies IM, Baxter JM, Diele K, Sanderson WG (2020) Missing the full story: first estimates of carbon deposition rates for the European flat oyster, *Ostrea edulis*. Aquat Conserv Mar Freshwat Ecosyst 30:2076–2086. https://doi.org/10.1002/aqc.3402

Lenton TM (2014) The global potential for carbon dioxide removal. In: Hester RE, Harrison RM (eds) Geoengineering of the climate system, pp 52–79. From the book series Issues in environmental science and technology. Published by the Royal Society of Chemistry [www.rsc.org]. Book https://pubs.rsc.org/en/content/ebook/978-1-84973-953-5. Chapter https://doi.org/10.1039/9781782621225-00052

Liu L-C, Li Q, Zhang J-T, Cao D (2016) Toward a framework of environmental risk management for CO_2 geological storage in China: gaps and suggestions for future regulations. Mitig Adapt Strat Glob Change 21:191–207. https://doi.org/10.1007/s11027-014-9589-9

Lorenz K, Lal R (2005) The depth distribution of soil organic carbon in relation to land use and management and the potential of carbon sequestration in subsoil horizons. Adv Agron 88:35–66. https://doi.org/10.1016/S0065-2113(05)88002-2

Lorenz K, Lal R (2014) Soil organic carbon sequestration in agroforestry systems. A review. Agron Sustain Dev 34:443–454. https://doi.org/10.1007/s13593-014-0212-y

Lorenz K, Lal R, Shipitalo MJ (2011) Stabilized soil organic carbon pools in subsoils under forest are potential sinks for atmospheric CO_2. For Sci 57:19–25. https://doi.org/10.1093/forestscience/57.1.19

Maack M, Davidsdottir B (2015) Five capital impact assessment: appraisal framework based on theory of sustainable well-being. Renew Sustain Energy Rev 50:1338–1351. https://doi.org/10.1016/j.rser.2015.04.132

Mace MJ, Fyson CL, Schaeffer M, Hare WL (2021) Governing large-scale carbon dioxide removal: are we ready? - an update. New York, USA: Carnegie Climate Governance Initiative (C2G), February 2021. https://climateanalytics.org/media/c2g2-2018-cdr-governance-1_1_1.pdf

Macreadie PI, Anton A, Raven JA, Beaumont N, Connolly RM et al (2019) The future of Blue Carbon science. Nat Commun 10:article 3998. https://doi.org/10.1038/s41467-019-11693-w

Mandal BP, Guha M, Biswas AK, Bandyopadhyayc SS (2001) Removal of carbon dioxide by absorption in mixed amines: modeling of absorption in aqueous MDEA/MEA and AMP/MEA solutions. Chem Eng Sci 56:6217–6224. https://doi.org/10.1016/S0009-2509(01)00279-2

Marshall S, Hoffman N, Hardng A, Zapantis A, Temple-Smith L (2020) CarbonNet project: a hub for climate change action and economic growth. A webinar produced by the Global CCS Institute. https://www.globalccsinstitute.com/wp-content/uploads/2020/07/CCS-Talks-CarbonNet-Webinar-Presentation_23-July.pdf

Masson-Delmotte V, Zhai P, Pörtner H-O., Roberts D, Skea J et al (eds) (2019) IPCC, 2018: Global Warming of 1.5°C. An IPCC Special Report on the impacts of global warming of 1.5°C above pre-industrial levels and related global greenhouse gas emission pathways, in the context of strengthening the global response to the threat of climate change, sustainable development, and efforts to eradicate poverty. Open access at this https://www.ipcc.ch/sr15/ and for PDF download: https://www.ipcc.ch/sr15/download/

Meadowcroft J, Langhelle O (eds) (2011) Caching the carbon: the politics and policy of carbon capture and storage. Edward Elgar Publishing Ltd., Cheltenham, UK, pp 320. ISBN: 9780857933874

Miocic JM, Gilfillan SMV, Frank N, Schroeder-Ritzrau A, Burnside NM, Haszeldine RS (2019) 420,000 year assessment of fault leakage rates shows geological carbon storage is secure. Sci Rep 9:1–9. https://doi.org/10.1038/s41598-018-36974-0

Moore D (2020) A biotechnological expansion of shellfish cultivation could permanently remove carbon dioxide from the atmosphere/Una ampliación biotecnológica del cultivo de moluscos bivalvos podría eliminar permanentemente el dióxido de carbono de la atmósfera. Mex J Biotechnol 5:1–10. https://doi.org/10.29267/mxjb.2020.5.1.1

Moore D (2021) Saving the planet with appropriate biotechnology: 4. Coccolithophore cultivation and deployment/Salvando el planeta con biotecnología apropiada: 4. Cultivo de cocolitóforos e implementación. Mex J Biotechnol 6:129–155. https://doi.org/10.29267/mxjb.2021.6.1.129

Moore D, Heilweck M, Petros P (2021a). Saving the planet with appropriate biotechnology: 1. Diagnosing the problems/Salvando el planeta con biotecnología apropiada: 1. Diagnóstico de los problemas. Mex J Biotechnol 6:1–30. https://doi.org/10.29267/mxjb.2021.6.1.1

Moore D, Heilweck M, Petros P (2021b). Saving the planet with appropriate biotechnology: 2. Cultivate shellfish to remediate the atmosphere/Salvando el planeta con biotecnología apropiada: 2. Cultivar mariscos para remediar la atmósfera. Mex J Biotechnol 6:31–91. https://doi.org/10.29267/mxjb.2021.6.1.31

Mori AS (2020) Advancing nature-based approaches to address the biodiversity and climate emergency. Ecol Lett 23:1729–1732. https://doi.org/10.1111/ele.13594

Müller LJ, Kätelhön A, Bachmann M, Zimmermann A, Sternberg A, Bardow A (2020) A guideline for life cycle assessment of carbon capture and utilization. Front Energy Res 8:article 15. https://doi.org/10.3389/fenrg.2020.00015

NASEM (2019) Negative emissions technologies and reliable sequestration: a research agenda. National academies of sciences, engineering, and medicine of the United States of America. The National Academies Press, Washington, DC:. https://doi.org/10.17226/25259

National Research Council (2010) Ecosystem services of bivalves: implications for restoration. In: Ecosystem concepts for sustainable bivalve mariculture. Report of the committee on best practices for shellfish mariculture. The National Academies Press, Washington, DC, pp 123–132, 190 p. ISBN: 9780309146951. https://doi.org/10.17226/12802

Natural Capital Committee (2020) Natural capital committee advice on reaching net zero by 2050: nature based interventions. https://www.gov.uk/government/publications/a-natural-capital-approach-to-attaining-net-zero-nature-based-interventions

Natural England (2010) England's peatlands: carbon storage and greenhouse gases. Report NE257. Open access and PDF download at http://publications.naturalengland.org.uk/search?q=englands+peatlands&num=100

Nellemann C, Corcoran E, Duarte CM, Valdés L, De Young C, Fonseca L, Grimsditch G (eds) (2009) Blue carbon: the role of healthy oceans in binding carbon. A Rapid Response Assessment for the United Nations Environment Programme. Published by GRID-Arendal (www.grida.no), 80 pp. ISBN: 9788277010601. https://www.grida.no/publications/145

Neori A, Nobre AM (2012) Relationship between trophic level and economics in aquaculture. Aquac Econ Manag 16:40–67. https://doi.org/10.1080/13657305.2012.649046

Nguyen HM, Ralph PJ, Marín-Guirao L, Pernice M, Procaccini G (2021) Seagrasses in an era of ocean warming: a review. Biol Rev; Online Version before Inclusion in an Issue. https://doi.org/10.1111/brv.12736

ONS (2016) UK natural capital: experimental carbon stock accounts, preliminary estimates. https://www.ons.gov.uk/economy/environmentalaccounts/bulletins/uknaturalcapital/experimentalcarbonstockaccountspreliminaryestimates

Pearse R, Böhm S (2014) Ten reasons why carbon markets will not bring about radical emissions reduction. Carbon Manag 5:325–337. https://doi.org/10.1080/17583004.2014.990679

Pendleton L, Donato DC, Murray BC, Crooks S, Jenkins WA, Sifleet S (2012) Estimating global "blue carbon" emissions from conversion and degradation of vegetated coastal ecosystems. PLoS One 7:article e43542. https://doi.org/10.1371/journal.pone.0043542

Pines L (2020) Environmental commodities: what are they & how can you trade them? The Essential Guide. Commodity.com website. https://commodity.com/environmental/

Primavera JH, Friess DA, Van Lavieren H, Lee SY (2019) The mangrove ecosystem. In: Sheppard C (ed) World seas: an environmental evaluation (Second Edition), Volume III: ecological issues and environmental impacts. Academic Press an imprint of Elsevier BV, Cambridge, MA, US, pp 1–34. ISBN 9780128050521. https://doi.org/10.1016/B978-0-12-805052-1.00001-2

Pugh TAM, Lindeskog M, Smith B, Poulter B, Arneth A, Haverd V, Calle L (2019) Role of forest regrowth in global carbon sink dynamics. Proc Natl Acad Sci USA 116:4382–4387. https://doi.org/10.1073/pnas.1810512116

Purvis B, Mao Y, Robinson D (2019) Three pillars of sustainability: in search of conceptual origins. Sustain Sci 14:681–695. https://doi.org/10.1007/s11625-018-0627-5

Rizvi S, Pagnutti C, Fraser E, Bauch CT, Anand M (2018) Global land use implications of dietary trends. PLoS ONE 13:article e0200781. https://doi.org/10.1371/journal.pone.0200781

Röhr ME, Holmer M, Baum JK, Björk M, Boyer K et al (2018) Blue carbon storage capacity of temperate Eelgrass (*Zostera marina*) meadows. Glob Biogeochem Cycles 32:1457–1475. https://doi.org/10.1029/2018GB005941

Sadiq MM, Konstas K, Falcaro P, Hill AJ, Suzuki K, Hill MR (2020) Engineered porous nanocomposites that deliver remarkably low carbon capture energy costs. Cell Rep Phys Sci 1:article 100070. https://doi.org/10.1016/j.xcrp.2020.100070

Secretariat of the Convention on Biological Diversity (2009) Scientific synthesis of the impacts of ocean fertilization on marine biodiversity. CBD Technical Series No. 45, 53 pp. Published by the Secretariat of the Convention on Biological Diversity: Montreal, Canada. ISBN: 929225166X. https://www.cbd.int/doc/publications/cbd-ts-45-en.pdf

Singh J, Dhar DW (2019) Overview of carbon capture technology: microalgal biorefinery concept and state-of-the-art. Front Mar Sci 6:article 29. https://doi.org/10.3389/fmars.2019.00029

Smith P, Adams J, Beerling DJ, Beringer T, Calvin KV et al (2019) Land-management options for greenhouse gas removal and their impacts on ecosystem services and the sustainable development goals. Annu Rev Environ Resour 44:255–286. https://doi.org/10.1146/annurev-environ-101718-033129

Soudzilovskaia NA, van Bodegom PM, Terrer C, Zelfde MV, McCallum I, McCormack ML, Fisher JB, Brundrett MC, de Sá NC, Tedersoo L (2019) Global mycorrhizal plant distribution linked to terrestrial carbon stocks. Nat Commun 10:article number 5077. https://doi.org/10.1038/s41467-019-13019-2

Stern N (2007) The economics of climate change: the stern review. Cambridge University Press,.Cambridge, UK, 712 pp. ISBN: 9780521700801

Stern N (2015) Economic development, climate and values: making policy. Proc R Soc Ser B, Biol Sci 282:article 20150820. https://doi.org/10.1098/rspb.2015.0820

The Climate Question (2021a) Does big money really believe green is good? BBC World Service podcast [TheClimateQuestion-20210214-DoesBigMoneyReallyBelieveGreenIsGood.mp3] broadcast 15 February 2021. Download from https://www.bbc.co.uk/programmes/w3ct0xbd

The Climate Question (2021b) Have we planted too much faith in trees? BBC World Service podcast [TheClimateQuestion-20210221-HaveWePlantedTooMuchFaithInTrees.mp3] broadcast 22 February 2021. https://www.bbc.co.uk/programmes/w3ct0xbf

Trujillo P (2008) Using a mariculture sustainability index to rank countries' performance. In: Alder J, Pauly D. (eds) A comparative assessment of biodiversity, fisheries and aquaculture in 53 countries' exclusive economic zones. University of British Columbia, Vancouver, Canada; Fisheries Centre Research Reports 16(7):28–56 https://open.library.ubc.ca/soa/cIRcle/collections/facultyresearchandpublications/52383/items/1.0074742

Tsuji Y, Yoshida M (2017) Biology of Haptophytes: complicated cellular processes driving the global carbon cycle. In: Hirakawa Y (ed) Advances in botanical research, vol 84, pp 219–261. https://doi.org/10.1016/bs.abr.2017.07.002

Turolla E, Castaldelli G, Fano EA, Tamburini E (2020) Life cycle assessment (LCA) proves that Manila Clam farming (*Ruditapes philippinarum*) is a fully sustainable aquaculture practice and a carbon sink. Sustainability 2020 12:article 5252. https://doi.org/10.3390/su12135252

Tussenbroek BI (1989) The life-span and survival of fronds of *Macrocystis pyrifera* (Laminariales *Phaeophyta*) in the Falkland Islands. Eur J Phycol 24:137–141. https://doi.org/10.1080/00071618900650131

UN (2016) Transforming our world: the 2030 agenda forsustainable development. https://sdgs.un.org/2030agenda

UNDP Thailand (2019) Blue carbon society and UNDP team up on a 'masterplan' for Phetchaburi's mangroves and its coastal ecosystems. Posting on the *United Nations Development Programme - Thailand* website. https://tinyurl.com/2m8cse7x

Watson RT, Noble IR, Bolin B, Ravindranath NH, Verardo DJ, Dokken DJ (eds) (2000) IPCC special report: land use, land-use change and forestry. Cambridge University Press for IPCC, Geneva, Switzerland, Cambridge, UK, 375 pp. https://www.ipcc.ch/report/land-use-land-use-change-and-forestry/

Wernberg T, Krumhansl K, Filbee-Dexter K, Pedersen MF (2019) Status and trends for the world's kelp forests. In: Sheppard C (ed) World seas: an environmental evaluation (Second Edition), Volume III: Ecological Issues and Environmental Impacts. Academic Press an imprint of Elsevier BV, Cambridge, MA, US, pp 57–78. ISBN 9780128050521. https://doi.org/10.1016/B978-0-12-805052-1.00003-6

WMO (2020) State of the global climate 2020: provisional report. Published by the world meteorological organization, 38 pp. https://library.wmo.int/index.php?lvl=notice_display&id=21880

Xu G, Liang F, Yang Y, Hu Y, Zhang K, Liu W (2014) An improved CO_2 separation and purification system based on cryogenic separation and distillation theory. Energies 7:3484–3502. https://doi.org/10.3390/en7053484

Zapantis A (2017) Benefits of CCS presented at significant conferences. Global CCS Institute Insight website news item. https://www.globalccsinstitute.com/news-media/insights/benefits-of-ccs-presented-at-significant-conferences/

Zevenhoven R, Fagerlund J (2010) Mineralisation of CO_2. In: Maroto-Valer MM (ed) Developments and innovation in carbon dioxide (CO_2) Capture and Storage Technology, pp 433–462. Woodhead Publishing Ltd., Cambridge, UK; an imprint of Elsevier B. V. ISBN: 9781845697976. https://doi.org/10.1533/9781845699581.4.433

What Should Be Done

8

David Moore, Matthias Heilweck,
and Peter Petros

8.1 In this Chapter...

We present our Executive Summary of Chaps. 1–7 and Summary of Recommendations, which includes our Action Plan, and ending with a section of FAQs: Frequently Asked Questions and Online Comments.

It's encouraging that more international attention has been paid to the plight of the oceans in recent years, culminating in the ***UN Decade of Ocean Science for Sustainable Development 2021–2030*** (https://www.oceandecade.org/wp-content/uploads//2021/09/337521-Ocean%20Decade%20Implementation%20Plan:%20Summary). Although much of the attention has been occasioned by, on the one hand, the desire to exploit the ocean deeps with mining operations, or on the other hand a drive to clean up past mistakes made by the casual discard of plastics articles, it does at least show that 'the powers that be' are beginning to recognise the ocean's role in a healthy planet. An intergovernmental conference (IGC) is developing an international legally binding instrument under the United Nations Convention on the Law of the Sea on the conservation and sustainable use of *Marine Biodiversity of Areas Beyond National Jurisdiction* (BBNJ) (https://www.iucn.org/theme/environmental-law/our-work/oceans-and-coasts/marine-biodiversity-areas-beyond-national-jurisdiction-bbnj).

Our suggestions go further than regulating mining or tidying up discarded wastes, with plans for a programme of sustainable and eco-friendly activities, on a global scale, that will provide vast amounts of human food and animal feed, while at the same time removing and sequestering permanently massive quantities of carbon dioxide from the atmosphere.

Our overriding message is that we must properly harness the ability of calcifying marine organisms (molluscs, crustacea and coccolithophore algae) to **remove permanently CO_2** from the atmosphere into solid (crystalline) $CaCO_3$. Emphasising that this $CaCO_3$ not only sequesters atmospheric carbon, but has the bio-circular economic potential for use as a sustainable biomaterial, with calcifying organisms producing many other ecosystem benefits.

Atmospheric CO_2 sequestered by shellfish is indigestible, crystalline and chemically stable calcium and calcium-magnesium carbonates; when the animal dies the shell remains for geological periods of time. Effectively, the CO_2 is permanently removed from the atmosphere. That's the animal's generous legacy and our inheritance. It is the certainty and permanence of the removal of CO_2 from the atmosphere that makes biotechnology using calcifying organisms so attractive as a means to ameliorate climate change.

The crucial first step is acceptance of the proposition that the present-day paradigm of current aquaculture, which is to cultivate shellfish for food, should be changed to ***cultivate shellfish for their shells***. This **paradigm change** places the **value** of the cultivation exercise on the

D. Moore et al., *Aquaculture: Ocean Blue Carbon Meets UN-SDGS*,
Sustainable Development Goals Series, https://doi.org/10.1007/978-3-030-94846-7_8

production of shell and its removal of carbon from the atmosphere. This allows us to take the monetary value of the food that results as a by-product, and effectively the earned interest on the capital invested in the shell-cultivation exercise. This ***new generation shellfish farming*** is aimed at whole-planet ecosystem repair and restoration. Take the food represented by shellfish meat as a by-product from the production of shell, and leave or return the shell to the seabed from which it was harvested. Bivalve molluscs have been described as ecosystem engineers because the shells of earlier generations create their own reef habitats, which are of such significant size that they become important to general marine biodiversity. By providing habitats at different depths they support and enhance entire ecosystems.

If we do amplify farming and harvesting greatly, we will start to produce shellfish meat in excess of that required for the 'shellfish-as-a-delicacy' market. Then we could start thinking about processed shellfish meat as an alternative to meat products produced from terrestrial farm-reared animals, in the expectation that reduced husbandry of farm animals for meat-eaters will release pastures for afforestation and reduce further destruction of existing natural forests. Pseudo-beef-burgers made from shellfish meat are likely to be more readily acceptable than those made from the insects or cultured animal cells that some food technologists are keen to promote. Another positive characteristic of shellfish farming is that it presents no conflict between using land to grow food crops and using land to grow trees, or, for that matter, using land for pasture animals. There is no need for irrigation, food or fertiliser. Farming shellfish can be combined with restoration and conservation of overfished fisheries and usually involves so little intervention (beyond provision of habitats and, where necessary, protection of larvae and juveniles from predation (in 'nurseries') that there is no inevitable conflict with other activities. About 70 per cent of the Earth's surface is covered by water. **We might as well use it sustainably to rescue our atmosphere, our planet and ourselves**.

Several recent publications have concluded that marine restoration projects are undervalued despite their ability to help maintain species survival and ecosystem services, ultimately providing humanity with the breathing space needed to stabilise the climate. Rebuilding marine life has been proposed as a doable **Grand Challenge for humanity**, an ethical obligation and a smart economic objective to achieve a sustainable future.

Securing that future for marine ecosystems suffering the effects of climate change is evidently a political challenge as much as an ecological or social one. The political limitations of conventional ecosystem governance have been recognised, but the immense promise of blue carbon science is so strikingly evident that it must be taken more seriously. But more than anything else it requires the recognition that cultivation of coccolithophores, corals, crustacea and molluscs ***on a massive scale*** would have the effect of removing a massive amount of CO_2 directly from the atmosphere; ***here, now and permanently***, making a ***continued*** contribution to the health of the whole planetary ecosystem.

It would be a criminal dereliction of duty if humanity failed to grasp this last opportunity to carry out this 'doable Grand Challenge'. And the sentence for such a criminal act is extinction.

8.2 Executive Summary of Chaps. 1–7

8.2.1 Chapter 1. Diagnosing the Problem by David Moore, Matthias Heilweck and Peter Petros

In this chapter we give a plain language guide to the Earth's carbon cycle by briefly summarising the observations and origins of increased levels of greenhouse gases, mainly CO_2 but including CH_4 and N_2O, in our present-day atmosphere. They are increased in the sense that they have not occurred naturally in the Earth's atmosphere at any time during the past 420,000 years. The only tenable explanation for our atmosphere's present state is that it is the consequence of mankind's excessive use of fossil fuels since the Industrial Revolution onwards. Something that has been described as a *planetary-scale experiment* in which humans return to the atmosphere and oceans the concentrated organic carbon that had previously been stored in sedimentary rocks for many hundreds of millions of years. We deal with the arguments that deny the truth of anthropogenic CO_2-driven climate change, then illustrate the Earth's global carbon cycle. Explaining how it was almost exactly in equilibrium for several thousand years while humans were evolving, before industrial humans intervened. We describe how the excess greenhouse gas emissions are projected to change the global climate over this century and beyond, and discuss '*dangerous anthropogenic interference*' (DAI), '*reasons for concern*' (RFCs) and *climate tipping points*. We give a short account of the various improved management, engineering and natural climate solutions advocated to increase carbon storage (sequestration) and/or avoid greenhouse gas emissions across global forests, wetlands, grasslands, agricultural lands and industry, and indicate how they are discussed in our later chapters. Finally, we outline the alternative natural carbon sink we propose that is currently so greatly undervalued and underdiscussed.

8.2.2 Chapter 2. Cultivate Shellfish to Remediate the Atmosphere by David Moore, Matthias Heilweck and Peter Petros

In this chapter the very recent research that indicates that massive tree planting is not the panacea that many believe, is discussed. Photosynthetic carbon capture by trees and other green plants is widely thought to be our most effective strategy to limit the rise of CO_2 concentrations in the atmosphere by pulling carbon from the atmosphere into the sinks represented by the plant body and the soil. However, practical experience indicates that putting such plans into effect could do more harm than good to our environment. Planting trees can release more carbon from the soil sink than the plants sequester into their biomass. And, ***in all cases***, the plant biomass sink is only ever a ***temporary sequestration*** because when the plant dies its biomass rots, and its sequestered carbon is returned to the atmosphere. Forests should be planted for the intrinsic values of forests; for clean, oxygenated air, natural biodiversity and restorative conservation of terrestrial ecosystems, rather than tree planting as a means to sequester atmospheric CO_2. This chapter describes the basic message of the book, which is that ***shellfish cultivation as a carbon sequestration strategy*** is both more immediately rewarding and more helpful ***in the very long term***. A considerable proportion of shellfish biomass is represented by the shells of the animals, and shellfish shell is made by converting atmospheric CO_2 into crystalline calcium carbonate which is stable for geological periods of time. The essentials of habitat conservation, ecosystem balance and carbon sequestration for carbon-offsetting programmes are also introduced; topics developed in chapters that follow.

8.2.3 Chapter 3. Aquaculture: Prehistoric to Traditional to Modern by David Moore and Matthias Heilweck

In this chapter it is pointed out that the human tradition of eating shellfish goes back to the time when *Homo sapiens* first started to migrate out of Africa, between 200,000 and 100,000 years ago. Archaeological finds of ancient meals of shellfish and ancient middens of shellfish shells track human migrations around the world. Middens do more than track migrations. They show that ***wooden artefacts and plant residues do not survive, but shells do***. Illustrating the truth of our fundamental claim that shellfish shells sequester atmospheric carbon permanently. The coastal migrations of early humans continued across the Bering Strait to North America. Along the northwest coast of North America, early humans, referred to as First Nations in Canada, actively, and sympathetically, managed the resources of their shoreline habitats, engineering intertidal rock-walled terraces as ***clam gardens***, ancient sustainable mariculture technologies. When we reach recorded history, we enter a phase of increasing exploitation of marine resources for an ever-growing human population. ***By the end of the nineteenth century oysters had become a cheap staple food*** on both sides of the Atlantic. The working man could get a decent meal of oysters at any street corner for a few cents in New York or a penny or two in London. The real price we all paid for this was that ***oyster dredging*** on both sides of the Atlantic ***destroyed 85% of the world's oyster beds***. New Yorkers in the 1800s ate about 600 oysters a year each; the average American today eats about 3 oysters each year. Farmed oysters account for 95% of the world's total present-day oyster consumption. The animal, which has been described as an ***ecosystem engineer*** for its reef-building abilities, is one of those that we have driven to the verge of extinction in the wild. In the twenty-first century, the oyster deserves to have the same vigour applied to its ***restoration and conservation*** as was applied to dredging it from the seabed during the nineteenth century.

8.2.4 Chapter 4. The High Seas Solution by Matthias Heilweck

In this chapter the case is made for greater use of the High Seas ***to replace forage fish with mussels in the diet of farmed fish*** and produce the increasing amounts of food that will be required by the growing human population, while at the same time pulling down carbon from the atmosphere with bivalve cultivation. The vision is to preserve the oceans as a healthy and sustainable food source for mankind by emphasising conservation and ecosystem balance beyond coastal waters. The plans are for huge (centralised) bivalve mollusc farming facilities on the high seas, using factory ships and offshore factory rigs (re-purposed disused oil rigs?) located on seamounts outside Exclusive Economic Zones and employing Perpetual Salt Fountains on the flanks of the seamount to bring nutrients to the farms. If properly designed (and the design and building capabilities exist throughout the offshore industries around the world), this will immediately provide (i) feed for animals and food for humans, (ii) sustainable marine ecosystems and (iii) permanent atmospheric carbon sequestration in the form of reefs of bivalve shells.

8.2.5 Chapter 5. Farming Giant Clams in 2020: A Great Future for the 'Blue Economy' of Tropical Islands by David Moore

In this chapter a specific and dramatic example for the tropics is detailed to avoid too much attention being diverted to Northern Hemisphere shellfish cultivation. There is nothing more dramatic than Giant Clams, which have been fished to extinction in many Indian Ocean and Pacific waters, but elsewhere contribute to a still thriving industry, though clam dredging is now doing immense damage to coral reefs in many areas. The topic of giant clam cultivation covers conservation and restocking of clams, but with the potential bonus of ***rehabilitating coral reefs*** degraded by bleaching induced by climate change, as well as food production, and development of remunerative local industry for local Pacific Island communities. It's not just the food value of the animal; the shells are used for carving (large!) ornaments, and several species are traded around the world for marine aquariums. Work towards 'seeding' and recolonising has been going on in the Pacific region for more than 30 years. Much of this work has been published and many of the faults in approach and problems of governance identified. In addition, though, several local enterprises have developed methods to produce economically large numbers of young giant clams for restocking tropical seas. The conservation and educational programmes that have resulted deserve ***wider and more prolonged attention and greater investment*** as they tie-in well with our call to 'cultivate shellfish to remediate the atmosphere'.

8.2.6 Chapter 6. Coccolithophore Cultivation and Deployment by David Moore

In this chapter the potential for cultivation of coccolithophore golden-brown algae for carbon sequestration is addressed. Coccolithophores have been major calcium carbonate producers in the world's oceans for about 250 million years. Today they account for about a third of the total marine $CaCO_3$ production by coating their single cells externally with delicately sculptured plates of *microcrystalline $CaCO_3$*. The possibility that these algae could be used to trap atmospheric CO_2 with existing technology has not been widely recognised. There is scope, however, for both high technology cultivation in bioreactors and low technology cultivation on the High Seas or in terraced raceway ponds or lagoons on tropical coastal sites. The latter could produce a sludge of pure $CaCO_3$ that could be harvested as a ***feedstock for cement production in place of the fossiliferous limestone that is currently used*** (cement production accounts for around 8% of industrial fossil CO_2 emissions). Bioreactor cultivation of genetically engineered coccolithophores could produce ***customised calcite crystals for nanotechnology industries***. On the high seas coccolithophores naturally produce extensive ***blooms***, and the blooms emit a volatile gas (dimethyl sulphide) to the atmosphere, where it ***promotes formation of clouds that block solar radiation***. Imagine aquaculture nurseries onboard factory ships, cultivating both coccolithophores and bivalve molluscs. During their open ocean cruises the ships could produce biodegradable floats already spawned with fixed juvenile bivalve molluscs and streams of coccolithophore algae that could be released into the ocean currents and ocean gyres nourished by artificial upwelling of nutrient-rich waters when the ship deploys its perpetual salt fountains. The dual aim is to be creating and maintaining *blooms of coccolithophores in the oceanic high seas to sequester carbon from the atmosphere, and generation of cloud cover* to cool the immediate environment.

8.2.7 Chapter 7. Comparing Industrial and Biotechnological Solutions for Carbon Capture and Storage by Peter Petros and David Moore

In this chapter we deal with the current artificial/industrial Carbon Dioxide Capture, Utilisation and Storage (CCUS) solutions and shows their power and potential in curtailing greenhouse gas (GHG) emissions. Key valuation models of sustainability for current carbon capture and storage (CCS) infrastructure will be used to explain what problems could arise and potential ways to avoid the likely risks through drastic changes in fundamental attitudes. The shortfalls of each industrial solution are also presented in the context that all activities should be carried out with due regard for long-term human and environmental well-being, rather than economic growth alone. Overall, we discuss: solutions for atmospheric carbon reduction; the carbon market; industrial/artificial carbon dioxide capture, utilisation and storage systems; carbon emissions reduction targets. We make comparisons between 'soft' nature-based biotechnological solutions, including coastal blue carbon and the ultimate blue carbon, which is the ocean's calcifiers. Following a discussion of sustainability assessment of CCUS methods we conclude that changing the paradigm of shellfish farming from 'shellfish as food' to 'shellfish for carbon sequestration' places the value of the exercise of shellfish cultivation onto the production of shell, taking the food value of the animal protein as one of the several ecosystem services that bivalve molluscs supply. We calculate that this paradigm shift makes mussel farming, and by default other ***bivalve mollusc farming enterprises, viable alternatives to all the CCUS industrial technologies in use today***.

8.2.8 Summary of Recommendations

In reading this section you should bear in mind that the key objective we wish to achieve is to enable the world's oceans to produce the increasing amounts of food that will be required by the growing human population in a sustainable manner, while at the same time permanently removing carbon from the atmosphere with ecologically friendly bivalve cultivation. To this we couple the determined use of coccolithophore algae cultivation, in the High Seas and in raceway lagoons on land, to extract permanently more carbon from the atmosphere and make further contributions to the amelioration of the dangerous anthropogenic interference that our industrial society has inflicted on the atmosphere.

In order to carry out our recommendations we need:

- planetary-scale funding, and
- central management with global authority to initiate, fund and maintain projects over several decades as necessary.

Most important of all, though, is that we (meaning humanity as a whole) must develop the determination to make the changes in human activity and human behaviour that are essential if we are to meet the challenge of climate change.

Importantly, this means not only all the widely discussed matters involved in reducing fossil fuel usage but serious changes in the attitudes of the world's scientific communities in respect of the solutions they promote.

Most of today's scientists would recommend Negative Emissions Technologies, or **NETs**, which are technologies that remove and sequester CO_2 from the atmosphere with the intention of mitigating climate change. NETs that are currently most widely ***expected*** to be of value are:

- biological processes to increase carbon stocks in soils, forests and wetlands,
- generate energy from biomass, and capture and store the resulting CO_2 emissions,
- capture CO_2 directly from the air with chemical processes and sequester it in geological reservoirs,
- formal consideration has only been given to near-shore coastal ***Blue Carbon***, namely, mangroves, tidal marshlands, and other tidal or saltwater wetlands, seagrass beds, and kelp 'forests'. However, these Blue Carbon options are, like terrestrial forests, reversible if the carbon sequestering practices are not maintained, because they depend on sequestering carbon in the biomass of ***living*** plants; when the plants die they are digested by microorganisms and their carbon is returned to the atmosphere as respiratory CO_2.

Focussing exclusively on ***near-shore coastal*** NETs wilfully ignores the **oceanic options for CO_2 removal and sequestration** that are offered by the 70% of the Earth's surface covered by the high seas.

We wish to remedy this exclusion. The central thrust of our argument is that the physiological chemistry of a few types of aquatic creatures, the ***calcifiers of the coasts and open seas***, (coccolithophore algae, corals, crustacea and molluscs) enables them to extract CO_2 from the atmosphere and sequester it **permanently** as crystalline $CaCO_3$.

8.2.9 The Action Plan

Our suggestions for a realistic action plan would fall into three levels of activity:

- Immediate activity.
- Infrastructural activity designed to change the paradigm.

8.2.10 Immediate Activity (Assuming Global Funding and Programme Management Are Both in Place)

- As the shellfish cultivation industry is the only industry on the planet that can expand without damaging the atmosphere, we want shellfish producers to greatly expand their production specifically to generate more shell.

- Central funding and management (a development foundation?) should be available to invest cash *immediately* in every existing aquaculture enterprise with the aim of doubling their production each season for the next five to ten seasons.
- Central funding should guarantee farm gate prices as the markets react and adapt to successively greater production volumes.

- From the project launch date, the ability of shellfish to sequester carbon permanently should be used in promotional and advertising materials at all levels of the shellfish food supply chain to encourage enhanced sales [‘*Eat more shellfish. SAVE the atmosphere*’].
- Carbon-offsetting programmes, those used by the **general public** to offset the carbon emissions of their transport and other domestic activities, should include projects to fund shellfish cultivation ***because of its ability to offer a permanent removal of atmospheric carbon***. There is a wide variety of potential projects, ranging from support for developing/expanding local subsistence fisheries as a means to employ and feed communities in need, through to supplementing the funding of local aquaculture programmes to enable them to expand their activities continually for several to many years.
- Primary **CO_2 emitter industries** might be encouraged to **sponsor** a different kind of help to balance their carbon footprints by funding the larger scale infrastructural activities which are anticipated, which include industrial scale installations offshore and ocean-going factory ships. The high-energy industries that most need to compensate their heavy carbon footprints have all the necessary skills and experience to take such large-scale efforts forward.
- Central governments should be persuaded and encouraged to fund shellfish cultivation to sequester atmospheric carbon as a contribution to their carbon neutrality goals. As well as making significant financial input to the projects most appropriate to them, their responsibilities could include political, legal and administrative facilitation of the anticipated projects.

8.2.11 Waste Shells?

If mollusc aquaculture is to play an increasingly significant role in the global provision of protein foods and feed, then it can be expected that there will be a diversification of mollusc products, with more sold in processed form, where shells are removed during processing. In such a scenario, shell waste valorisation will be of increasing concern. In areas of high shellfish production, such as China, Europe and the Americas, shell waste is already an issue, with shell dumps providing an unsightly and odorous nuisance. This is completely unjustified because far from being a nuisance waste product shellfish shells are an environmentally and economically valuable commodity. By far the best thing to do with waste shells is return them to the seabed where the scraps of flesh that remain can feed scavengers and detritus feeders and the shells contribute to reef formation. On the other hand, we have mentioned before some of the extra value that has been found by exploiting the ‘waste’ shells of dead calcifiers (Section headed *Additional benefits* in Chap. 2). Such uses for waste shells that have been published include:

- Calcium supplementation in poultry farming.
- Acidity regulation in hobbyist aquarium systems.
- Use of crushed mollusc shells as a replacement for more commonly used mined limestone for addition to agricultural land to adjust soil pH and/or drainage.
- For use in paper whitening.
- As an eco-friendly road de-icer.
- Calcination of waste shells produces quicklime (CaO) which also has many uses, most notably as far as release of fossil carbon to the atmosphere is concerned, in cement manufacture.

Clearly, using shells like this will return their carbon to the global carbon cycle, instead of sequestering it. The advantageous point here, though, is that all the applications mentioned above (including, most dramatically, cement manufacture) normally use ***fossil limestone*** for their purposes. Consequently, the ‘additional benefit' gained by using 'waste' shells is that it is **today's CO_2** that is returned to the carbon cycle rather than the inevitable return to the atmosphere being an **emission of fossil CO_2.**

8.2.12 Infrastructural Activity Designed to Change the Paradigm

This phase assumes that (a) an administrative, legal and political secretariat is in place. This could be an authority formed under the United Nations Convention on the Law of the Sea (UNCLOS) which will fund, regulate, supervise and, where necessary, impose, activities aimed at sustainable atmosphere amelioration in both coastal and international waters. Let's call it ***The Ocean Decade Commission***; (b) Central government start-up funding and major energy-industry sponsorship-funds are secured. (c) All activities listed under 'Immediate activity', above, have been initiated.

We expect the ***Ocean Decade Commission*** to fund developmental research into high technology programmes. Biotechnological research on aquaculture is well-established around the world but we need to coordinate this varied activity towards the common goal of extracting CO_2 from the atmosphere and sequestering it permanently as crystalline $CaCO_3$.

- To provide the calcium carbonate for use as a feedstock for cement production, replacing the fossil limestone currently used to make quicklime, we need to exploit fully the potential for cultivation of coccolithophore algae on large scales:
  - in giant illuminated fermenters;
  - in 'rice-paddy-like' terraces flooded with flowing seawater.
- Coccolithophore cellular biomass will also provide lipids and biofuels to replace fossil fuel usage, as well as other bioactive substances with potential pharmaceutical uses.
- Genetically manipulated coccolithophores could provide tailor-made coccoliths for devices in the nanotechnology industries.
- We need to fund research programmes specifically to develop shellfish cultivation aimed primarily at farming shells (taking any food extracted as a by-product at a guaranteed price).
- We need to adapt existing aquaculture farming methods to a wider range of sites and locations, for example: a mussel farm or other bivalve farm on every offshore wind turbine, every oil and gas rig, every pier, wharf and jetty, every breakwater or harbour wall. This is to include standardising methods of creating clam gardens and bivalve shell-reefs as a contribution to shore protection and wave-calming measures. In fact, bivalve farming wherever possible, at low risk and low effort, taking any food extracted as a by-product at a guaranteed price.
- We need to develop new aquaculture farming methods to establish new organisms and new methods to enhance incorporation of atmospheric carbon into shells.

8.3 Specific Recommendations

8.3.1 Seamount Installations and Factory Ships

The **largest installations** we wish to build are factory rigs, either floating above 1000 m deep anchorages (or with dynamic positioning) or fixed to the flat tops of seamounts (guyots or table mounts) that rise close to the surface. These are extinct volcanoes rising up from the seafloor, sometimes almost to the surface, perfectly suited to support an infrastructure on its top, with any necessary pipework along its slopes. For the first such international installation we have located a suitable guyot in the Vitória-Trindade Chain, which is called Davis Bank, which is located off the central coast of Brazil. Starting 175 km off the coast of Espírito Santo State and extending for 950 km eastward, the seamounts of this chain are disposed almost linearly at 20° and 21 °S. There are many other seamounts in the world's oceans that we hope would be utilised once the value of the operation to the atmosphere has been demonstrated.

The **primary function** of these factory rigs will be to provide the infrastructure necessary for massive scale cultivation of mussels on long lines. Current mussel farms based on this method are yielding 150–300 metric tonnes of prime mussels, per hectare per year. To put these figures into perspective, beef production is only around 0.340 tonne per hectare per year, almost a thousand times less. We can reasonably expect production of between 3 and 6 million tonnes of mussel flesh in a square of 90,000 ha, like the flat top of Davis Bank. The seamount installations will be equipped with:

- The manufacturing facilities to establish, maintain and harvest 90,000 ha of long line mussel farm.
- The mariculture facilities for mussel hatcheries/nurseries, macroalgae cultivation and distribution (for the kelp forests), zooplankton (copepod) nurseries and coccolithophore cultivation.
- Pipe laying equipment for the illuminated perpetual salt fountains that will be created around the seamount.
- Industrial equipment for processing harvested mussels to make aquafeed from the mussel flesh.
- Equipment for the return of waste mussel shells to the seamount top and sides (some of this will be pre-attached with mussel spat to establish shell-reefs around and across the seamount.
- Wind (and possibly water) turbines for renewable energy generation to supplement a geothermal energy power plant.

The **purpose of these massive mussel farming installations** is to provide mussel meat intended as an ***aquafeed***, for fish farms elsewhere, as a replacement for the fishmeal that is currently derived from capture fishing of forage fish.

As fish farming expands to feed the growing human population, increasing quantities of wild captured forage fish are necessary to feed the farmed fish but forage fish catches are already declining. Current attempts to feed farmed carnivorous marine fish with fishmeal substitutes for from terrestrial agricultural resources are fundamentally flawed because it is illogical to use scarce agricultural land to feed a marine resource. A 'fishmeal' produced from mussel meat would be a natural and well-balanced diet for farmed fish. The expected production capacity mentioned above (3–6 million tonnes of mussel flesh in a square of 90,000 ha) can be compared to the world's largest capture fishery dedicated to the production of fishmeal which is currently the ±5.5 million tonne of anchovies caught in Peruvian waters. This kind of activity, occurring in all oceans, is environmentally destructive because these low trophic fish species are the subsistence food for around a billion people who live in coastal communities, not to mention food chains involving higher trophic animals, all of which depend on a healthy marine environment which is currently being jeopardised by overfishing.

Our concept is that as uncontrolled harvesting of forage fish as fish food is not sustainable, we need to establish extensive mussel farms as an alternative source of nutritious aquafeed for future generations of fish farms, especially in developing countries where the essential development of aquaculture is delayed by the lack of aquafeed. In addition, although the planned mussel production by seamount installation is centred on harvesting mussel meat, it also represents a sequestration programme for atmospheric carbon on a massive scale. If, say, the Davis Bank installation produces 200 tonnes of mussels on long lines per hectare per year (a modest production average), about 50% of that harvest will be shell, representing 100 t of calcium carbonate containing 12% of carbon, that is 12 t of carbon per hectare per year, being permanently removed each year (=about one million t y^{-1} across the fully operational 90,000 ha seamount farm). Terrestrial ecosystems retain about 4 t of carbon per hectare per year, and the carbon is sequestered only as long as the plants remain alive (see Chap. 4).

We also plan that the seamount installations will create small biodegradable floating devices, spawned with bivalve mollusc larvae, to be released from the facility into the passing Brazil current towards the South Atlantic gyre. There is no intention to harvest these, but to let them sink when the shells are heavy enough. This is a highly scalable simple technology to create a self-replicating carbon sink.

A final point is that the seamount installations are planned to be Integrated Multi-Trophic Aquaculture (IMTA) facilities, where the waste products of one species are recycled as feed for others. Mussel faeces cause pollution problems in most of today's monoculture farm locations. To avoid this, the soluble nutrients in faeces can be assimilated by macroalgae ('kelp forests'), and the solids can be assimilated by scavengers and detritus feeders on the sea bottom. IMTA establishes greatly improved biodiversity so that the 'mussel farm' becomes a self-sustaining ecological community or biotope/habitat.

8.3.2 Factory Ships

More mobile versions of the seamount installations will be a fleet of factory ships intended only to enhance shell production. These will be equipped with bivalve hatcheries and production facilities for biodegradable floatation devices that will be released, already spawned with fixed juvenile bivalve molluscs, into ocean currents and ocean gyres. There is no intention to harvest these self-replicating carbon sinks. In addition, the factory ships will be equipped with bioreactors to cultivate coccolithophore algae (derived from waters local to their operating zones) that will be used to establish and maintain extensive coccolithophore blooms in the open ocean well away from shipping lanes and fishing areas.

- Coccolithophore blooms produce the volatile gas dimethyl sulphide (DMS), which promotes cloud formation above the bloom. So, here is potential to stimulate formation of clouds that reflect solar radiation, which cools the ocean by altering the radiative energy budget, consequently reduces coccolithophore activity, thereby reducing levels of DMS in a classic, self-regulating feedback loop.

8.3.3 Coral Reef Restoration

For 40 years or more a wide range of academics and agencies have studied the decline of stocks of giant clams and their coral reef habitats due to commercial over-fishing, climate change and growth in demand for aquarium supplies and recreational (tourist) SCUBA fishing. Numerous well tested techniques and protocols exist that are able, within a reasonable timescale, to restore the biodiversity of coral reef systems in the wild to something close to normality. These include growth and reattachment of reef-building corals, coupled with distribution of captive bred, adult giant clam restorations, in which the giant clams share the role of ecosystem engineers with the corals, building the reef framework. Unfortunately, local efforts to implement these conservation schemes have in general been only partially successful for a mixture of reasons, among which are:

- Limited time and limited funding both contributing to limited scale of the operations.
- Conflicting demands between conservationists and local communities.
- Conflicting politics between local, regional and even national and international administrations.
- The high costs and lengths of time required to produce 'seed' clams have been problems for many operations.
- Lack of consistently committed involvement of local communities in the projects. In some cases, projects were not matched to what the local community needed or wanted.
- Poor survey and reporting protocols, together with poor funding for monitoring, have limited assessment of some reintroduction and restocking programmes even to the point of failing to report successful results.

We would expect the ***Ocean Decade Commission*** to intervene in these fragmented activities to unify the operations, supply generous funding for their expansion, and, probably most important, provide an over-arching secretariat offering consistent transnational activity over several decades and across the Indian and Pacific Oceans and into the South China Sea.

8.4 FAQs—Frequently Asked Questions and Online Comments

As with many documents these days, we have uploaded preprints of early versions of these documents to a variety of social networking sites and blog posts. These have given rise to questions and comments which we include below. The external questions/comments have been anonymised and are listed in date order.

> 17/11/19 R*******A writes: Yes, shellfish can help to reduce the CO_2 levels in the atmosphere, but scientists also have noted that during the climate change seawater has become more acidic and shellfish species are shrinking in size and the shells deform. What can the author comment on this? Are all the oceans becoming more acidic or are there regional differences? How could one also lower the acidification of the ocean? Shellfish are great purifiers of the ocean, but they also are great vectors for diseases and potent phycotoxins. The proposal of the authors is interesting, but it should go hand in hand with the monitoring of shellfish safety that in many regions worldwide is not done, especially in countries with less resources. Seaweed forests have also been proposed as atmospheric CO_2 removers, it should be mentioned briefly in the text.

REPLY: Only a very small percentage of shellfish you might eat are full of toxins or diseases. The shellfish farmer has the CHOICE of where and how to cultivate his shellfish. If the shellfish are farmed for food they can be grown in clean toxin-free waters and regularly monitored for food-safety; if the shellfish are grown ***for their shells and their permanent carbon sequestration ability***, as we suggest, it doesn't matter where or why you grow them. Grow them in polluted water to remove the pollutants, then ignore the meat. Grow them in exposed offshore positions to create wave-calming shell-reefs but ignore the meat. Grow them around offshore wind turbines and oil/gas rigs but don't bother trying to harvest them in such exposed places, just ignore the meat. The investment in growing them is worth it for the amount of carbon they can remove from the atmosphere. We want to '**change the paradigm**' to grow shellfish for their SHELLS, because the shells are a permanent atmospheric carbon sink. If you ***can*** grow them in clean waters, you can take the meat to sell as an extra return on your investment, BUT return the waste shells to the ocean, so they can continue to engineer their habitat as they have done for hundreds of millions of years.

Ocean acidification is caused by the uptake of carbon dioxide from the atmosphere. If we can take control of CO_2 emissions, we will also control ocean acidity. Thankfully, oceans have not yet been acidified to the extent that the water affects crystalline calcium carbonate. Experiments have indeed demonstrated significant changes in mussel (*Mytilus edulis*) shells cultured under acidification conditions, BUT these experiments used CO_2 concentrations that were 2½ times higher than the present natural levels. Adverse effects of present day ocean acidification impact the viability of symbiotic algae in the tissues of coral-forming animals (and giant clams). It is their death/expulsion due to elevated water temperature and/or acidification that kills ('bleaches') the coral.

> 17/11/19 R******B writes: [You] talk about the aquaculture of crustacean and mollusk … The question is, why combine crustaceans and mollusk?

REPLY: Crustacean shells are made of protein and calcium carbonate, and a lot of chitin. But they still sequester atmospheric carbon in the calcium carbonate.

> This idea … that mollusk shell is atmospheric solidified CO_2, is not correct.

REPLY: That depends where you start your analysis. The experiments you are thinking about analysed the origin of shell-carbonate-carbon from within the internal metabolism of the mollusc. They showed that some derived from bicarbonate in the water (which does come from dissolved atmospheric CO_2), but most was harvested from the animal's food.

Don't stop there, though, since all food chains start with a photosynthetic producer organism that makes its own food, whatever the shellfish animal eats depends ultimately on ***fixation of photosynthetic carbon from the atmosphere***. This is true of all the carbon in all the food of predators, scavengers, filter-feeders and detritus feeders alike, aquatic and terrestrial. **Globally, metabolic carbon is derived from photosynthetic fixation of *atmospheric carbon dioxide;* there is no other source**.

> Increase the invest[ment] cash in existing aquaculture enterprises is not the proper solution. Every aquaculture enterprise has a limited area (several acres) and have a [limited] carrying capacity.

REPLY: That's perfectly true. It's also true of animal pastures on land. But when the terrestrial farmer reaches the carrying capacity of his first field, he can burn down a forest to make a second field. The bivalve mollusc farmer can simply move a little further down the coast, or a little further away from the shore to lay his second farm. It's less destructive than farming on land. There's more ocean than land on this planet.

> 19 Feb 2020 J*** H******d Managing Director of a shellfish farm writes: I have read your paper with interest and I agree with the basic science you have presented … you need to remember that marketing and product development also costs money and we are competing for market in a country where, except for fish and chips, the average person never eats fish of any sort, let alone bivalve shellfish and the great majority of consumers do not really care about their food carbon footprint which is why you can see queues at '000s of drive-in Burger shops.

REPLY: Yes, I understand; at present we cultivate shellfish for the meat (the shell is food-waste) and the industry is scaled according to that market. And that's exactly why we say **we must change the paradigm**: ***cultivate the bivalves, and those other shellfish, to sequester permanently CO_2 from the atmosphere and accept the food as a sellable by-product***.

Don't lose sight of the fact that shellfish farming is unique. Bivalves and other shellfish are the only actively farmed organisms in which a third to a half of the weight of the harvest is crystalline, chemically-stable, calcium carbonate made from atmospheric CO_2. I reckon today's global aquaculture farming is removing about 5.5 million tonnes of CO_2 from the atmosphere each year. And that's permanent removal (calcium carbonate, limestone, is converted to quicklime, calcium oxide, at about 1,000 °C). We can easily increase the removal of carbon from the atmosphere

> We live in a real world where there are many other stakeholders with claims on marine space and we cannot unilaterally build mussel farms wherever we please, we have to get permits, licences, leases etc. This is a process which takes many years, costs money and often ends in failure but it is one we are continuing with and steadily making progress on.

REPLY: OK, if all of that is a problem, **CHANGE IT**. We need an international secretariat with the international political authority, the international funding, and a focussed administration to drive it all forward around the entire globe. Nobody expects it to be easy; but failure means that *Homo sapiens* stands a good chance of being just another species driven to extinction by humanity's folly.

15/4/2020: B** E***l writes: ... there is a major problem with large scale intensive bivalve culture ... The impact of any bivalve culture at the levels you describe would have a very profoundly damaging effect on the overall marine ecosystems ... because bivalves of most species filter out huge quantities of both plant and animal plankton (including many larval stages). When grown intensively they create single species monocultures ... [with] reduced growth rates, increased ... diseases and an environment dominated by bivalves.

REPLY: We discuss Integrated Multi-Trophic Aquaculture at some length in Chap. 4. Don't think about filling the ocean with mussel farms. Think about seeding the oceans with appropriately designed ***communities*** of organisms that will support the basic need: which is to cultivate shellfish for their ***shells*** and take whatever else they offer, human food (and/or animal feed), reef-building, pollution-filtration, coral reef reconstruction, as free by-products.

23/12/2020: B***S**R***** writes: This manuscript is outstanding. I have just seen calcifying organisms in a corner of the carbon sequestration diagram. I never realised their prominence in the elimination of CO_2. I would like to thank you for sharing this insight.

I have a few queries regarding this:

1. Is excess (more than normal) production of $CaCO_3$ good? Even though they are good alternatives to harness CO_2, might this have any negatives?
2. Is safe artificial upwelling feasible with current development? What about ocean nutrient removal as in case of Chinese coastal waters? I know the significance of artificial upwelling.
3. Is the release of dimethyl sulfide safe? (when grown in bioreactor). I think they have a few issues. (High vapor concentrations might be irritating to the eyes and respiratory tract, and may have various adverse effects on the central nervous system).

REPLY: Thank you for your kind words about our manuscript. Let me answer your questions in turn:

1. There's no such thing as 'excess $CaCO_3$' it's insoluble and present in vast quantities on Earth already. The fossilised stuff is called limestone and is the fossilised $CaCO_3$ silt of ancient shallow seas. The 'White Cliffs of Dover' are made out of it. We don't think there are any negatives.
2. In terms of the scale of the overall volume of the ocean, our upwelling plans are literally 'a drop in the ocean'. And we are not removing the nutrients, we are aiming to cycle them into bivalve and/or alga biomass; when they die the nutrients will be returned to the ocean one way or another.
3. The bioreactors are, by definition, closed vessels with gas outflows which can be safely monitored and controlled to ensure safety of operating staff. The release of DMS is a completely natural process and DMS has been released into the atmosphere by coccolithophore blooms throughout the 250 million years the things have been growing in our oceans. Well, you might query that last phrase, maybe they're not OUR oceans, maybe they're THEIRS. Maybe that's the point. Humans have messed up their ocean and atmosphere so now we have to enable them to solve the problem for us.

23/12/2020: William B. L***s writes: Here in NYC we're rebuilding the oyster beds that the Dutch ate. This is in response to hurricane Sandy, with its 13' storm surge that would've been resisted by those oysters. It's a lot of oysters! Supposedly one of the largest beds in the world. Still, I don't think it's that much carbon, honestly. Compared to a herd of cattle, I don't think oysters are going to offer an effective carbon sink. A single cow can sequester tones of carbon and other GHGs over the course of its life, I don't think oysters do that, even pound for pound.

REPLY: Hi William, thanks for your comment. Yes we know and appreciate the Billion Oyster Project in NYC harbor, but you SERIOUSLY underestimate the abilities of bivalve molluscs. Consider:

1. At least HALF the fresh weight of bivalve molluscs is made up of shell which is composed of calcium carbonate. This is stable for millions of years and represents atmospheric CO_2 PERMANENTLY removed from the atmosphere (we have fossil mollusc shells hundreds of millions of years old). You can't say that for a herd of cows because they sequester carbon only temporarily (when the cow dies the carbon in its biomass will be digested and returned to the atmosphere, only the skeleton remains and that's made from calcium phosphate).
2. There are old records of oyster beds over 200 km long and 30–40 km wide off the European coasts (that were fished out in the 1800s) and they must have been of similar size in US waters. You'd need a hell of a herd of cows to match the biomass in those beds. Pity we destroyed almost all of them in the 1800s. They need to be restored to their former glory. Go to it NYC!
3. I don't have production data for oysters to hand but we do know that long lines mussel farming is by far the world's most productive meat-production method, currently yielding 150–300 metric tonne per hectare per year. To put these figures into perspective, beef production is only around 0.34 tonne per hectare per year, almost a thousand times less!
4. A human subsistence diet requires about 180 kg of grain per person per year, and this can be produced on 0.045 ha of land. In contrast, an affluent high-meat diet requires at least four times more grain (and four times more land, 0.18 ha) because the animals are fed on grain and conversion of grain to meat is very inefficient. As it stands, the Earth does not have enough land for all its inhabitants to enjoy an affluent high-meat diet.
5. Bivalve molluscs and the other marine calcifiers (crustacea, corals, coccolithophore algae) don't need farmland, don't need fresh water irrigation, don't need supplementary fertilisers. They just use (a small part of) the 70% of the Earth's surface that's covered in ocean.
6. And finally, bivalve molluscs and the other marine calcifiers don't burp and fart methane into our poor misused atmosphere! They just take carbon out of it. Permanently (or have I said that before?).

28/12/20: William B. L***s writes again:
... ya I've heard about the seaweed but all of that is only relevant to CAFOs. When grazing on pasture the net effect of ruminants is to store tons of GHGs.
... you left out the soil and ecosystems that benefit from ruminants. even after they die. There was up to 14' of topsoil in the mission before the bison were eliminated and agriculture destroyed the land. Do you really think all those bison were causing global warming? Cmon.
... im surprised you aren't all more familiar with regenerative agriculture and ranching. Check out the Savory Institute.

REPLY: Hi William, You're missing a few important points, there, William. The total number of dairy cows in the world is around 0.3 billion, in addition there are about 1.5 billion beef cattle worldwide. In total, therefore, there are approximately 1.8 billion cattle worldwide, and that's just the global cattle herd, which is mostly fed intensively, so CAFOs are NOT irrelevant. Then there are about a billion sheep in the world, as well as a lot of antelopes, bison, buffalo, camels, deer, goats, oxen, wildebeest ... and we've not even mentioned the hind-gut fermenting horses, zebra, rhinoceros, elephants, tapirs, sloths, pigs, peccaries, guinea pigs, chinchillas…and rabbits! All of these are HUGE methane emitters (a gas 8 times stronger in greenhouse effect than carbon dioxide). So, YES, all those ruminants and hind-gut fermenters ARE CONTRIBUTING very significantly to global warming. And they do not STORE GHGs—they fart and burp them out as gas emissions all the time.

You are confusing pasture grazing in the wild with current farming practice. Those ancient herds of North American Bison grazed the open plains, perhaps for millions of years. When grass became scarce, they either died or moved on. It was a feedback self-regulated system, beneficial to both animals and their pastures. When a present-day cattle herder runs out of fodder he either (a) burns down a forest to make more grazing land; (b) adds fertiliser to his pastures to force them to grow more grass; or (c) buys in fodder from somebody else's pastures to feed his own animals.

It takes an area of cropland 7 times the size of the European Union to produce feed for the livestock animals of Europe (most of the additional cropland required to meet this demand is in China, the US, and Indonesia). Globally, we consume around 350 MILLION TONNES of meat a year. Meat has a much higher energy footprint than any other food. It takes 75 times more energy to produce meat than corn (https://www.theworldcounts.com/challenges/consumption/foods-and-beverages/world-consumption-of-meat/story). The Earth does not have enough land for all its human inhabitants to have a high-meat diet. Crops grown to feed farm animals are competing for land that might otherwise grow food for direct human consumption. This is not sustainable.

Bivalve molluscs and the other marine calcifiers (crustacea, corals, coccolithophore algae) don't need farmland, don't need freshwater irrigation, don't need supplementary fertilisers. If we put that amount of effort into producing 350 million tons of ***shellfish*** meat (rather than livestock meat), then the shells of the annual shellfish harvest would be ***removing*** 42 million tonnes of carbon from the atmosphere *EACH YEAR*: and using just a small part of the 70% of the Earth's surface that's covered in ocean. Have a Happy and shellfish-eating New Year.

24/12/2020: E***n B*****d B****y II writes: Very interesting. Perhaps fungi and algae work together in symbiosis? ... The principal organisms involved are bacteria, particularly cyanobacteria, small algae and fungi, that participate in the growth of microbial biofilms and mats…

REPLY: Hi E***n, thank you for your comments. I'm absolutely sure that fungi and algae must cooperate somewhere in the ocean; after all, the first photosynthetic terrestrial organisms were lichens—the archetypical fungus-cyanobacterium-bacterium-alga organised communities. Also, I know that fungal exoenzymes are capable of extracting nutrients from most terrestrial rocks (check out the chapters on geomycology and lichens in Moore, Robson & Trinci, 2020. *21st Century Guidebook to Fungi*, Second Edition. Cambridge, UK: Cambridge University Press

But, while I fully appreciate the importance of biofilms (see for example Moore, 2013. *Fungal Biology in the Origin and Emergence of Life*. Cambridge, UK: Cambridge University Press. 230 pp. ISBN-10: 1107652774, ISBN-13: 978-1107652774), we are talking here about coccolithophores which are PLANKTONIC. So, their coccoliths will rain down on the sediments from the massive populations of the algae in the photic zone above. And they will be in sufficient quantity to form the limestone strata to which they've been contributing for 250 million years.

28/12/2020: E***n B*****d B****y II writes again: Thanks … my understanding is lichens were not the first photosynthetic terrestrial life and are younger than previously thought. They came from the ocean preadapted for symbiosis. How might the role of diatoms as silicon transporters affect the coccolithophore calcification process? How might the differing requirements for silicon among the coccolithophores affect the process?

REPLY: Hi E***n, Generally speaking, a low silicate to nitrogen and phosphorus ratio in the surface waters allows coccolithophores to outcompete diatoms, when silicate to phosphorus and nitrogen ratios are high coccolithophores are outcompeted by diatoms.

Because their 'shell' is siliceous, diatoms sequester carbon only as long as they remain alive; when they die the carbon in their biomass is digested by something or other and the biomass carbon is respired as CO_2 (it is not included in their fossils, any more than dinosaur meat is included in their fossils). On the other hand, coccolithophores sequester carbon in their extracellular $CaCO_3$ 'shells' (they're called coccoliths) as well as in their biomass carbon. Some of them regularly shed their coccoliths during normal life, but all of them, when they die and however they die, leave behind the indigestible, microcrystalline, chemically-stable $CaCO_3$ coccoliths. These contribute to the seabed silt and, given a few million years, will form a layer of limestone. It was the co-existence of coccolithophores with diatoms of the day that caused the "significant removal of carbon dioxide from the atmosphere" long ago, while diatoms (and a lot of cyanobacteria) contributed to increasing oxygen levels.

It's perfectly true that calcifiers release carbon dioxide as a by-product of the calcification reaction. That reaction scheme is: $2HCO_3^- + Ca^{2+} = CaCO_3 + CO_2 + H_2O$. So, you see it starts with TWO bicarbonate ions from the environment ONE is precipitated as $CaCO_3$; ONE is returned to the environment as CO_2. So, while it is true that "it releases CO_2 as a by-product during shell formation" this is only HALF the story, you should add that the other one of those two carbons on the left of the reaction scheme is precipitated as $CaCO_3$ and permanently removed from the atmosphere.

Diatoms are very good at precipitating silica, but diatomaceous particles are NOT composed of carbon. And as industrial humans are not destroying the atmosphere with fossil silicate emissions, they're not doing our atmosphere any favours.

4/1/2021: Erlon B*****d B****y II writes yet again: "Calcium carbonate is essentially insoluble in sea surface waters today. Shells of dead calcareous plankton sinking to deeper waters are practically unaltered until reaching the lysocline, the point about 3.5 km deep past which the solubility increases dramatically with depth and pressure. By the time the CCD is reached all calcium carbonate has dissolved…". "...Calcareous plankton and sediment particles can be found in the water column above the CCD. If the sea bed is above the CCD, bottom sediments can consist of calcareous sediments called calcareous ooze, which is essentially a type of limestone or chalk. If the exposed sea bed is below the CCD tiny shells of $CaCO_3$ will dissolve before reaching this level, preventing deposition of carbonate sediment. As the sea floor spreads, thermal subsidence of the plate, which has the effect of increasing depth, may bring the carbonate layer below the CCD; the carbonate layer may be prevented from chemically interacting with the sea water by overlying sediments such as a layer of siliceous ooze or abyssal clay deposited on top of the carbonate layer.[1]" https://en.wikipedia.org/wiki/Carbonate_compensation_depth. Accessed 28 Dec 2020. "Results from this global study confirm that fast-sinking mechanisms can transport fresh organic carbon down into the deep ocean. By keeping carbon in the ocean and preventing it from re-entering the atmosphere as CO_2, the ocean acts to mediate Earth's climate". https://www.oceanbites.org/carbon-sinks-diatoms-in-the-deep-sea/

REPLY: Hi Erlon. Yes, great research. Of course, you could have got all of that from our preprints - check out number 2 [*Cultivate Shellfish to Remediate the Atmosphere*] and (mostly) number 3 [*The High Seas Solution*]. Still, pat on the back for 'due diligence'. I hope you appreciate what this research is telling you: when the calcifiers make carbonate from CO_2, the carbonate is STABLE. Even at the pressures of the deep ocean, only its solubility alters. The carbonate ion is absolutely stable and no CO_2 is released. When the solubilised $CaCO_3$ flows into coastal waters; it recrystallises! You might like to investigate what set of circumstances will release CO_2 from the limestone (that the $CaCO_3$ eventually becomes)... David

24/12/2020: CA***********d writes: There's no clear indication that the carbon taken up by coccolithophores would actually be sequestered. It is likely to be degraded in the surface ocean by heterotrophic bacteria and return to the atmosphere. Coccolithophore blooms typically end in viral lysis which releases cellular material into surface waters.
Also, there's a long history ecological dysfunction resulting from introducing non-native species. The second order effects of introducing large numbers of coccolithophores could be worse for the climate.

REPLY: Hi CA, No clear indication of carbon sequestration??? What about all those limestone strata all around the world? The one on my doorstep forms the well-known White Cliffs of Dover. Coccolithophores have been sequestering vast quantities of CO_2 from the atmosphere for at least 250,000,000 years.

With regard to viruses, and the giant DNA-viruses known as EhVs are possibly the largest viruses that exist anywhere, then yes, they would be something that would have to be avoided (by standard axenic culture) in any cultivation of coccolithophores. But they do specifically attack the diploid phase of the life cycle, so the haploids, that can also be cultivated, will be disease-free. However, the point about coccolithophores is that the $CaCO_3$ plates they make (which is where the CO_2 is sequestered) are extracellular. So viral lysis simply releases them into the sea where they are still chemically stable and indigestible, forming silts on the seabed which eventually fossilise into those limestone strata.

Finally, who mentioned non-native species? Only you. Many coccolithophores have a global oceanic distribution. *Emiliania huxleyi* is most commonly encountered, but there are several more. These planktonic algae are native to the planet. We don't plan to introduce them anywhere else.

30/12/2020: CA***********d writes again: White Cliffs of Dover are not necessarily representative of modern oceanographic conditions—note that they are a very localised phenomenon. please point me to a study showing the percentage of carbon that is degraded versus Buried in situ, eg using sediment traps. Also, the ocean is acidifying—how will projected

changes in pH affect the solubility of coccoliths? Finally what percent of cellular carbon is in coccoliths, rather than in more labile forms?
Yes, E hux is globally distributed, but that doesn't mean that it is wise to shift the balance away from other phytoplankton taxa by introducing an unprecedented amount to a natural system—ecology is a nonlinear dynamic system and the consequences are unpredictable eg dead zones.

REPLY: Hi CA, Well, Dover might be localised, but it IS representative of the MANY limestone-forming environments on Earth today. "Limestone is forming in the Caribbean Sea, Indian Ocean, Persian Gulf, Gulf of Mexico, around Pacific Ocean islands, and within the Indonesian archipelago" quoted from https://www.geology.com/rocks/limestone.shtml.

You don't need to worry about phytoplankton community structure; coccolithophores have been a dominant part of it for 250-million years. And you don't need to worry about "what percent of cellular carbon is in coccoliths, rather than in more labile forms" because the POINT is that the labile carbon is just part of the normal carbon cycle (like you breathing out CO_2); the coccolith is totally stable, crystallised, precipitated carbon that is REMOVED from the normal carbon cycle. You should get what you need to know about coccolithophores from this paper: Rigual-Hernández, A.S., Trull, T.W., Flores, J. A., Nodder, S.D., Eriksen, R. and nine others. (2020). Full annual monitoring of Subantarctic Emiliania huxleyi populations reveals highly calcified morphotypes in high-CO_2 winter conditions. *Scientific Reports*, **10**: article number 2594. https://doi.org/10.1038/s41598-020-59375-8.

25 December 2020: J***-P***** G*****o writes: Shellfish are key animals in marine ecosystems, sometimes ecosystem engineers and key resources for food security. There are therefore many reasons to conserve them and develop a sustainable aquaculture. However, not for their role in climate change mitigation.
The contribution by Moore et al. is highly misleading and demonstrates a profound misunderstanding of simple, but admittedly counterintuitive, biogeochemical principles. The precipitation of calcium carbonate is a source of carbon dioxide (CO_2):

$$2HCO_3^- + Ca^{2+} = CaCO_3 + CO_2 + H_2O$$

This reaction is valid in freshwater. In seawater the molar ratio $CO_2/CaCO_3$ is about 0.6 rather than 1 due to the buffering effect (Frankignoulle et al. 1994).
As pointed out by Berner et al. (1983), precipitation of $CaCO_3$ is an important process and the major way by which CO_2 is returned to the atmosphere.
Berner R. A., Lasaga A. C. & Garrels R. M., 1983. The carbonate-silicate geochemical cycle and its effect on atmospheric carbon dioxide over the past 100 million years. *American Journal of Science* **283**: 641–683.
Frankignoulle M., Canon C. & Gattuso J.-P., 1994. Marine calcification as a source of carbon dioxide- Positive feedback of increasing atmospheric CO_2. *Limnology and Oceanography* **39**:458–462.

REPLY: Sorry, but your comment "is highly misleading and demonstrates a profound misunderstanding of simple" arithmetic. AS YOU SHOW, the reaction removes TWO bicarbonate ions from the environment ONE is precipitated as $CaCO_3$; ONE is returned to the environment as CO_2. How can that be a " major way by which CO_2 is returned to the atmosphere"?

Very simple and well-known of every chemist: precipitation of $CaCO_3$ consumes 2 bicarbonate ions and releases 1 CO_2, which makes it a source of CO_2 (in addition to respiratory CO_2 of course. Do not take my word for it and check marine chemistry textbooks as well as the references I provided.

REPLY: YES they are the FACTS; it's your INTERPRETATION of them that is misguided. As my first comment pointed out (and you have just repeated), the calcifying reaction scheme shows that two bicarbonate ions (which originally were derived from the atmosphere) react with Ca ions and one of them is precipitated as $CaCO_3$, and the other released as CO_2. So, while it is true that "precipitation of calcium carbonate is a source of carbon dioxide (CO_2)" it is illogical to claim that returning one out of two carbons to the environment is a "major way by which CO_2 is returned to the atmosphere".

BUT, if you go one step further and ADD the consideration that there are a great many calcifying organisms in the oceans, which are all cycling through this reaction 24/7, then you can rightly claim that this is a major way by which CO_2 is returned to the atmosphere; PROVIDING you remember that the other one of those two carbons on the left of your reaction scheme is precipitated as $CaCO_3$ and ADMIT the matching claim that if this IS a major way by which CO_2 is RETURNED to the atmosphere then it is ALSO a major way by which carbon is REMOVED FROM the atmosphere.

3/1/2021: J***-P***** G*****o writes again:
Enough said, the evidence that dissolution of Ca-Mg rocks is a sink of CO_2, that calcification is a source of CO_2, and that the balance between the two is a sink for CO_2 is in the literature. I look forward to reading a peer-reviewed paper of yours demonstrating that the literature is wrong.

REPLY: ... and the end result is that the $CaCO_3$ carbon sink made by calcifiers lasts for hundreds of millions of years. Happy New Year JP, Happy New Year.

23 April 2021: M***** S****t of Aquafish Solutions Ltd, UK added this comment to the above conversation: Not my area, but...If bicarbonate is formed initially from CO_2 dissolving in seawater, and CO_2 released through shell formation is then redissolved into seawater (presumably?) wouldn't the net overall affect of shell production be less CO_2 in the environment? Is it a matter of looking at overall CO_2 budgets rather than individual reactions/components?
Does shell formation help stop ocean acidification through this process?

REPLY: Thanks for your contribution to this discussion Martin. Yes, you've got to the essential truth very succinctly. I might well adopt your phrase "a matter of looking at overall CO_2 budgets rather than individual reactions/components" because the best I've managed so far is this diatribe:

- The calcifying reaction scheme shows that two bicarbonate ions (which originally were both derived from the atmosphere, photosynthetic fixation of atmospheric CO_2 being the only source of metabolic carbon) react with a Ca + ion and one of them is precipitated as $CaCO_3$, and the other released as CO_2.
- So, the ***meaningful*** reaction scheme is: two atmospheric $CO_2 \rightarrow 2$ metabolic bicarbonates $\rightarrow 1$ insoluble carbonate + 1 potentially atmospheric CO_2, and during the calcification process, half of the involved atmospheric CO_2 is sequestered permanently and the other half can potentially return to the atmosphere. Though in photosynthetic algae (e.g. coccolithophores) this CO_2 is more likely to be used

immediately for photosynthesis and in animals without photosynthetic symbionts, is more likely to form carbonic acid and contribute to another round of calcification.

- This does NOT amount to a net production and release of CO_2 to the atmosphere. Rather, the produced (respired) CO_2 is just part of the normal carbon cycle and biological calcification is a major way by which carbon is removed permanently from the atmosphere.

As far as acidification is concerned, the 'usual' calcification reaction scheme [$2HCO_3^- + Ca^{2+} = CaCO_3 + CO_2 + H_2O$] suggests that shell formation ***can*** help stop ocean acidification. But by far the biggest ***cause*** of ocean acidification is the ***excess*** atmospheric CO_2 our industrial activities have generated by burning fossil fuels. Until we ***greatly amplify*** our oceanic farming of calcifying organisms (bivalves and coccolithophores alike), as we advocate in our publications, the formation of calcifier shells is unlikely to make much of an impact on ocean acidification. Check out our 'master-plan' in https://doi.org/10.29267/mxjb.2021.6.2.1.

25 December 2020: T***a M***a R****u writes: All great in theory, but these shellfish and other biota need food sources and a livable habitat. And I vaguely remember reading something about acidic ocean water making shell formation more difficult (but don't quote me on that). I'm just wary of solutions that require the proliferation of a small number of species within ecosystems, that will have other consequences).

REPLY: T***a, we have already fished out so many shellfish fisheries world-wide that many are close to extinction. For example, 85% OF THE WORLD'S oyster beds were fished out during the 19th Century by the end of which oysters were a cheap staple food for New Yorkers and Londoners alike. We are suggesting restoring balances that have been unbalanced by past human activities. Please read Chap. 3 '*Aquaculture—Prehistoric to Traditional to Modern*' for the full story about our history of eating shellfish. Thankfully, oceans have not been acidified to the extent that the water affects calcium carbonate. Acidification is more damaging to symbiotic algae in the tissues of coral-forming animals. It is their death/expulsion due to elevated water temperature and/or acidification that kills the coral. If you look at Chap. 4 '*High Seas Solution*' you will see that we are fully aware of the need to maintain ecosystem balance with multi-trophic solutions.

25 December 2020: H*t C****y writes: Shellfish are the cleaners of the sea. They take in the water, filter it is flush out cleaner water holding the toxins in their body with no harm to themselves.
Of course if you eat them and they are holding heaps of toxins ...

REPLY: H*t: that's very true (but only applies to a small percentage of shellfish you might eat). However, the shellfish farmer has the CHOICE of where and how to cultivate his shellfish. If the shellfish are farmed for food they can be grown in clean toxin-free waters; if the shellfish are grown for their shells (and their permanent carbon sequestration ability) it doesn't matter where or why you grow them. Grow them in polluted water to remove the pollutants, then ignore the meat. Grow them in exposed offshore positions to create wave-calming shell-reefs but ignore the meat. Grow them around offshore wind turbines and oil/gas rigs but don't bother trying to harvest them in such exposed places, just ignore the meat. The investment in growing them is worth it ***for the amount of carbon they can remove from the atmosphere***. We want to 'change the paradigm' to grow shellfish for their SHELLS, because the shells are a permanent atmospheric carbon sink. If you can grow them in clean waters, you can take the meat to sell as an extra return on your investment, BUT return the waste shells to the ocean, so they can continue to engineer their habitat as they have done for hundreds of millions of years.

29/12/2020: D***a D*******a of G*******r writes: Can coccolithophores be cultivated in a closed system only using ambient air for CO_2 supply? Can the CO_2 produced in the formation of the coccoliths be trapped and recycled? Can the closed system be fed with seawater? There is a desalination plant in Gibraltar for drinkable water, can the brine be used diluted with some seawater? Are there any existing patents that you may have come across that are worth investigating into with regards to the subject?

REPLY: Dear Diana, Thank you for your interest in, and kind words about, our preprints. These four preprints will be published in the *Mexican Journal of Biotechnology* on 1st January 2021 and you will be able to download all of them (free) from the journal using these DOIs

Moore D., Heilweck M. & Petros, P. 2021. Saving the Planet with Appropriate Biotechnology: 1. Diagnosing the Problems/Salvando el planeta con biotecnología apropiada: 1. Diagnóstico de los problemas. *Mexican Journal of Biotechnology*. **6**(1): 1–30. https://doi.org/10.29267/mxjb.2021.6.1.1.

Moore D., Heilweck M. & Petros, P. 2021. Saving the Planet with Appropriate Biotechnology: 2. Cultivate Shellfish to Remediate the Atmosphere/Salvando el planeta con biotecnología apropiada: 2. Cultivar mariscos para remediar la atmósfera. *Mexican Journal of Biotechnology*. **6** (1): 31–91. https://doi.org/10.29267/mxjb.2021.6.1.31.

Heilweck M. & Moore D. 2021. Saving the Planet with Appropriate Biotechnology: 3. The High Seas Solution/Salvando el planeta con biotecnología apropiada: 3. La solución de alta mar. *Mexican Journal of Biotechnology*. **6**(1): 92–128. https://doi.org/10.29267/mxjb.2021.6.1.92.

Moore D. 2021. Saving the Planet with Appropriate Biotechnology: 4. Coccolithophore cultivation and deployment/Salvando el planeta con biotecnología apropiada: 4. Cultivo de cocolitóforos e implementación. *Mexican Journal of Biotechnology*. **6** (1):129–155. https://doi.org/10.29267/mxjb.2021.6.1.129.

In April 2021 we published the fifth paper in this series:

Petros, P., Heilweck, M. & Moore, D. (2021). Saving the planet with appropriate biotechnology: 5. An action plan/Salvando el planeta con biotecnología apropiada: 5. Un plan de acción. *Mexican Journal of Biotechnology*, **6**(2): 1–60. https://doi.org/10.29267/mxjb.2021.6.2.1.

All five are extracted from an 8-chapter book manuscript we are finishing off at the moment. Just yesterday I finished the first draft of this chapter, which is our 'executive summary', so it gives a good idea of what the book contains. I attach it here but please treat it as private personal communication of a first draft which has not yet been seen by my co-authors.

To your specific questions:

First. if you haven't found it yet, I suggest you look at our chapter about fermenter technology, which you can view here: http://www.davidmoore.org.uk/21st_Century_Guidebook_to_Fungi_PLATINUM/Ch17_00.htm. This deals with fungi, but the basic principles apply to other microbes including algae (with provision for illumination). Trying to go straight to industrial level fermentation would be extremely expensive, and with algae, access to light is a major engineering complication. The following paper is the best I have found so far: Xu, L., Weathers, P. J., Xiong, X.-R. & Liu, C.-Z. (2009). Microalgal bioreactors: challenges and opportunities. Engineering in Life Sciences, 9: 178–189. https://doi.org/10.1002/elsc.200800111.

My personal view is that modest-scale bioreactors would be used to prepare inoculum for open-air raceways (or specialist purposes like genome manipulations). My thoughts about 'industrial scale' production would concentrate on raceways or the open sea. Check out:

Moheimani, N.R. & Borowitzka, M.A. (2006). The long-term culture of the coccolithophore *Pleurochrysis carterae* (Haptophyta) in outdoor raceway ponds. Journal of Applied Phycology, 18: 703–712. https://doi.org/10.1007/s10811-006-9075-1.

Second. This notion that calcification is a major emitter of CO_2 to the atmosphere is a half-truth which has been grasped by marine biologists in what I consider a misguided and almost mischievous manner. My response to it is this: The calcifying reaction scheme shows that two bicarbonate ions (which originally were derived from the atmosphere) react with Ca ions and one of them is precipitated as $CaCO_3$, and the other released as CO_2. So, while it is true that "precipitation of calcium carbonate is a source of carbon dioxide (CO_2)" it is illogical to claim that returning one out of two carbons to the environment is a "major way by which CO_2 is returned to the atmosphere". BUT, if you go one step further and ADD the consideration that there are a great many calcifying organisms in the oceans, which are all cycling through this reaction 24/7, then you can rightly claim that this is a major way by which CO_2 is returned to the atmosphere; **PROVIDING** you remember that the other one of those two carbons on the left of your reaction scheme is precipitated as $CaCO_3$ and **ADMIT** the matching claim that if this ***IS*** a major way by which CO_2 is **RETURNED** to the atmosphere then it is ALSO a major way by which **carbon is REMOVED FROM the atmosphere**.

In raceways or bioreactors, you don't have to worry about coccolithophores 'returning CO_2 to the atmosphere' because it's todays CO_2 being recycled (as it has been for hundreds of millions of years) it is not a damaging net-additional emission like that that comes from fossil fuel usage. And anyway, chances are the algae in your raceways or bioreactors will transport the CO_2 into their chloroplasts and fix it photosynthetically into sugar almost as soon as the calcifying reaction emits it! This notion that the cell is likely to allow it to escape to the atmosphere is ludicrous.

29/12/2020: G*****y B*****r of G*******r writes: I found your recent paper fascinating. You won't believe how fascinating. I have to say that your conclusions (built on more scientific research that our work) were almost identical to ours. I.e. Your (a) Tiered systems of running seawater downhill (possibly in desert areas), (b) bioreactors to produce $CaCO_3$, (c) anchoring ships near to seamounts to produce $CaCO_3$. Indeed, I even have a list of possible seamounts that one of my team identified on one of my desks.

Some decades ago, I started off trying to produce products sustainably for my vegetarian products company. I installed wind and solar energy generation (amongst other innovations) a quarter of a century ago.

Some years after that I expanded into other sectors, however I am concerned that I are not producing things sustainably … I know we do produce very large volumes of CO_2 in the construction of our resorts. I have spent several years asking our team to research how we reverse all the CO_2 that we have and do currently produce.

… I congratulate you on a Tour de Force in your paper relating to coccolithophores. … Because the production of coccolithophores leads to the release of a molecule of CO_2 we looked to sequester this permanently by either turning some of the algae biomass into bio-char or by introducing fungi of the type that produce calcium oxalate. Thereby ensuring that there was no overall release of CO_2. We have also looked at trying to infuse more CO_2 back into the run-off seawater, than is produced, in order again, not to be producing CO_2 from the production.…

REPLY: Dear G**g, Thank you for your kind comments about our coccolithophore preprint. I am delighted to hear that we are thinking along similar lines and very happy to offer comments and advice. … Taking your queries in turn:

1. Most importantly you don't have to worry about coccolithophores 'returning CO_2 to the atmosphere' the notion that any photosynthetic cell is likely to allow it to escape to the atmosphere is ludicrous.
2. If you want to gain experience cultivating interesting micro-'algae' I would suggest looking into the cultivation of the product known as Spirulina, which is actually the dried biomass of a photosynthetic bacterium (cyanobacterium) *Arthrospira platensis*. you are undoubtedly aware of this as a health food (which has the distinction of being advocated by both NASA and ESA for cultivation on long-term space missions as food for astronauts travelling to Mars). There is plenty of advice on the Internet for 'home grown' cultivation of Spirulina. An example is the YouTube video Smart Microfarms—Algae Growing Systems for Home and Backyard (https://www.youtube.com/results?search_query=XZW0NpvxTH8); and compare this with the commercial production process which is shown at this URL: http://www.aurospirul.com/production-process.html.
   Much more basic methods of cultivating home-grown algae aimed at aquarists is shown at: https://www.wikihow.com/Grow-Algae and https://www.bitesizebio.com/27998/open-closed-two-ways-grow-algae/; and you can buy your Spirulina starter cultures from Amazon (https://www.amazon.co.uk/HealthAlgae-Spirulina-platensis-living-culture/dp/B07F93L1C7). If you have a system that grows commercial quantities of Spirulina, you could use the profits to create a parallel production line of coccolithophores (but don't forget the likely profits there - in biofuel potential and even omega-3-fatty acids) which you can take while leaving the calcium carbonate they also make behind (it also has value as a very fine chalk in several processes).
3. I will end by making a plea that you also remember bivalve molluscs. There's plenty of 'how-to' cultivation advice referenced in our second paper: Moore D., Heilweck M. & Petros, P. 2021. Saving the Planet with Appropriate Biotechnology: 2. Cultivate Shellfish to Remediate the Atmosphere/Salvando el planeta con biotecnología apropiada: 2. Cultivar mariscos para remediar la atmósfera. *Mexican Journal of Biotechnology*. **6** (1): 31–91. https://doi.org/10.29267/mxjb.2021.6.1.31.
   Cultivate mussels, clams or oysters and you have a farmed-product that is 50% shell (made of crystalline $CaCO_3$); sell the shellfish meat as a valuable and nutritious food (sorry, I know you're vegetarian) and take the profits as return on investment, but make sure the 'waste' shells go back to the farm. That's how to recreate the shellfish reefs that were fished out in the 19th century.
4. And finally. What about Giant Clams? (see Chap. 5). This paper: Schmidt-Roach, S., Duarte, C.M., Hauser, C.A.E. & Aranda, M. (2020). Beyond reef restoration: next-generation techniques for coral gardening, landscaping, and outreach. Frontiers in Marine Science, 7: article 672. Open access. https://doi.org/10.3389/fmars.2020.00672 reviewed proposals to restore coral reef populations, pointing out that the scale of the work required to implement these concepts in ecosystem-wide habitats is a major limitation for logistical and, more importantly, financial reasons. Their solution is to suggest implementation by including land-based coral gardening into architectural elements to enhance and beautify coastal development sites. I hear you've got one or two of those?

30/12/2020: V******e R**h writes: I'm so excited to read this! I always wanted to learn more about CO_2 sequestering and why algae is not used as a biofuel.

REPLY: Hi V******e, thank you for your kind comments. Microalgae are indeed used as biofuels see: Borowitzka, M.A. & Moheimani, N.R. (eds) (2013). Algae for Biofuels and Energy. Developments in *Applied Phycology* **5**. Dordrecht: Springer Science+Business Media, pp. 288. ISBN: 9789400754782. https://doi.org/10.1007/978-94-007-5479-9.

Biofuels replace fossil fuels; they still emit CO_2 but it's today's CO_2 not fossil CO_2 so it is not a net addition to the atmospheric load of carbon.

30/12/2020: J**n S. W***h writes: Very interesting and it covered most of the questions I had. I do have a couple of questions though; (1) When the calcium carbonate is formed, it take the calcium from the ocean. Once it dies, it sinks to become limestone. Does sequestering the Calcium exacerbate Ocean acidification? (2) Have you looked into using Calcium Carbonate as an additive in biochar process to sequester the pure, inert, carbon? This sequesters the carbon for thousands of years. There have been studies on using calcium carbonate in the pyrolysis of sewage to make a higher BTU fuel, and create more carbon as a waste product. This is the first I've read on this and am genuinely interested to see how far this can be taken.

REPLY: Hi J**n. Calcium is the fifth most abundant metal in the Earth's crust (=about 4.1% of the crust) and occurs abundantly as limestone (calcium carbonate) and a few other salts. It has no bearing on acidification which is due to anthropogenic CO_2 dissolving in the ocean and becoming carbonic acid. There are fossil limestones made by coccolithophores (i.e. with recognisable coccoliths in them) dating back to the Triassic (250,000,000 years ago) so the few thousand years of biochar is no advantage at all in regards to sequestration, but likely helps plant growth in other ways.

30/12/2020: I******g L******a writes: I've been reading about this. It is great to be able to read actual research and not the articles about it. Thank you!

REPLY: Hi I******g. Thank you for your interest. after the 1st of January, you can download the published paper (free): Moore D. 2021. Saving the Planet with Appropriate Biotechnology: 4. Coccolithophore cultivation and deployment/Salvando el planeta con biotecnología apropiada: 4. Cultivo de cocolitóforos e implementación. *Mexican Journal of Biotechnology*. **6** (1):129–155. https://doi.org/10.29267/mxjb.2021.6.1.129.

6/1/2021: K*m F******n writes: Reading your work and wonder how you overcome the issue of shell making being a CO_2 production and not a sink?

REPLY: Hi K*m. Thank you for your interest in our work. My usual response to this question is as follows: The calcifying reaction scheme shows that two bicarbonate ions (which ultimately were derived from the atmosphere) react with Ca ions and one of them is precipitated as $CaCO_3$, and the other released as CO_2. So, while it is true that "precipitation of calcium carbonate is a source of carbon dioxide (CO_2)" it is illogical to claim that returning one out of two carbons to the environment is a "major way by which CO_2 is returned to the atmosphere" as some have put it to me. BUT, if you go one step further and ADD the consideration that there are a great many calcifying organisms in the oceans, which are all cycling through this reaction 24/7, then you can rightly claim that this is a major way by which CO_2 is returned to the atmosphere; PROVIDING you remember that the other one of those two carbons on the left of your reaction scheme is precipitated as $CaCO_3$ and ADMIT the matching claim that if this IS a major way by which CO_2 is RETURNED to the atmosphere then it is ALSO a major way by which carbon is REMOVED FROM the atmosphere.

13/2/2021
S****a E. S*****y passed on the following message from an anonymous 'highly respected scientist' as a 'heads up' "re: https://www.thefishsite.com/articles/can-bivalve-aquaculture-prevent-the-widespread-institutional-failure-of-our-attempts-to-tackle-climate-change do you know this guy? he does know that you generate a molecule of CO_2 every time you form a molecule of $CaCO_3$ right? and the animal respires when it does the work of making that molecule… right? I love the thought of people growing shellfish - but there's a problem with the stoichiometrics. Is he a real scientist?".

REPLY: Hi S***y, Thanks for that and yes, rattled cages have messaged me before. My usual response is: The calcifying reaction scheme shows that two bicarbonate ions (which originally were derived from the atmosphere, photosynthetic fixation of atmospheric CO_2 being the only source of metabolic carbon) react with Ca ions and one of them is precipitated as $CaCO_3$, and the other released as CO_2. So, the stoichiometry is two atmospheric CO_2 => 2 metabolic bicarbonate => 1 insoluble carbonate + 1 potentially atmospheric CO_2. That's how a real scientist looks at it. A real scientist also appreciates that respired CO_2 is just part of the normal carbon cycle. Insoluble carbonates remove CO_2 from that cycle and could therefore compensate for the fossil CO_2 emitted by burning fossil fuel. You could ask your respected scientist what he thinks that solid shell-stuff is made of that's left over after his moules marinière. It's solidified atmosphere lad, that's what it is.

10/04/2021
Balamurali Sreedhar of Midland, MI USA, asked: "I am a climate enthusiast who happens to be worried about the role of CO_2 in atmosphere, just like many others. … One more idea could be to combine mollusk farming + agriculture … One of the challenges facing agriculture across the world is the presence of soil acidity (pH <= 5). It is well known that soil acidification reduces net primary productivity and carbon sequestration by accelerating leaching of nutrients The idea is to combine mollusk farming + soil fixation + agriculture to bring about a comprehensive, synergistic ocean and land based farming strategy to sequester carbon … Central theme of the idea would be to carry out shellfish (or any other commercially viable mollusk) farming and in the process producing calcium carbonate from its shells. Then, using it to fix soil acidity in severely affected areas. This would be followed by planting commercially viable, fast growing crops that would sequester carbon as biomass in the soil. When combined, the above approach would give nutrition to the masses as well as fix carbon".

REPLY: Hi Bala, Thanks for your interest in our work. The bottom line of all of our publications to date is that we believe that shellfish aquaculture can readily sequester far more carbon from the atmosphere (returning the anthropogenically released fossil CO_2 BACK into 'fossil' shell form) than any terrestrial forest or even coastal (kelp/mangrove) forest, AND we calculate that this paradigm shift (from 'shellfish as food' to 'shellfish for carbon sequestration') makes bivalve mollusc farming and microalgal farming enterprises, viable, profitable, and sustainable, alternatives to **all** CCUS industrial technologies and terrestrial biotechnologies in use today.

By far the best thing to do with waste mollusc shells is return them to the seabed where the scraps of flesh that remain can feed scavengers and detritus-feeders and the shells contribute to reef formation. We realise that other uses for waste shells have been published, including pH adjustment.

Our problem with using mollusc shells this way in soils is that the pH-adjustment necessarily causes release of CO_2 from the shells back into the atmosphere. Matthias Heilweck describes it like this: "Mollusc farming + soil fixation + agriculture is like using steam to produce electricity to boil water". It only makes logical sense if the shells ***replace*** use of fossilised limestone. Then at least the CO_2 released is today's CO_2 rather than yet more fossil CO_2.

Come all you no-hopers, you jokers and rogues

Awash on a sea of our own vanity,
We should rejoice in our individuality.
Though it's gale force, lets steer a course for sanity.
Come all you no-hopers, you jokers and rogues,
We're on the road to nowhere, lets find out
where it goes.
It might be a ladder to the stars, who knows?
Come all you no-hopers, you jokers and rogues.
Watch ***Port Isaac's Fisherman's Friends***
sing the original at this URL:
https://www.youtube.com/watch?v=AYA_0R7Vw1s

Index

A
Acanthostrongylophora ingens, 120
Acidification, 9, 25, 30, 122, 123, 156, 164, 203, 227, 236, 237, 242, 243
Acmaea spp., 69
Action Plan, 48, 88, 217, 222, 239
Aerobic, 6, 36, 52, 102, 112
Afforestation, 15, 20, 26, 27, 35, 36, 41, 44, 48, 49, 190, 197, 200, 201, 204, 218
Agar production, 75
Aggregation approach, 208
Aggressive emission reduction, 36, 41
Agricultural land, 1, 18, 19, 36, 49, 100, 125, 186, 204, 219, 223, 225
Agricultural practices, 27, 29, 197
Agricultural soil, 20, 27, 50, 100, 197
Agroforestry, 40, 103, 200
Airlift photobioreactors, 169
Alaska pollock, 76
Algal lipids, 167, 169
Algal oils, 102, 168
Alkenones, 161
Alkylamines, 187
Amazon river, 75
American whelk tingle, 79, 84
Ammonites, 24, 46, 163, 164
Amphidinium klebsii, 143
Anchovies, 97, 111, 225
Antarctic, 2, 10, 105
Antarctic Intermediate Water (AIW), 105–107, 110, 112, 120
Anthropocene, 107, 119
Anthropogenic, 1, 3, 5, 6, 8, 9, 10, 13, 14, 19, 25, 26, 40, 41, 107, 110, 124, 138, 156, 159, 164, 179, 193, 219, 222, 242, 243
Antioxidant, 58, 115, 120, 127, 168
Aoshan Bay, 106
Apalachicola Bay, 83
Aphanomyces astaci, 77
Aphotic zone, 105
Applecross Peninsula, 67
Aquaculture, 24, 25, 28–30, 41–45, 47, 48, 50, 51, 53, 55–58, 70–78, 80, 85, 88–91, 97–103, 110–112, 115, 118, 124, 126, 141, 142
Aquaculture management, 103
Aquafeed, 99, 101, 107, 111, 119, 125, 225
Aquarium fish, 74
Aragonite, 24, 25, 46, 163
Arctic, 2, 13, 16, 45, 202
Artemia, 159
Arthropod, 44, 164
Arthrospira platensis, 167, 240
Assessment models, 206
Astrea gibgerosa, 69
Atmospheric carbon, 5, 23, 27, 35, 39, 44, 49, 51, 55, 58, 65, 70, 97, 121, 132, 138, 140, 141, 159, 177, 197, 199, 203, 204, 217, 219, 220, 221, 223–225, 227, 228, 235, 238
Atmospheric CO2, 3–5, 8–11, 23, 25, 35, 39, 43, 46, 49, 53, 70, 71, 86, 155, 156, 164, 166, 178, 183, 193, 196, 198, 203, 204, 210, 217, 219, 221, 227, 228, 229, 230, 235, 237, 243
Australia, 66, 90, 102, 134, 135, 137, 149, 185, 187
Autotrophic, 138, 167
Aztec, 74

B
Bacillariophyta, 109
Baitfish, 76
Balanus spp., 69
Baltic Sea, 45, 114
Bangka Island, 120
Barnacles, 66, 69
Barrier reef, 137, 141–143, 149
Bass, 77, 100
Bass Strait, 185
Bedeque Bay, 83
Bedrock, 70, 181
Benin, 72
Bering Sea, 76
Betula pubescens, 36, 37, 199
Bicarbonate (hydrogencarbonate), 23, 24, 44, 121, 124, 162, 163, 205, 228, 233, 236, 237, 239, 242, 243
Bilayer membranes, 25
Billingsgate Market, 29
Biodegradable, 125, 155, 172, 221, 226
Biodeposition, 86, 88, 202, 208, 210
Biodiesel, 58, 167

D. Moore et al., *Aquaculture: Ocean Blue Carbon Meets UN-SDGS*, Sustainable Development Goals Series, https://doi.org/10.1007/978-3-030-94846-7

Biodiversity, 18–21, 30, 36, 39, 41, 45, 50, 86, 103, 109, 112, 114, 117, 138, 139, 146, 191, 193, 196, 199, 200, 206, 207, 210, 217–219, 226
Biodiversity Beyond areas of National Jurisdiction (BBNJ), 117, 217
Bioenergy, 20, 27, 186, 190, 194, 195, 210
Bioenergy with Carbon Capture & Storage (BECCS), 20, 27, 190, 194, 195
Biofilm reactors, 169
Biogenic reef habitats, 58, 86
Biolipids, 161
Biomass yield, 169, 170
Biomaterial, 49, 217
Biomineral, 44
Biomineralization, 123
Biopolymer, 58
Bioreactors, 155, 165, 168–172, 221, 226, 230, 239, 240
Biosphere sink, 41
Birch tree, 36, 199
Bivalve, 25, 42, 44–46, 48, 49, 51, 53–55, 57–59, 66, 67, 69, 76, 80, 82, 86, 88–90, 97, 110, 114, 121, 124, 133, 134, 136, 155, 172, 177, 178, 204, 208–210, 218, 220–222, 224, 226, 228–232, 237, 241, 243
Bivalve ecosystem services, 88, 208
Bivalve farming, 88, 208
Black pacu, 75
Black Soldier Fly (BSF), 102
Blackwater river, 80
Bleaching, 131, 138, 141, 145, 220
Blooms, 103, 155–157, 159–161, 170–172, 221, 226, 230, 234
Blue carbon, 21–23, 27, 29, 50, 54, 56, 57, 140, 177, 190, 200–202, 204, 205, 218, 221, 222
Blue crabs, 86
Blue economy, 179, 194, 220
Bog, 36, 38, 51
Boreal forests, 13, 200
Borneo, 131
Bottlenose dolphins, 120
Bottom culture, 47, 90, 110
Bouchot culture, 47, 110
Brachiopods, 24, 46
Brassica napus, 102
Brazil, 30, 74, 105, 108, 109, 117, 118, 120, 125, 224, 226
Brazil Current, 108, 125, 228
Bream, 74, 78, 120
Brevoortia patronus, 76
Brightlingsea, 79–81
Brine shrimp, 159
British Broadcasting Corporation (BBC), 6, 12, 16, 148
British Columbia Shellfish Growers' Association (BCSGA), 89
Broodstock, 73, 142
Brown algae, 155, 202, 221
Brown Mussels, 66
Bryozoa, 24, 46
Butter clams, 69
Byssus, 136

C
Cachama, 75
Calanus, 114–116
Calcareous sediments, 23, 159, 233
Calcidiscus, 158, 159, 165
Calcification, 23–25, 44, 51, 59, 121, 123, 124, 143, 155–157, 159, 162–164, 203, 205, 232, 233, 235–237, 239
Calcifier, 18, 22, 23, 25, 28, 44, 48, 122, 123, 156, 161, 162, 164, 177, 178, 204, 209, 221–223, 233, 234, 236, 237
Calcination, 58, 223
Calciosolenia murrayi, 158, 162
Calcite, 23, 25, 46, 58, 155, 157, 159, 163, 165, 166, 203, 221
Calcite Compensation Depth (CCD), 23, 122, 125, 126, 233
Calcium Oxide (CaO), 58, 223, 229
Caliciviruses, 82
California, 66, 67, 90
California mussel, 66
Callinectes sapidus, 86
Calluna vulgaris, 36, 37, 199
Cambrian, 25, 123, 133, 163
Cambrian fossil, 133
Camulodunum, 79
Canada, 6, 14, 17, 57, 65, 68, 69, 83, 90, 91, 113, 219
Cane molasses, 102
Capelin, 97
Capture CO_2, 19, 27, 41, 187, 194, 197, 222
Capture fisheries, 70, 74–76, 78, 100, 120, 209, 225
CarbFix, 187
Carbon capture, 20, 27, 35, 36, 39, 41, 44, 48, 53, 170, 177, 180, 181, 183, 189, 190, 194, 197, 201, 206, 207, 219, 221
Carbon Corp, 195
Carbon credits, 178, 179, 193
CarbonCure Technologies, 195
Carbon cycle, 1, 6–9, 39, 40, 58, 121, 125, 170, 198, 202, 203, 218, 219, 223, 224, 235, 243
Carbon deposition, 1, 6–9, 39, 40, 58, 121, 125, 170, 198, 202, 203, 218, 219, 223, 224, 235, 237, 243
Carbon Dioxide (CO_2), 1–12, 14–20, 22–28, 35, 36, 38, 39, 41, 42–44, 46, 50–53, 55–59, 70, 71, 85, 86, 121, 122–124, 132, 138, 140, 155, 156, 162, 163–167, 169–171, 177, 178, 180, 181, 183–191, 193–199, 201, 203–206, 209, 210, 217–219, 221–224, 227–231, 233–244
Carbon dioxide Capture & Storage (CCS), 41, 53, 177, 178, 180–183, 185–193, 206, 221
Carbon dioxide Capture & Utilisation (CCU), 180, 181, 185, 206, 207
Carbon dioxide Capture, Utilisation & Storage (CCUS), 177, 180, 181, 185, 186, 188–194, 196, 209, 210, 221, 243
Carbon Dioxide Removal (CDR), 19, 22, 178, 194, 196
Carbon emissions, 2, 5, 28, 39, 40, 42, 56, 171, 177, 193, 194, 199, 221, 223
Carbon flux, 8, 121, 124, 159

Carbon market, 17, 57, 124, 177–180, 193, 204, 221
CarbonNet Project, 185, 187
Carbon neutral, 27, 185, 186
Carbon offsets, 56, 59, 121, 125, 179
Carbon offsetting, 56, 57, 223
Carbon reduction, 39, 177, 199, 206, 221
Carbon sequestration, 24, 27, 35, 37–39, 41, 46, 48, 51, 57, 97, 121, 125, 126, 155, 165, 166, 171, 177, 178, 180, 181, 185, 193, 194, 196, 197, 199–204, 209, 219–221, 227, 229, 234, 238, 243
Carbon sink, 1, 5, 16, 20, 24, 36, 38–41, 50, 59, 121, 125, 159, 161, 185, 198, 199, 203, 204, 219, 226, 227, 230, 236, 238
Carbon Upcycling, 195
Cardigan Bay, 81, 82
Cardium edule, 67
Caribbean Regional Fisheries Mechanism (CRFM), 75
Carnivorous, 97, 100, 101, 208, 225
Carp, 73–75, 77, 78, 100
Carp aquaculture, 77
Carrageenan, 75
Catfish, 73, 76, 209
Cement, 2, 8, 46, 83, 155, 164, 171, 172, 186, 187, 194, 195, 221, 223, 224
Center for Climate and Energy Solutions (C2ES), 186, 187
Cerebral cortex, 102
Channel catfish, 76
Chemotactic, 136
Chicken feed, 67
Chicxulub crater, 134
China, 14, 17, 39, 41, 52, 73, 74, 106, 112, 114, 131, 132, 135, 137, 149, 152, 197, 223, 227, 232
Chinese mitten-handed crab, 74
Chitin, 22, 44, 58, 228
Chitons, 69
Chitosan, 58
Chloromethane, 36
Chlorophyll-a, 104, 161, 162, 203
Chub, 74
Cipangopaludina chinensis, 74
Circular economic value, 178
Circular economy, 57, 124, 178, 180, 181, 193
Clam garden, 65, 68–70, 72, 78, 149, 219, 224
Clam garden habitats, 69, 70
Clam gardening, 68
Clean Development Mechanism (CDM), 180
Climate Action Tracker website (CAT), 16
Climate assessment, 11, 13
Climate change, 1, 5, 6, 8, 10–22, 26–29, 35, 37, 39, 40, 42, 45, 48, 50, 55, 56, 123–126, 131, 135, 139, 140, 141, 144, 149, 156, 161, 164, 170, 177–180, 183–186, 188, 189, 191, 193, 194, 196–199, 202, 203, 205–207, 209, 217–220, 222, 226, 227, 235
Climate mitigation, 8, 51, 125, 196, 199
Clinocardium nuttallii, 69
Closed systems, 169, 170, 238
Cloud brightening, 171
CO2CRC, 187
CO_2 injection, 185, 186
Coastal blue carbon, 22, 27, 140, 177, 200, 204, 221, 222
Coastal protection, 49, 58, 202
Coastline, 20, 50, 66, 68, 74, 79, 85, 137, 171, 202, 203
Cobia, 100, 101
Coccolith, 24, 157, 159, 162–165, 203, 235
Coccolithophore, 22, 43, 155–157, 159–162, 164, 166, 167, 170–172, 183, 204, 217, 221, 222, 224–226, 230–232, 234, 239, 240, 242
Coccolithus, 158, 159, 160, 167
Coccolithus braarudi, 167
Coccosphere, 158–160, 165, 170
Cockle culture, 76
Cockles, 42, 43, 56, 58, 67, 69, 76
Cod, 97
Coelenterates, 98
Colchester, 79, 83
Colossoma macropomum, 75
Combustion, 2, 3, 5, 10, 36, 39, 194
Commission on the Limits of the Continental Shelf (CLCS), 117
Common carp, 73, 77, 78
Common good, 117, 118
Connecticut, 83
Conservation, 18–21, 26, 30, 35, 36, 38, 40, 42, 50, 53, 56, 65, 75, 82, 86, 88, 97, 109, 111, 119, 125, 131, 134, 135, 137, 139–141, 145–147, 149, 150, 179, 197, 202, 217–220, 226
Conserve, 22, 25, 30, 76, 117, 127, 140, 149, 191, 209, 235
Continental shelf, 50, 117, 118
Continuous forest, 27, 35
Convention on International Trade in Endangered Species (CITES), 135
Cook Islands, 138
COP26, 15–17
Copepods, 97, 114, 116, 225
Coral, 22, 24, 28, 46, 48, 53, 55, 67, 114, 123, 131, 134–141, 143, 144, 146, 147, 149, 156, 163, 204, 209, 218, 220, 222, 226, 227, 229, 231, 232, 237, 241
Coral gardening, 139, 146, 241
Coral reef, 53, 55, 67, 131, 134–141, 144, 146, 147, 149, 163, 220, 226, 229, 241
Coral reef bleaching, 138
Coral reef restoration, 139, 144, 226
Corn sugar, 102
Council for the Protection of Rural England (CPRE), 39, 199
Countryside Charity, 39, 199
Crab, 42, 44, 57, 58, 66, 67, 74, 86, 89, 90, 111, 202
Crassostrea gigas, 87
Crassostrea rhizophorae, 75
Crawfish, 76, 77, 90
Crayfish, 76, 77
Crayfish farming, 76
Crayfish plague, 77
Crepidula fornicate, 79
Cretaceous, 24, 108, 134, 160, 163, 164

Cretaceous chalk, 159
Crinoid, 24, 46
Cropland, 52, 200, 232
Crouch river, 80
Crustacea, 22, 28, 48, 55, 58, 70, 88, 89, 124, 140, 204, 208, 209, 217, 218, 222, 231, 232
Cryogenic distillation, 187
Crystalline calcium carbonate, 35, 44, 86, 125, 219, 227
Ctenopharyngodon idellus, 75
Cultivation, 18, 22, 35, 42, 43, 46–49, 52, 53, 55–57, 70, 72–74, 76, 77, 84, 87–89, 91, 97, 110–112, 120, 121, 124–126, 131, 135, 139–141, 145, 147, 149, 155, 166–171, 177, 209, 210, 217–225, 234, 239, 240–242
Cultivation economics, 47
Cultured pearls, 74
Cyanobacteria, 109, 167, 169, 232, 233, 240
Cyprinus carpio, 73, 77
Cytochrome oxidase subunit I (COI), 137

D
Damariscotta River, 67
Dangerous Anthropogenic Interference (DAI), 1, 13, 219, 222
Davis Bank, 108, 109, 111, 114–117, 120, 121, 125, 126, 224, 225
Deep Chlorophyll maximum (DCM), 161, 162
Deepwater biomass, 162
Deforestation, 3, 4, 17, 178
Delaware Bay, 83
Delmarva Peninsula, 83
Department of Energy (DOE), 6, 123
Detoxifying, 164, 171
Detritus feeder, 44, 53, 223, 226, 228, 244
Devonian, 164
Diamonds, 116, 187
Diatom, 109, 111, 143, 156, 160, 203, 232, 233
Dictyopteris membranacea, 120
Diesel fuel, 25, 52
Diethanolamine (DEA), 187
Dimethyl sulphide (DMS)), 155, 161, 171, 226, 230
Dinoflagellates, 109, 138, 143
Dinophyta, 109
Direct Air Capture (DAC), 20, 27, 180, 185, 190, 194, 196
Discosphaera tubifera, 158, 161, 165
Disused oil rigs, 97, 220
Diversity, 18, 45, 66, 69, 71, 88, 109, 117, 134, 137, 158, 159, 161, 164, 190, 203, 208
Docosahexaenoic Acid (DHA), 102, 111
Dog whelk, 45, 67, 79
Dolphins, 103, 120
Dongsha Atoll, 137
Drought, 11, 13, 15, 19, 29, 40, 48, 198

E
Early humans, 24, 46, 65–68, 219
Eastern soft shell clams, 69
Economic, 5, 13, 18, 20, 26, 27, 30, 40, 45, 47, 49, 55, 74–76, 78, 85, 88, 97, 102, 108, 117–119, 140, 141, 147, 168, 177–179, 185, 187, 188, 196, 197, 204–209, 217, 218, 220, 221
Ecosystem engineer, 58, 65, 86, 89, 144, 218, 220, 226, 235
Ecosystem goods, 71
Ecosystem services, 18, 26, 54, 55, 58, 59, 71, 88, 140, 141, 171, 177, 178, 180, 191, 193, 196, 197, 199, 201, 203, 205, 207–210, 218, 221
Ecosystem stewardship, 18, 179, 196
Edible red alga, 74
Eelgrass, 69, 202
Effluent, 82
Egypt, 18, 73, 74
EhVs, 120, 234
Eicosapentaenoic acid (EPA), 102, 111
El Niño, 111
Emiliania, 159, 160
Emiliania huxleyi, 156–161, 164–167, 170, 234, 235
Emissions, 1–6, 8, 9, 11, 12, 14–22, 26–28, 36, 39–42, 44, 46, 50, 56, 59, 85, 139, 140, 155, 156, 171, 177–181, 183, 185–188, 191, 193–199, 202, 205–207, 209, 219, 221–224, 227, 231, 233, 240
Emissions market, 177, 178
Emission Trading System (ETS), 178
Energy from biomass, 19, 196
Energy penalty, 53, 189
Enhanced weathering, 190, 195
Ensis sp., 67
Environmental Commodity, 178, 179
Environmental credits, 193
Environmental health, 179, 191, 193
Environmental impact, 13, 29, 52, 134, 180, 206, 208, 209
Eocene, 160
Epibionts, 136
Eriocheirn sinensis, 74
Erosion, 13, 38, 50, 54
Essex dredgermen, 80
Eucheuma spp., 75
Euphotic zone, 108, 160
European Environment Agency (EEA), 20, 21
European Union (EU), 17, 47, 49, 52, 56, 88, 124, 208, 210, 232
Eutrophication, 98, 114, 135, 136
Exclusive Economic Zone (EEZ), 108, 109, 117, 118
Executive summary, 217, 218, 239
Exoskeleton, 44, 58, 157, 164
Exploration, 66
Extinction, 9, 13, 14, 49, 56, 65, 66, 78, 84, 86, 88, 98, 131, 149, 163, 164, 193, 218, 220, 229, 237
Extreme weather, 11, 14, 20, 26, 29, 48

F
Factory rigs, 97, 220, 224, 225
Factory ships, 57, 97, 125, 155, 172, 220, 221, 223, 224, 226
Farasan Islands, 67
Farmed fish, 74, 97, 98, 101, 102, 107, 118, 220, 225
Farm hatcheries, 90
Farmed oysters, 65, 85, 220
Farming activities, 71, 72
Farming shellfish, 50, 218
Fatty acids, 85, 101, 102, 111, 112, 159, 167, 241
Feedback loop, 161, 171, 226
Fermenters, 43, 102, 169, 171, 224, 231, 239
Filter-feeder, 44, 53, 110, 113, 114, 134, 228
Filter feeding, 45, 58, 82, 86, 110, 136, 143, 208
Finfish, 71, 73, 78, 113, 124, 202, 208, 209
Finfish aquaculture, 124, 208
Finfish farms, 73
First Nations, 65, 68–70, 72, 78, 219
Fished out, 53, 57, 86, 97, 131, 137, 149, 230, 237, 241
Fish farming, 70, 72, 73, 77, 78
Fish food, 98, 225
Fish holes, 72
Fishing smack, 79, 80, 81
Fishing vessel, 25, 28, 52, 80, 114, 115
Fishmeal, 71, 98, 101, 107, 111, 225
Fish oil, 71, 98, 101, 102, 111, 116, 159
Fish stocks, 78, 97, 98
Fish weirs, 68, 70, 72
Flagella, 157, 203
Flat-plate photobioreactors, 169
Floatation devices, 125, 172, 226
Floating cages, 73
Flood, 15, 18, 48, 50, 74, 77, 196
Florida, 83, 91
Florisphaera profunda, 158, 162
Flue gas, 41, 177, 186, 194, 195
Flupsy, 90
Food and Agriculture Organization of the United Nations (FAO), 40, 42, 43, 70, 73–77, 87, 89, 100, 190, 197
Food value, 46–48, 72, 88, 131, 177, 208, 209, 220, 221
Food-waste, 46, 58, 228
Food webs, 98, 160
Forage fish, 97–99, 101–103, 107, 220, 225
Foraminifera, 24, 25, 28, 48, 164
Forest conservation, 35, 36
Forest ecosystem, 9, 19, 40, 49, 51, 198, 199, 201
Forest management, 20, 27, 40, 195, 197, 198
Forest of Arden, 78
Fossil fuel, 1–5, 8, 9, 19, 27, 29, 36, 39, 41, 42, 49, 51, 56, 107, 163, 167, 171, 177, 178, 183, 185, 186, 189, 191–194, 196, 198, 218, 222, 224, 237, 240, 241, 243
Fossiliferous limestone, 46, 155, 171, 221
Fossils, 24, 46, 66, 134, 163, 164, 167, 171, 172, 177, 178, 183, 185, 186, 188, 189, 191–194, 196, 198, 218, 221, 222–224, 230, 233, 234, 237, 240–244
Framework approach, 208
France, 78, 98, 103, 120, 122
French Polynesia, 141, 144–146
Frequently Asked Questions (FAQ's), 217, 227
Fresh-water mussel, 74
Fresh-water shrimp, 74
Fronds, 203, 204
Fungal chloromethane, 36

G
Gadus chalcogrammus, 76
Gas liquefaction, 187
Gastroenteritis, 82
Gastropods, 25, 45, 79, 133
Gene-editing, 172
Genetically modified canola, 102
Genetically Modified Organism (GMO), 103
Genome variability, 161
Geoengineering, 171, 178, 195, 196
Gephyrocapsa, 159, 160
Gephyrocapsa oceanica, 158, 161
Germ theory, 82
GHG emissions, 16, 26, 85, 139, 140, 177–181, 185, 193, 221
Giant clam farming, 139, 141, 145, 149
Giant clams, 123, 131–139, 141–149, 156, 220, 226, 227, 241
Giant DNA-containing viruses, 170
Giant river freshwater prawn, 73
Gibbula cineraria, 67
Gigatonne, 2, 35
Giving Pledge, 119
Glacial ice, 2
Glasgow Financial Alliance for Net Zero (GFANZ), 17
Glidden midden, 67
Global Carbon Cycle (GCC), 1, 7–9, 125, 202, 203, 219, 223
Global Carbon Project (GCP), 3, 6, 186
Global CCS Institute, 185, 191
Global health, 179, 180
Global pollution, 99, 179
Global warming, 8, 10, 12, 14–18, 27, 35, 36, 40, 50, 59, 99, 124, 137, 138, 155, 159, 170, 178, 193, 197–199, 206, 231
Gondwana, 134
Goods and services, 58, 202, 206, 207
Governance, 19, 22, 26, 28, 41, 55, 70, 117, 131, 140, 141, 146, 149, 179, 192, 194, 218, 220
Gracelaria spp., 75
Gracilaria debilis, 75
Grand Terre, 137
Gravitactic, 136
Great Barrier Reef, 141–143, 149
Great Lakes, 13, 73, 76
Great scallop, 52
Green carbon, 27, 200
Green diesel, 167, 168
Greenhouse effect, 4, 231

Greenhouse gas (GHG), 1–7, 9, 11–16, 18, 21, 26, 27, 36, 39, 41, 50, 85, 139, 171, 177–181, 185, 188, 193, 198, 199, 209, 218, 219, 221
Greenling, 69
Green-lipped mussel, 111
Green turtle, 75
Grouper, 75, 76, 97
Growout phase, 90
Gulf menhaden, 76
Gulf of Mexico, 76, 83, 235
Guyot, 109–109, 117, 224

H
Hacrobia, 157
Haliotis kamtschatkana, 69
Hampton Roads, 83
Haptonema, 157
Haptophyte, 157, 160, 161, 170, 203
Hatchery, 72, 76, 89, 90, 125, 142, 145, 149, 225, 226
HCO3 , 44, 122
Heart cockles, 69
Heather moorland, 36, 199
Heatwaves, 14, 15, 20
Heavy metals, 107, 120
Hellisheidi, 187
Hepatitis A virus, 82
Hermatypic corals, 144
Herring, 97
Heterotrophic microalgae, 102
Heterotrophs, 108
Hexagrammidae, 69
H. floresiensis, 66
High Seas, 21, 22, 76, 78, 97, 99, 100, 103, 108, 116–119, 125, 155, 171, 172, 201, 205, 220–222, 234, 237, 238
High trophic level fish, 97
Hippopus hippopus, 134, 135, 137, 142–143
Historical aquaculture, 71
Holdfast, 203, 204
Holocene, 2, 6
Hominid species, 66
Homo sapiens, 65, 66, 164, 219, 229
Horse clams, 69
H. porcellanus, 134, 136
Human capital, 207
Hurricane, 12, 103, 104, 230
Hybrid cultivation system, 169
Hydrate-based processing, 41
Hydrodynamic stress, 170
Hydrofluorocarbon (HFC), 178
Hydrogen peroxide, 142
Hymenomonas, 159
Hypnea sp., 75
Hypophthalmichthys molitrix, 75
Hypophthalmichthys nobilis, 75

I
Ice age, 2–4, 7, 9
Ice core, 2, 3, 10
Iceland, 186, 187
Ice sheet, 2, 13, 16
Ictalurus punctatus, 76
Immunomodulator, 58, 168
Inclusive governance, 179
Indian ocean, 131, 138, 144, 220, 235
Indicator set, 208
Indigenous population, 66
Indonesia, 6, 17, 52, 73, 120, 139, 232
Indo-Pacific region, 131, 133, 135, 144, 146
Induced spawning, 142, 145
Industrial humans, 1, 7, 22
Industrial revolution, 1, 5, 22, 59, 156, 218
Industrial technologies, 177, 185, 210, 221, 243
Infrastructure, 13, 20, 103, 107, 110, 117, 119, 125, 177, 188, 189, 191, 196, 206, 207, 221, 224, 225
Inoceramus, 133, 134
Insect meal, 102
Integrated Multi-Trophic Aquaculture (IMTA), 53, 112, 113, 226, 208, 226, 229
Intergovernmental Conference (IGC), 217
Intergovernmental Panel on Climate Change (IPCC), 3, 11, 13, 15, 16, 27, 35, 50, 140, 170, 177, 178, 180, 181, 183, 184, 186, 188–190, 197, 200
Intergovernmental Science-Policy Platform on Biodiversity and Ecosystem Services (IPBES), 18, 191
International Energy Agency (IEA), 180, 188
International Seabed Authority (ISA), 117
International Union for Conservation of Nature (IUCN), 20, 21, 135
Intertidal, 45, 47, 65, 66, 68–70, 90, 110, 219
Inundation, 13
Invertebrate, 66, 69, 86, 89, 124, 143, 163, 166
Irrigation water, 71, 99, 100
Isle of Sheppy, 78
Isotherm, 107

J
James River, 83
Jellyfish, 98, 143
Jersey fishery, 80
Jomonlithus spp., 159
Jurassic, 81, 134, 160, 163
Jurassic Coast, 81
Jutland, 114
Juvenile clams, 138, 143

K
Kelp forest, 36, 39, 41, 51, 69, 112, 202–204, 225, 226
Krill, 58, 97
Kyoto Protocol, 121, 178–180

L
Lagoon reef, 137
Laguna, Brazil, 120
Laminariales, 202
Larvae, 19, 40, 50, 89, 90, 124, 125, 136, 142, 143, 218, 226
Laurasia, 134
Laver seaweed, 74
LC-PUFA, 101, 102, 114, 116, 159
Leakage, 181, 183, 185, 193
Lecithotrophic, 136
Leukoma staminea, 69
Life cycle analysis, 52, 59
Life Cycle Assessment (LCA), 24, 52, 206, 207, 210
Light Emitting Diode (LED), 104, 169
Limestone strata, 7, 232, 234
Limpet, 66, 67, 69, 79–81, 84
Lincoln County, 67
Lincolnshire, 80
Lingcod, 69
Liquid CO_2, 187
Littleneck clams, 69
Littorina littoralis, 67
Littorina spp., 67, 69
Livestock animal, 52, 232
Lobster, 44, 57, 58, 89, 90
Local weather, 103
London, 6, 12, 56, 65, 78–80, 83, 85, 88, 165, 220, 237
Long chain omega-3 polyunsaturated fatty acids, 101, 159
Long Island Sound, 83
Longline culture, 111
Lottiidae, 69
Louisiana, 77, 90
Louisiana Estuaries, 83
Louisiana wetlands, 83
Low trophic level fish, 97
Low-carbon technologies, 19, 185
Lumpfish, 101

M
Macoma clams, 69
Macoma spp., 69
Macrobrachium rosenbergii, 73, 74
Maine, 67
Mandarin fish, 74
Mangrove oyster, 75, 89
Mangrove oyster farming, 75
Mangroves, 22, 28, 75, 87, 89, 140, 200–202, 222, 243
Manila Clam, 24, 210
Mantle colours, 134
Mantle tissues, 134
Manufactured capital, 207
Manzamine A, 120
Margaritifera margaritifera, 74
Mariana Trench, 106
Mariculture, 65, 68, 70, 75, 88, 132, 133, 135, 138, 141, 145, 147–150, 208, 209, 220, 225
Maricultured clams, 144
Mariculture of seaweeds, 75, 76
Mariculture Sustainability Index (MSI), 139, 208
Marine calcifiers, 28, 122, 164, 231, 232
Marine ecosystems, 28, 53, 55, 87, 88, 97, 103, 140, 218, 220, 229, 235
Marine Protected Area (MPA), 117, 138
Marine snail, 69
Market, 17, 29, 46, 49, 57, 73–75, 78–80, 83–85, 87, 98, 102, 111, 117–119, 124, 132, 134, 135, 167, 177–180, 185, 193, 204, 218, 221, 223, 228, 229
Market value, 111, 179
Marshall Islands, 138
Mauritius, 131, 134, 138
Meagre, 73
Medieval Europe, 73
Mediterranean, 77, 134, 162
Mediterranean mussel, 51, 52
Mesoamerican, 74
Mesolithic, 67
Mesozoic Era, 134, 163
Methane (CH_4), 85, 178, 187
Methanol, 58, 167
Methyldiethanolamine (MDEA), 187
Miasma, 82
Michaelsarsia adriaticus, 162
Microalgae, 102, 165–170, 177, 203, 241
Microalgal farming, 177, 243
Microcrystalline $CaCO_3$, 43, 155, 221, 233
Middens, 24, 46, 65, 67, 70, 219
Middle Pleistocene, 66
Migrations, 12, 65–68, 72, 107, 219
Million years ago (Mya), 25, 69, 163, 164
Mill ponds, 77
Mineralisation, 27, 181, 185, 194
Mineral nutrients, 103, 110, 167
Mire, 36
Mississippi River delta, 83
Mitigation, 8, 18–21, 29, 37, 51, 54, 55, 125, 140, 180, 181, 191, 196, 199, 204, 206, 207, 235
Mixotrophs, 108
Molluscan shell, 44
Mollusc farming, 97, 210, 220, 221, 243, 244
Molluscs, 22, 28, 42–44, 48, 53, 55, 58, 66, 70, 71, 76, 77, 88, 89, 97, 110, 114, 124, 133, 140, 155, 172, 177, 178, 202, 204, 208–210, 217, 218, 220–223, 226, 228, 230–232, 241, 243, 244
Monoethanolamine (MEA), 187
Monterey Bay, 84
Moor, 36
Moss, 38
Mullet, 73, 77
Multi-trophic aquaculture, 53, 112, 113, 208, 226, 229
Muskeg, 36
Mussel, 24, 41, 42, 44, 45, 47, 48, 51–54, 56–59, 66, 67, 69, 74, 76, 77, 80, 82, 87, 89, 90, 97, 110–114, 117, 120, 121, 123–125, 142, 148, 156, 177, 201, 208–210, 220, 221, 224–227, 229, 231, 241
Mya arenaria, 69
Mytilus californianus, 66

Mytilus edulis, 45, 46, 67, 76, 112, 123, 156, 227
Mytilus galloprovincialis, 51, 52

N
Nanodevices, 165, 172
National Academies of Sciences, Engineering, and Medicine, 19, 21, 22, 27, 185, 190, 196, 197, 200, 201, 204
National Aeronautics and Space Administration (NASA), 12, 56, 104, 157, 167, 240
National Marine Fisheries Service (NMFS), 77
National Oceanic & Atmospheric Administration (NOAA), 5, 6, 77
Native Americans, 67, 88
Native Oyster Restoration Alliance (NORA), 89
Natural capital, 39, 40, 180, 191, 194, 196, 198, 199, 207
Natural Climate Solution (NCS), 1, 18, 196, 219
Nature's Contributions to People (NCP), 191
Nature-Based Negative Emissions Technology (NB-NET), 196
Nature-based solution (Nbs), 20, 21, 179
Nature-friendly, 41
Negative emissions, 19, 20, 185, 196, 197, 222
Negative Emissions Technology (NET), 19, 20, 185, 196, 197, 222
Net zero, 14–17, 39, 44, 50, 183, 191, 199
Neuse River, 86
New Caledonia, 137
New food resources, 99
New generation shellfish farming, 218
New Haven Harbor, 83
New York, 8, 10, 56, 65, 82–85, 88, 220, 237
Nile Delta, 73
Ninety Mile Beach, 185
Nitrous Oxide (N_2O), 1, 3, 36, 85, 178, 198, 218
Nordic Blue Carbon Project, 202
Norovirus, 82
North Carolina, 86
Northern abalone, 69
Northern Lights project, 187
North Sea, 79, 80, 86, 98, 187
Nucella spp., 69
Nucellus lapillus, 67
Nudibranchs, 143
Nursery system, 147, 148
Nutricline, 162

O
Ocean fertilisation, 190, 195
Ocean floor, 90, 108, 125, 133, 159, 196
Oceanic Anoxic Event (OAE), 160
Ocenebra erinacea, 79
Ochrophyta, 109
Office for National Statistics (ONS), 38, 199
Offsetting, 35, 56, 57, 219, 223
Offshore factory rigs, 97, 220
Offshore industries, 97, 220
Omega-3 fatty acids, 85, 101, 102, 111, 112, 241
Oncorhynchus mykiss, 76
Online Comments, 217, 227
Ooids, 24, 46
Oomycota, 77
Open ponds, 168
Ophiodon elongates, 69
Ordovician, 163
Oreochromis spp., 75
Orinoco river, 75
Ornamental fish, 76
Ostrea edulis, 67, 85
Oyster aquaculture, 85, 88, 91
Oyster bed, 65, 80, 82–88, 201, 202, 220, 230, 237
Oyster Day, 78, 79
Oyster dredging, 65, 81, 83, 85–87, 220
Oysters, 50–52, 54, 57, 58, 65, 67, 74–91, 111, 124, 133, 142, 144, 148, 201, 202, 208, 209, 220, 230, 231, 237, 241
Oyster saloons, 83, 84
Oyster shell midden, 67
Oyster spat, 89

P
Pacific, 56, 68, 69, 76, 83, 87, 88, 126, 131, 132, 137, 139, 141, 145, 146, 147, 171, 220, 227, 235
Pacific Islands, 137, 141
Pacific oysters, 87, 88
Pacific salmon, 68
Palaeocene, 163
Palaeocene/Eocene Thermal Maximum, 163
Palaeolithic migration, 67
Palaeozoic, 163
Palau, 137, 138, 142, 147
Paleogene, 160, 164
Paradigm, 46, 49, 88, 141, 177, 192, 209, 217, 221, 222, 224, 227, 229, 238, 243
Paris Agreement, 16, 17, 19, 121, 185
Patella vulgata, 67
PBR, 168
Pearls, 74, 133, 147
Peatland, 36–39, 50, 198, 199, 200, 204
Pecten maximus, 52, 67
Pediveliger, 143
Pelican Site CCS, 185
Penaeid shrimp, 73, 75, 76
Penaeus chinensis, 74
Penaeus spp., 73–75
Pentacyclic alkaloid, 120
Perch, 78
Perfluorocarbon, 178
Periodicity, 136, 142
Periwinkle, 67
Permaculture, 103
Permafrost, 13, 16, 36, 198, 200
Permanence, 27, 179, 204, 217
Permian, 160, 163, 164
Perna perna, 66

Persistent Organic Pollutant, 107
Peruvian anchovies, 111
Petagram, 2
PETM, 160, 163
PFC, 178
Phaeodactylum tricornutum, 143
Philippines Sea, 106
Phosphate, 22, 124, 163, 166, 230
Phospholipid, 25
Photic zone, 103, 105, 108, 160, 162, 165, 232
Photobioreactor, 165, 168, 169, 170, 171, 172
Photoperiod, 142
Photosynthesis, 4, 7, 24, 40, 44, 52, 105, 108, 122, 134, 136, 143, 157, 159, 162, 163, 166, 167, 203, 237
Photosynthetic bacterium, 167, 240
Phototrophy, 136, 143
PhycoCosm, 161
Physiological tools, 25, 164
Phytoplankton, 55, 101, 103, 104, 110, 112, 114, 121, 124, 143, 156, 159, 160, 161, 162, 166, 195, 196, 203, 235
Phytoplankton farming, 53
Pinnacle Point, 66
Pinus sylvestris, 37
Pipeline, 187, 188, 189, 191
Planetary ecosystem, 22, 55, 140, 218
Plankton, 23, 24, 44, 53, 71, 107, 108, 111, 124, 136, 143, 159, 164, 229, 233
Planktotrophic, 136
Plant biomass, 27, 35, 194, 219
Pleurochrysis carterae, 165, 167, 170, 239
Pleurochrysis spp., 159, 165, 167, 170, 239
Po delta, 51, 52
Podzol, 37
Pollutant content, 107
Pollution-filtration, 28, 53, 229
Polyplacophora, 69
Polysaccharide, 58
Pond based aquaculture, 76
POP, 107
Porphyra sp., 74
Potable water, 100, 164
Poultry feeds, 57
Power plants, 41, 53, 183, 188, 189, 195, 225
Precambrian, 123, 163
Precipitation, 11, 15, 16, 24, 44, 46, 54, 124, 163, 181, 205, 235, 236, 239, 242
Predator, 43–46, 72, 79, 89, 91, 114, 136, 148, 159, 169, 228
Primary productivity, 109, 161, 162, 243
Prince Edward Island, 83
Procambarus clarkii, 76
Procambarus zonangulus, 77
Production cost, 47, 111, 210
Protandric hermaphrodites, 142
Protection of larvae, 50, 218
Proton transporters, 25
Perpetual Salt Fountain (PSA), 71, 97, 106, 107, 116, 120, 155, 172, 220, 221, 225
Pteria, 134
Puget Sound, 83

Q
Quadra Island, 69
Queen conch, 76

R
Raceways, 76, 142, 168, 171, 239, 240
Rachycentron canadum, 101
Raft culture, 47, 74, 111
Rainbow , 76
Ranching, 72, 75, 89, 231
Rapeseed, 102
Razor shell, 67
Reason For Concern (RFC), 1, 13, 219
Recirculating Aquaculture System (RAS), 101
Recommendations, 21, 28, 75, 89, 118, 217, 222, 224
Recreational fishing, 137
RED List, 135
Red Sea, 66, 67, 194
Red swamp crawfish, 76
Reef, 25, 49–51, 55, 58, 67, 69, 82, 85–88, 97, 114, 131, 134, 135-138, 140, 141–147, 149, 163, 208, 210, 218, 220, 223, 226, 241, 244
Reef associated fish, 69
Reef-building corals, 144, 226
Reef reconstruction, 53, 229
Reef restoration, 139, 144–146, 226, 241
Reforestation, 20, 27, 121, 125, 197, 199, 202
Regulatory policy, 179
Reliable sequestration, 19, 185, 196
Renewable energy, 28, 56, 171, 180, 225
Reservoir leaks, 183
Respiratory CO2, 39, 51, 52, 222, 236
Restocking, 57, 75, 131, 135, 137–139, 141, 144, 147, 149, 220, 226
Restoration, 18, 20, 21, 26, 27, 28, 30, 35, 36, 39, 40, 50, 55, 65, 69, 76, 82, 87, 88, 89, 90, 91, 125, 131, 138, 139, 140, 141, 144, 145, 146, 197, 199, 200, 202, 204, 205, 208, 218, 220, 226, 241
Restorative Shellfish Mariculture (RSM), 88
Rhabdosphaera clavigera, 158, 162
Rice paddies, 71, 74, 224
Río de la Plata, 108
River Colne, 79, 81
River Po, 51, 52
River-snail, 74
Rochester, 78
Rock fish, 69
Rope culture, 111, 112
Rudista, 134
Ruditapes, 24, 210

S

Salmon, 68, 72, 77, 97, 100, 107, 113, 115, 118, 209
Salted fish, 77
Salt-water wetlands, 22, 200
Sardines, 97
Satellite images, 157
Saudi Arabia, 56, 67
Saxidomus gigantea, 69
Scallop dredging, 78, 81
Scallops, 52, 57, 67, 74, 78, 80, 81, 89, 90, 113, 133, 142, 208
Scavenger, 44, 113, 136, 223, 226, 228, 244
Schizochytrium sp., 102
Scotland, 37, 50, 51, 67, 81
Seabass, 73
Seabream, 73, 120
Sea cucumbers, 113, 149
Seagrass beds, 22, 50, 134, 200, 222
Seagrasses, 87, 140, 201, 202
Sea ice, 13, 16
Sea level, 10, 12, 13, 109, 134
Sea lice, 101
Seals, 103, 120
Seamount, 97, 107–109, 114, 117, 118, 126, 220, 224, 225, 226, 240
Seamount installations, 57, 224, 225, 226
Sea turtle farms, 75
Sea urchin, 66, 69, 113
Sea-Viewing Wide Field-of-View Sensor, 104
Seaweed farming, 202, 203
SeaWIFS, 104
Sebastes spp., 69
Self-Contained Underwater Breathing Apparatus (SCUBA), 137, 144, 149, 226
Self-regulating, 171, 226
Semibalanus spp., 69
Sequester, 19, 22, 35, 36, 39, 40, 44, 46, 49–51, 65, 70, 71, 121, 140, 155, 164, 185, 193, 196, 197, 200, 201, 203, 204, 206, 217, 219, 221–223, 228, 229, 230, 233, 240, 242, 243
Sequestering carbon, 27, 36, 197, 202, 222
Serotonin, 142
Sewer networks, 82
Seychelles, 144
Shallow Coastal Ecosystems (SCEs), 201
Shared Socio-economic Pathway, 16
Shellfish as food, 177, 208, 209, 221, 243
Shellfish cultivation, 35, 42, 46, 49, 52, 53, 56, 121, 125, 126, 131, 177, 209, 219, 220, 221, 222, 223, 224
Shellfish farm, 51, 52, 56, 78, 89, 126, 177, 218, 221, 227–229, 238
Shellfish for carbon, 24, 50, 177, 209, 221, 243
Shellfish-associated infections, 82
Shellfish-eating, 65, 232
Shipworm, 133
Shoreline, 50, 65, 68, 69, 70, 88, 203, 208, 210, 219
Shrimp, 44, 58, 66, 73, 74, 75, 76, 77, 89, 100, 159, 202, 209
Silkworm cultivation, 73, 112
Silurian, 163
Single-use plastics, 58
Sink persistence, 39, 198
Siphon, 136, 142, 143
Slipper limpets, 79, 80, 81, 84
Slugs, 133
Snails, 67, 69, 74, 133
Snapper, 76
Social capital, 141, 207
Soft-shelled turtle, 74
Soil organic carbon, 36, 37, 199, 200, 204
Solar radiation, 5, 155, 161, 171, 178, 221, 226
Solar-energy, 171
Solomon Islands, 137, 147
South Atlantic Gyre, 105, 125, 226
South China Sea, 131, 132, 135, 137, 227
South Pacific, 131, 137, 147, 162, 166
Spawning, 72, 74, 87, 110, 114, 136, 142, 145
Species complex, 45, 161
Species richness, 160, 161
Sphagnum, 37, 38
Spirulina, 167, 168, 240, 241
Sprats, 97
SSP, 16
Stabilisation, 37, 88, 181, 208, 210
Staple food, 65, 84, 85, 132, 220, 237
Storage ponds, 71
Strombus gigas, 76
Strongylocentrotus spp., 69
Summary for Policymakers (SPM), 15, 16
Super trawlers, 97
Superfood, 168
Supplement, 49, 57, 71, 101, 102, 115, 142, 159, 168, 223, 225, 231, 232
Sustainability, 18, 19, 49, 52, 67, 70, 114, 116, 137, 177, 191, 197, 206, 207, 208, 210, 221
Sustainability Assessment, 177, 206, 210, 221
Sustainability Index, 208
Sustainable, 12, 18, 20, 21, 24, 26, 27, 28, 29, 30, 40, 49, 55, 65, 70, 72, 84, 85, 88, 97, 99, 118, 119, 125, 126, 134, 135, 139, 140, 145, 146, 147, 149, 166, 178, 179, 196, 198, 205, 206, 208, 209, 210, 217, 218, 220, 222, 224, 235, 243
Sustainable development, 19, 21, 25, 28–30, 48, 58, 126, 127, 139, 140, 190, 191, 207–209, 217
Swansea Bay, 81
Swordfish, 97
Symbiodinium spp., 134, 136, 143
Symbiotic, 19, 40, 134, 136, 138, 143, 200
Symbiotic algae, 123, 143, 156, 227
Symbiotic dinoflagellate, 143
Synchiropus splendidus, 74

T

Taiwan, 132, 137
T. crocea, 134, 138, 142–144
T. derasa, 132, 134, 137, 138, 142, 148

Temperature, 3, 5, 7, 10, 11, 12, 15, 16, 17, 18, 22, 27, 28, 45, 46, 54, 75, 101, 107, 110, 111, 116, 122, 123, 124, 125, 134, 137, 138, 142, 156, 157, 159, 161, 162, 165, 169, 194, 227, 237
Teredo navalis, 133
Terraces, 65, 68, 171, 219, 224
Terrestrial ecosystems, 4, 27, 35, 51, 121, 219, 225
Tethys Ocean, 134
Tetraselmis, 169
Thames estuary, 79
The Commons, 117
The Philippines, 5, 66, 106, 131, 137, 139, 148
Thermal Maximum, 160, 163
Three-pillar concept, 208
Tidal marshlands, 22, 200, 222
Tilapia, 73, 75, 100
Tingle borers, 79
T. maxima, 134, 136–138, 141–144
TMDC, 147, 149
T-OAE, 160
Toarcian Oceanic Anoxic Event, 160
Tonga, 137
Topshell, 67
Toxicity, 159
Transporters, 25, 44
Tree coverage, 27, 35
Tree planting, 35, 36, 37, 39, 40, 41, 197, 198, 199, 219
Tresus spp., 69
Triacylglycerol, 167
Tridacna gigas, 122, 134, 135, 139, 142, 143, 145, 149
Tridacna Mariculture Development Center, 132, 133, 135, 147, 148, 149, 150
Tridacnidae, 133, 145
Trillion Tree, 40, 41, 197
Trilobites, 24, 46, 163, 164
Trindade Archipelago, 108
Trochophore, 143
Trout, 76, 78
T. squamosa, 134, 137, 138, 142, 143, 144
T. tevora, 134
Tuamotu atolls, 141, 145
Tubuai, 144
Tuna, 76, 97
Turbulent flow, 170
Turtle, 74–76
Turtle farming, 75

U
UCLA CarbonBuilt, 195
UK National Ecosystem Assessment (UKNEA), 54
Umbellosphaera foliosa, 162
Umbellosphaera irregularis, 158, 161
UN Convention on the Law Of the Sea (UNCLOS), 30, 117, 224
UN Educational, Scientific and Cultural Organization (UNESCO), 141
UN Environment Programme (UNEP), 17, 40, 197
UN Framework Convention on Climate Change (UNFCCC), 22, 50, 178, 180
United Nations Conference of the Parties, Glasgow 2021, 15, 16, 17, 20, 21, 26
United Nations Conference of the Parties, Paris 2015, 16
United Nations Conference of the Parties, Sharm al-Sheikh 2022, 18
UN Sustainable Development Goals (UN-SDGs), 21, 28, 190, 191
Upper Chesapeake Bay, 83
Urosalpinx cinerea, 79
USA aquaculture, 76, 77

V
Vanuatu, 137
Vegetation, 9, 20, 36, 39, 40, 199, 200, 201
Veliger, 143
Vibrio spp., 82
Vibrio vulnificus, 82
Virginia, 83
Vitória-Trindade, 105, 108, 109, 114, 117, 118, 224
Volcanic, 5, 9, 14, 25, 46, 105, 108, 123, 163

W
Warwickshire, 78
Wave-calming, 49, 227
Wester Ross, 67
Wetland, 1, 18, 19, 22, 27, 72, 73, 183, 190, 196, 199, 200, 204, 219, 222
Wetland agriculture-with-aquaculture, 74
Whaleback midden, 67
Whedos, 72
White Cliffs of Dover, 156, 159, 203, 230, 234
Whitefish, 72
Whitstable, 78, 83
Wildfire, 13, 14, 19, 40, 48, 198
Wildlife Conservation Society (WCS), 40, 197
Willapa Bay, 83
Wood-decaying fungi, 36
Wood fuel, 39, 198
World Economic Forum (WEF), 20, 40, 197
World Food Program (WFP), 117
World Meteorological Organization (WMO), 4, 6, 194, 205
World Oceans Day, 99
World Wide Fund for Nature (WWF), 40, 197

X
XPRIZE, 195

Z
Zooplankton cultivation, 53
Zooxanthellae, 134, 136, 138, 143, 144

GPSR Compliance

The European Union's (EU) General Product Safety Regulation (GPSR) is a set of rules that requires consumer products to be safe and our obligations to ensure this.

If you have any concerns about our products, you can contact us on ProductSafety@springernature.com

In case Publisher is established outside the EU, the EU authorized representative is:

Springer Nature Customer Service Center GmbH
Europaplatz 3
69115 Heidelberg, Germany

Batch number: 10731385

Printed by Printforce, the Netherlands